Engineering Materials 2

An Introduction to Microstructures, Processing and Design

Learning Resource Centre

Park Road, Uxbridge Middlesex UB8 1NQ
Renewals: 01895 853326 Enquiries: 01895 853344

Please return this item to the Learning Centre on or before this last date
Stamped below:

1 6 JUN 2011

3-12 OCT 2012

- 2 OCT 2012

0 7 DEC 2017

620. 11

Engineering Materials 2

*An Introduction to Microstructures,
Processing and Design*

Third Edition

Michael F. Ashby

and

David R. H. Jones
Department of Engineering, Cambridge University, UK

ELSEVIER

AMSTERDAM • BOSTON • HEIDELBERG • LONDON • NEW YORK • OXFORD
PARIS • SAN DIEGO • SAN FRANCISCO • SINGAPORE • SYDNEY • TOKYO

Butterworth-Heinemann is an imprint of Elsevier

Butterworth-Heinemann is an imprint of Elsevier
The Boulevard, Langford Lane, Kidlington, Oxford, OX5 1GB
30 Corporate Drive, Suite 400, Burlington, MA 01803, USA

First edition 1986
Reprinted with Corrections 1988
Reprinted 1989, 1992
Second edition 1998
Reprinted 1999, 2000, 2001
Third edition 2006
Reprinted 2006, 2007, 2008, 2009

Copyright © 2006, Michael F. Ashby and David R. H. Jones. Published by Elsevier Ltd.
All rights reserved

The right of Michael F. Ashby and David R. H. Jones to be identified as the authors of
this work has been asserted in accordance with the Copyright, Designs and Patents Act 1988

No part of this publication may be reproduced, stored in a retrieval system
or transmitted in any form or by any means electronic, mechanical, photocopying,
recording or otherwise without the prior written permission of the publisher

Permissions may be sought directly from Elsevier's Science & Technology Rights
Department in Oxford, UK: phone (+44) (0) 1865 843830; fax (+44) (0) 1865 853333;
email: permissions@elsevier.com. Alternatively you can submit your request online by
visiting the Elsevier web site at http://elsevier.com/locate/permissions, and selecting
Obtaining permission to use Elsevier material

Notice
No responsibility is assumed by the publisher for any injury and/or damage to persons
or property as a matter of products liability, negligence or otherwise, or from any use
or operation of any methods, products, instructions or ideas contained in the material
herein. Because of rapid advances in the medical sciences, in particular, independent
verification of diagnoses and drug dosages should be made

British Library Cataloguing in Publication Data
A catalogue record for this book is available from the British Library

Library of Congress Cataloging-in-Publication Data
A catalog record for this book is available from the Library of Congress

ISBN: 978-0-7506-6381-6

For information on all Butterworth-Heinemann publications
visit our website at www.elsevierdirect.com

Printed and bound in *Great Britain*

09 10 11 12   12 11 10 9 8 7 6 5

Working together to grow
libraries in developing countries

www.elsevier.com | www.bookaid.org | www.sabre.org

ELSEVIER   BOOK AID
             International   Sabre Foundation

Contents

General introduction

Materials are evolving today faster than at any time in history. Industrial nations regard the development of new and improved materials as an "underpinning technology" – one which can stimulate innovation in all branches of engineering, making possible new designs for structures, appliances, engines, electrical and electronic devices, processing and energy conservation equipment, and much more. Many of these nations have promoted government-backed initiatives to promote the development and exploitation of new materials: their lists generally include "high-performance" composites, new engineering ceramics, high-strength polymers, glassy metals, and new high-temperature alloys for gas turbines. These initiatives are now being felt throughout engineering, and have already stimulated design of a new and innovative range of consumer products.

So the engineer must be more aware of materials and their potential than ever before. Innovation, often, takes the form of replacing a component made of one material (a metal, say) with one made of another (a polymer, perhaps), and then redesigning the product to exploit, to the maximum, the potential offered by the change. The engineer must compare and weigh the properties of competing materials with precision: the balance, often, is a delicate one. It involves an understanding of the basic properties of materials; of how these are controlled by processing; of how materials are formed, joined and finished; and of the chain of reasoning that leads to a successful choice.

This book aims to provide this understanding. It complements our other book on the properties and applications of engineering materials,* but it is not necessary to have read that to understand this. In it, we group materials into four classes: Metals, Ceramics, Polymers and Composites, and we examine each in turn. In any one class there are common underlying structural features (the long-chain molecules in polymers, the intrinsic brittleness of ceramics, or the mixed materials of composites) which, ultimately, determine the strengths and weaknesses (the "design-limiting" properties) of each in the engineering context.

And so, as you can see from the Contents list, the chapters are arranged in *groups*, with a group of chapters to describe each of the four *classes* of materials. In each group we first introduce the major families of materials that go to make up each materials class. We then outline the main microstructural features of the class, and show how to process or treat them to get the structures (really, in the end, the *properties*) that we want. Each group of chapters is illustrated by

---

* M. F. Ashby and D. R. H. Jones, *Engineering Materials 1: An Introduction to their Properties and Applications*, 2nd edition, Butterworth-Heinemann, 1996.

Case Studies designed to help you understand the basic material. And finally we look at the role of materials in the design of engineering devices, mechanisms or structures, and develop a methodology for materials selection. One subject – *Phase Diagrams* – can be heavy going. We have tried to overcome this by giving a short programmed-learning course on phase diagrams. If you work through this when you come to the chapter on phase diagrams you will know all you need to about the subject. It will take you about 4 hours.

At the end of each chapter you will find a set of problems: try to do them while the topic is still fresh in your mind – in this way you will be able to consolidate, and develop, your ideas as you go along.

To the lecturer

This book has been written as a second-level course for engineering students. It provides a concise introduction to the *microstructures and processing of materials* (metals, ceramics, polymers and composites) and shows how these are related to the properties required in engineering design. It is designed to follow on from our first-level text on the properties and applications of engineering materials,[*] but it is completely self-contained and can be used by itself.

Each chapter is designed to provide the content of a 50-minute lecture. Each block of four or so chapters is backed up by a set of *Case Studies*, which illustrate and consolidate the material they contain. There are special sections on design, and on such materials as wood, cement and concrete. And there are problems for the student at the end of each chapter for which worked solutions can be obtained separately, from the publisher. In order to ease the teaching of phase diagrams (often a difficult topic for engineering students) we have included a programmed-learning text which has proved helpful for our own students.

We have tried to present the material in an uncomplicated way, and to make the examples entertaining, while establishing basic physical concepts and their application to materials processing. We found that the best way to do this was to identify a small set of "generic" materials of each class (of metals, of ceramics, etc.) which broadly typified the class, and to base the development on these; they provide the pegs on which the discussion and examples are hung. But the lecturer who wishes to draw other materials into the discussion should not find this difficult.

Acknowledgements

We wish to thank Prof. G. A. Chadwick for permission to reprint Fig. A1.34 (p. 417) and K. J. Pascoe and Van Nostrand Reinhold Co. for permission to reprint Fig. A1.41 (p. 422).

---

[*] M. F. Ashby and D. R. H. Jones, *Engineering Materials 1: An Introduction to their Properties and Applications*, 3rd edition, Butterworth-Heinemann, 2005.

Figures 29.1 and 29.4 are reprinted courtesy of University of St Andrews.

Figure 29.6 is reprinted courtesy of the Dundee Central Library.

Figure 29.2 is reprinted from *Engineering Failure Analysis*, Jones, D. R. H., The Tay Bridge disaster, with permission from Elsevier.

Figures 29.11 and 29.14 are reprinted from *Engineering Failure Analysis* 12, Richard *et al.*, Fracture in a rubber-sprung railway wheel, 986–999, © 2005, with permission from Elsevier.

Figure 29.13 is reprinted from *Engineering Failure Analysis* 11, Esslinger *et al.*, The railway accident at Eschede, 515–535, © 2004, with permission from Elsevier.

Figures 29.15–29.19 are reprinted from *Engineering Failure Analysis* 11, Jones, D. R. H., Analysis of a fatal bungee-jumping accident, 857–872, © 2004, with permission from Elsevier.

Accompanying Resources

The following accompanying web-based resources are available to teachers and lecturers who adopt or recommend this text for class use. For further details and access to these resources please go to http://textbooks.elsevier.com

Instructor's Manual

A full Solutions Manual with worked answers to the exercises in the main text is available for downloading.

Image bank

An image bank of downloadable PDF versions of the figures from the book is available for use in lecture slides and class presentations.

Online Materials Science Tutorials

A series of online materials science tutorials accompanies *Engineering Materials 1* and *2*. These were developed by Alan Crosky, Mark Hoffman, Paul Munroe and Belinda Allen at the University of New South Wales (UNSW) Australia, based upon earlier editions of the books. The group is particularly interested in the effective and innovative use of technology in teaching. They realised the potential of the material for the teaching of Materials Engineering to their students in an online environment and have developed and then used these very popular tutorials for a number of years at UNSW. The results of this work have also been published and presented extensively.

The tutorials are designed for students of materials science as well as for those studying materials as a related or elective subject, for example mechanical or civil engineering students. They are ideal for use as ancillaries to formal teaching programs, and may also be used as the basis for quick refresher courses for more advanced materials science students. By picking selectively from the range of tutorials available they will also make ideal subject primers for students from related faculties.

The software has been developed as a self-paced learning tool, separated into learning modules based around key materials science concepts. For further information on accessing the tutorials, and the conditions for their use, please go to http://textbooks.elsevier.com

About the authors of the Tutorials

Alan Crosky is an Associate Professor in the School of Materials Science and Engineering, UNSW. His teaching specialties include metallurgy, composites and fractography.

Belinda Allen is an Educational Graphics Manager and Educational Designer at the Educational Development and Technology Centre, UNSW. She provides consultation and production support for the academic community and designs and presents workshops and online resources on image production and web design.

Mark Hoffman is an Associate Professor in the School of Materials Science and Engineering, UNSW. His teaching specialties include fracture, numerical modelling, mechanical behaviour of materials and engineering management.

Paul Munroe has a joint appointment as Professor in the School of Materials Science and Engineering and Director of the Electron Microscope Unit, UNSW. His teaching specialties are the deformation and strengthening mechanisms of materials and crystallographic and microstructural characterisation.

A. Metals

Chapter 1
Metals

Introduction

This first group of chapters looks at metals. There are so many different metals – literally hundreds of them – that it is impossible to remember them all. It isn't necessary – nearly all have evolved from a few "generic" metals and are simply tuned-up modifications of the basic recipes. If you know about the generic metals, you know most of what you need.

This chapter introduces the generic metals. But rather than bore you with a catalogue we introduce them through three real engineering examples. They allow us not only to find examples of the uses of the main generic metals but also to introduce the all-important business of how the characteristics of each metal determine how it is used in practice.

Metals for a model traction engine

Model-making has become big business. The testing of scale models provides a cheap way of getting critical design information for things from Olympic yacht hulls to tidal barrages. Architects sell their newest creations with the help of miniature versions correct to the nearest door-handle and garden shrub. And in an age of increasing leisure time, many people find an outlet for their energies in making models – perhaps putting together a miniature aircraft from a kit of plastic parts or, at the other extreme, building a fully working model of a steam engine from the basic raw materials in their own "garden-shed" machine shop.

Figure 1.1 shows a model of a nineteenth-century steam traction engine built in a home workshop from plans published in a well-known modellers' magazine. Everything works just as it did in the original – the boiler even burns the same type of coal to raise steam – and the model is capable of towing an automobile! But what interests us here is the large range of metals that were used in its construction, and the way in which their selection was dictated by the requirements of design. We begin by looking at metals based on *iron* (*ferrous* metals). Table 1.1 lists the generic iron-based metals.

How are these metals used in the traction engine? The design loads in components like the wheels and frames are sufficiently low that *mild steel*, with a yield strength σ_y of around 220 MPa, is more than strong enough. It is also easy to cut, bend or machine to shape. And last, but not least, it is cheap.

The stresses in the machinery – like the gear-wheel teeth or the drive shafts – are a good deal higher, and these parts are made from either *medium-carbon,*

Figure 1.1   A fully working model, one-sixth full size, of a steam traction engine of the type used on many farms a hundred years ago. The model can pull an automobile on a few litres of water and a handful of coal. But it is also a nice example of materials selection and design.

Table 1.1   Generic iron-based metals

Metal	Typical composition (wt%)	Typical uses
Low-carbon ("mild") steel	Fe + 0.04 to 0.3 C (+ ≈ 0.8 Mn)	Low-stress uses. General constructional steel, suitable for welding.
Medium-carbon steel	Fe + 0.3 to 0.7 C (+ ≈ 0.8 Mn)	Medium-stress uses: machinery parts − nuts and bolts, shafts, gears.
High-carbon steel	Fe + 0.7 to 1.7 C (+ ≈ 0.8 Mn)	High-stress uses: springs, cutting tools, dies.
Low-alloy steel	Fe + 0.2 C 0.8 Mn 1 Cr 2 Ni	High-stress uses: pressure vessels, aircraft parts.
High-alloy ("stainless") steel	Fe + 0.1 C 0.5 Mn 18 Cr 8 Ni	High-temperature or anti-corrosion uses: chemical or steam plants.
Cast iron	Fe + 1.8 to 4 C (+ ≈ 0.8 Mn 2 Si)	Low-stress uses: cylinder blocks, drain pipes.

Figure 1.2   A close-up of the mechanical lubricator on the traction engine. Unless the bore of the steam cylinder is kept oiled it will become worn and scored. The lubricator pumps small metered quantities of steam oil into the cylinder to stop this happening. The drive is taken from the piston rod by the ratchet and pawl arrangement.

high-carbon or *low-alloy steels* to give extra strength. However, there are a few components where even the strength of high-carbon steels as delivered "off the shelf" ($\sigma_y \approx 400\,\text{MPa}$) is not enough. We can see a good example in the mechanical lubricator, shown in Fig. 1.2, which is essentially a high-pressure oil metering pump. This is driven by a ratchet and pawl. These have sharp teeth which would quickly wear if they were made of a soft alloy. But how do we raise the hardness above that of ordinary high-carbon steel? Well, medium- and high-carbon steels can be hardened to give a yield strength of up to 1000 MPa by heating them to bright red heat and then quenching them into cold water. Although the quench makes the hardened steel brittle, we can make it tough again (though still hard) by *tempering* it – a process that involves heating the steel again, but to a much lower temperature. And so the ratchet and pawls are made from a quenched and tempered high-carbon steel.

Stainless steel is used in several places. Figure 1.3 shows the fire grate – the metal bars which carry the burning coals inside the firebox. When the engine is working hard the coal is white hot; then, both oxidation and creep are problems. Mild steel bars can burn out in a season, but stainless steel bars last indefinitely.

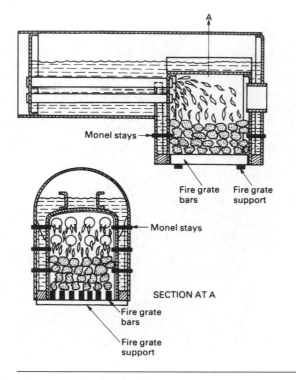

Monel stays

Fire grate bars   Fire grate support

Monel stays

SECTION AT A

Fire grate bars

Fire grate support

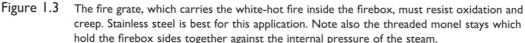

Figure 1.3   The fire grate, which carries the white-hot fire inside the firebox, must resist oxidation and creep. Stainless steel is best for this application. Note also the threaded monel stays which hold the firebox sides together against the internal pressure of the steam.

Finally, what about *cast iron*? Although this is rather brittle, it is fine for low-stressed components like the cylinder block. In fact, because cast iron has a lot of carbon it has several advantages over mild steel. Complicated components like the cylinder block are best produced by casting. Now cast iron melts much more easily than steel (adding carbon reduces the melting point in just the way that adding anti-freeze works with water) and this makes the pouring of the castings much easier. During casting, the carbon can be made to separate out as tiny particles of graphite, distributed throughout the iron, which make an ideal boundary lubricant. Cylinders and pistons made from cast iron wear very well; look inside the cylinders of your car engine next time the head has to come off, and you will be amazed by the polished, almost glazed look of the bores – and this after perhaps 10^8 piston strokes.

These, then, are the basic classes of ferrous alloys. Their compositions and uses are summarised in Table 1.1, and you will learn more about them in Chapters 11 and 12, but let us now look at the other generic alloy groups.

An important group of alloys are those based on copper (Table 1.2).

The most notable part of the traction engine made from copper is the boiler and its firetubes (see Fig. 1.1). In full size this would have been made from

Table 1.2   Generic copper-based metals

Metal	Typical composition (wt%)	Typical uses
Copper	100 Cu	Ductile, corrosion resistant and a good electrical conductor: water pipes, electrical wiring.
Brass	Zn	Stronger than copper, machinable, reasonable corrosion resistance: water fittings, screws, electrical components.
Bronze	Cu + 10–30 Sn	Good corrosion resistance: bearings, ships' propellers, bells.
Cupronickel	Cu + 30 Ni	Good corrosion resistance, coinage.

mild steel, and the use of copper in the model is a nice example of how the choice of material can depend on the *scale* of the structure. The boiler plates of the full-size engine are about 10 mm thick, of which perhaps only 6 mm is needed to stand the load from the pressurised steam safely – the other 4 mm is an allowance for corrosion. Although a model steel boiler would stand the pressure with plates only 1 mm thick, it would still need the same corrosion allowance of 4 mm, totalling 5 mm altogether. This would mean a very heavy boiler, and a lot of water space would be taken up by thick plates and firetubes. Because copper hardly corrodes in clean water, this is the obvious material to use. Although weaker than steel, copper plates 2.5 mm thick are strong enough to resist the working pressure, and there is no need for a separate corrosion allowance. Of course, copper is expensive – it would be prohibitive in full size – but this is balanced by its ductility (it is very easy to bend and flange to shape) and by its high thermal conductivity (which means that the boiler steams very freely).

Brass is stronger than copper, is much easier to machine, and is fairly corrosion-proof (although it can "dezincify" in water after a long time). A good example of its use in the engine is for steam valves and other boiler fittings (see Fig. 1.4). These are intricate, and must be easy to machine; dezincification is a long-term possibility, so occasional inspection is needed. Alternatively, corrosion can be avoided altogether by using the more expensive *bronzes*, although some are hard to machine.

Nickel and its alloys form another important class of non-ferrous metals (Table 1.3). The superb creep resistance of the nickel-based superalloys is a key factor in designing the modern gas-turbine aero-engine. But nickel alloys even appear in a model steam engine. The flat plates in the firebox must be stayed together to resist the internal steam pressure (see Fig. 1.3). Some model-builders make these stays from pieces of monel rod because it is much stronger than copper, takes threads much better and is very corrosion resistant.

Figure 1.4   Miniature boiler fittings made from brass: a water-level gauge, a steam valve, a pressure gauge, and a feed-water injector. Brass is so easy to machine that it is good for intricate parts like these.

Table 1.3   Generic nickel-based metals

Metals	Typical composition (wt%)	Typical uses
Monels	Ni + 30 Cu 1Fe 1Mn	Strong, corrosion resistant: heat-exchanger tubes.
Superalloys	Ni + 30 Cr 30 Fe 0.5 Ti 0.5 Al	Creep and oxidation resistant: furnace parts.
	Ni + 10 Co 10 W 9 Cr 5 Al2 Ti	Highly creep resistant: turbine blades and discs.

Metals for drinks cans

Few people would think that the humble drink can (Fig. 1.5) was anything special. But to a materials engineer it is high technology. Look at the requirements. As far as possible we want to avoid seams. The can must not leak, should use as little metal as possible and be recyclable. We have to choose a metal that is ductile to the point that it can be drawn into a single-piece can

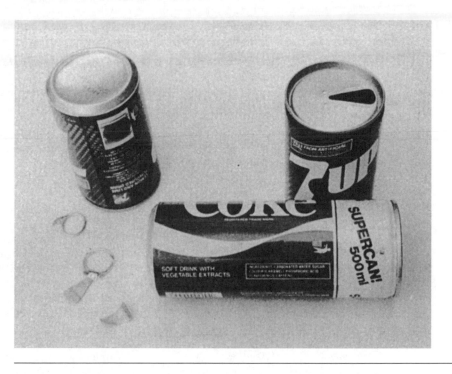

Figure 1.5   The aluminium drink can is an innovative product. The body is made from a single slug of a 3000 series aluminium alloy. The can top is a separate pressing which is fastened to the body by a rolled seam once the can has been filled. There are limits to one-piece construction.

body from one small slug of metal. It must not corrode in beer or coke and, of course, it must be non-toxic. And it must be light and must cost almost nothing.

Aluminium-based metals are the obvious choice* (Table 1.4) – they are light, corrosion resistant and non-toxic. But it took several years to develop the process for forming the can and the alloy to go with it. The end product is a big advance from the days when drinks only came in glass bottles, and has created a new market for aluminium (now threatened, as we shall see in Chapter 21, by polymers). Because aluminium is lighter than most other metals it is also the obvious choice for transportation: aircraft, high-speed trains, cars, even. Most of the alloys listed in Table 1.4 are designed with these uses in mind. We will discuss the origin of their strength, and their uses, in more detail in Chapter 10.

---

* One thinks of aluminium as a cheap material – aluminium spoons are so cheap that they are thrown away. It was not always so. Napoleon had a set of cutlery specially made from the then-new material. It cost him more than a set of solid silver.

Table 1.4   Generic aluminium-based metals

Metal	Typical composition (wt%)	Typical uses
1000 Series unalloyed Al	>99 Al	Weak but ductile and a good electrical conductor: power transmission lines, cooking foil.
2000 Series major additive Cu	Al + 4 Cu + Mg, Si, Mn	Strong age-hardening alloy: aircraft skins, spars, forgings, rivets.
3000 Series major additive Mn	Al + 1 Mn	Moderate strength, ductile, excellent corrosion resistance: roofing sheet, cooking pans, drinks can bodies.
5000 Series major additive Mg	Al + 3 Mg 0.5 Mn	Strong work-hardening weldable plate: pressure vessels, ship superstructures.
6000 Series major additives Mg + Si	Al + 0.5 Mg 0.5 Si	Moderate-strength age-hardening alloy: anodised extruded sections, e.g. window frames.
7000 Series major additives Zn + Mg	Al + 6 Zn + Mg, Cu, Mn	Strong age-hardening alloy: aircraft forgings, spars, lightweight railway carriage shells.
Casting alloys	Al + 11 Si	Sand and die castings.
Aluminium– lithium alloys	Al + 3 Li	Low density and good strength: aircraft skins and spars.

Metals for artificial hip joints

As a last example we turn to the world of medicine. Osteo-arthritis is an illness that affects many people as they get older. The disease affects the joints between different bones in the body and makes it hard – and painful – to move them. The problem is caused by small lumps of bone which grow on the rubbing surfaces of the joints and which prevent them sliding properly. The problem can only be cured by removing the bad joints and putting artificial joints in their place. The first recorded hip-joint replacement was done as far back as 1897 – when it must have been a pretty hazardous business – but the operation is now a routine piece of orthopaedic surgery. In fact half a million hip joints are replaced world-wide every year.

Figure 1.6 shows the implant for a replacement hip joint. In the operation, the head of the femur is cut off and the soft marrow is taken out to make a hole down the centre of the bone. Into the hole is glued a long metal shank which carries the artificial head. This fits into a high-density polythene socket which in turn is glued into the old bone socket. The requirements of the implant are stringent. It has to take large loads without bending. Every time the joint is used ($\approx 10^6$ times a year) the load on it fluctuates, giving us a

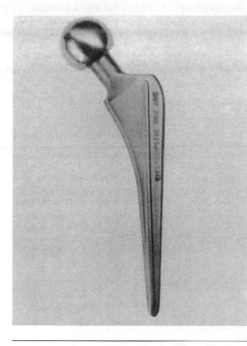

Figure 1.6   The titanium alloy implant for a replacement hip joint. The long shank is glued into the top of the femur. The spherical head engages in a high-density polythene socket which is glued into the pelvic socket.

high-cycle fatigue problem as well. Body fluids are as corrosive as sea water, so we must design against corrosion, stress corrosion and corrosion fatigue. The metal must be bio-compatible. And, ideally, it should be light as well.

The materials that best meet these tough requirements are based on *titanium*. The α–β alloy shown in Table 1.5 is as strong as a hardened and tempered high-carbon steel, is more corrosion resistant in body fluids than stainless steel, but is only half the weight. A disadvantage is that its modulus is only half that of steels, so that it tends to be "whippy" under load. But this can be overcome by using slightly stiffer sections. The same alloy is used in aircraft, both in the airframes and in the compressor stages of the gas turbines which drive them.

Table 1.5   Generic titanium-based metals

Metal	Typical composition (wt%)	Typical uses
α–β titanium alloy	Ti–6 Al4 V	Light, very strong, excellent corrosion resistance, high melting point, good creep resistance. The alloy workhorse: turbofans, airframes, chemical plant, surgical implants.

Table 1.6   Properties of the generic metals

Metal	Cost (UK£ (US$) tonne^{-1})	Density (Mg m^{-3})	Young's modulus (GPa)	Yield strength (MPa)	Tensile strength (MPa)	Ductility	Fracture toughness (MPa m$^{1/2}$)	Melting Temperature (K)	Specific heat (J kg^{-1}K^{-1})	Thermal conductivity (W m^{-1}K^{-1})	Thermal expansion coefficient (MK^{-1})
Iron	100 (140)	7.9	211	50	200	0.3	80	1809	456	78	12
Mild steel	200–230 (260–300)	7.9	210	220	430	0.21	140	1765	482	60	12
High-carbon steel	150 (200)	7.8	210	350–1600	650–2000	0.1–0.2	20–50	1570	460	40	12
Low-alloy steels	180–250 (230–330)	7.8	203	290–1600	420–2000	0.1–0.2	50–170	1750	460	40	12
High-alloy steels	1100–1400 (1400–1800)	7.8	215	170–1600	460–1700	0.1–0.5	50–170	1680	500	12–30	10–18
Cast irons	120 (160)	7.4	152	50–400	10–800	0–0.18	6–20	1403			
Copper	1020 (1330)	8.9	130	75	220	0.5–0.9	>100	1356	385	397	17
Brasses	750–1060 (980–1380)	8.4	105	200	350	0.5	30–100	1190		121	20
Bronzes	1500 (2000)	8.4	120	200	350	0.5	30–100	1120		35	19
Nickel	3200 (4200)	8.9	214	60	300	0.4	>100	1728	450	39	13
Monels	3000 (3900)	8.9	185	340	680	0.5	>100	1600	420	22	14
Superalloys	5000 (6500)	7.9	214	800	1300	0.2	>100	1550	450	11	12
Aluminium	910 (1180)	2.7	71	25–125	75–135	0.1–0.5	45	933	917	240	24
1000 Series	910 (1180)	2.7	71	28–165	75–180	0.1–0.45	45	915			24
2000 Series	1100 (1430)	2.8	71	200–500	300–600	0.1–0.25	10–50	860		180	24
5000 Series	1000 (1300)	2.7	71	40–300	120–430	0.1–0.35	30–40	890		130	22
7000 Series	1100 (1430)	2.8	71	350–600	500–670	0.1–0.17	20–70	890		150	24
Casting alloys	1100 (1430)	2.7	71	65–350	130–400	0.01–0.15	5–30	860		140	20
Titanium	4630 (6020)	4.5	120	170	240	0.25	50–80	1940	530	22	9
Ti–6 Al4 V	5780 (7510)	4.4	115	800–900	900–1000	0.1–0.2		1920	610	6	8
Zinc	330 (430)	7.1	105		120	0.4		693	390	120	31
Lead–tin solder	2000 (2600)	9.4	40					456			
Diecasting alloy	800 (1040)	6.7	105		280–330	0.07–0.15		650	420	110	27

Data for metals

When you select a metal for any design application you need *data* for the properties. Table 1.6 gives you *approximate* property data for the main generic metals, useful for the first phase of a design project. When you have narrowed down your choice you should turn to the more exhaustive data compilations given in Appendix 3. Finally, before making final design decisions you should get detailed material specifications from the supplier who will provide the materials you intend to use. And if the component is a critical one (meaning that its failure could precipitate a catastrophe) you should arrange to test it yourself.

There are, of course, many more metals available than those listed here. It is useful to know that some properties depend very little on microstructure: the density, modulus, thermal expansion and specific heat of *any* steel are pretty close to those listed in the table. (Look at the table and you will see that the variations in these properties are seldom more than ±5%.) These are the "*structure-insensitive*" properties. Other properties, though, vary greatly with the heat treatment and mechanical treatment, and the detailed alloy composition. These are the "*structure-sensitive*" properties: yield and tensile strength, ductility, fracture toughness, and creep and fatigue strength. They cannot be guessed from data for other alloys, even when the composition is almost the same. For these it is *essential* to consult manufacturers' data sheets listing the properties of the alloy you intend to use, with the same mechanical and heat treatment.

Examples

1.1 Explain what is meant by the following terms:

(a) structure-sensitive property;
(b) structure-insensitive property.

List five different structure-sensitive properties.
List four different structure-insensitive properties.

Answers

Structure-sensitive properties: yield strength, hardness, tensile strength, ductility, fracture toughness, fatigue strength, creep strength, corrosion resistance, wear resistance, thermal conductivity, electrical conductivity. Structure-insensitive properties: elastic moduli, Poisson's ratio, density, thermal expansion coefficient, specific heat.

1.2 What are the five main generic classes of metals? For each generic class:

(a) give one example of a specific component made from that class;
(b) indicate why that class was selected for the component.

Chapter 2
Metal structures

Introduction

At the end of Chapter 1 we noted that structure-sensitive properties like strength, ductility or toughness depend critically on things like the composition of the metal and on whether it has been heated, quenched or cold formed. Alloying or heat treating work by controlling the *structure* of the metal. Table 2.1 shows the large range over which a material has structure. The bracketed subset in the table can be controlled to give a wide choice of structure-sensitive properties.

Crystal and glass structures

We begin by looking at the smallest scale of controllable structural feature – the way in which the atoms in the metals are packed together to give either a crystalline or a glassy (amorphous) structure. Table 2.2 lists the crystal structures of the pure metals at room temperature. In nearly every case the metal atoms pack into the simple crystal structures of face-centred cubic (f.c.c), body-centred cubic (b.c.c.) or close-packed hexagonal (c.p.h.).

Metal atoms tend to behave like miniature ball-bearings and tend to pack together as tightly as possible. F.c.c. and c.p.h. give the highest possible packing density, with 74% of the volume of the metal taken up by the atomic spheres. However, in some metals, like iron or chromium, the metallic bond has some directionality and this makes the atoms pack into the more open b.c.c. structure with a packing density of 68%.

Table 2.1

Structural feature	Typical scale (m)	
Nuclear structure	10^{-15}	
Structure of atom	10^{-10}	
Crystal or glass structure	10^{-9}	
Structures of solutions and compounds	10^{-9}	Range that can be
Structures of grain and phase boundaries	10^{-8}	controlled to alter
Shapes of grains and phases	10^{-7} to 10^{-3}	properties
Aggregates of grains	10^{-5} to 10^{-2}	
Engineering structures	10^{-3} to 10^{3}	

Table 2.2   Crystal structures of pure metals at room temperature

Pure metal	Structure	Unit cell dimensions (nm)	
		a	c
Aluminium	f.c.c.	0.405	
Beryllium	c.p.h.	0.229	0.358
Cadmium	c.p.h.	0.298	0.562
Chromium	b.c.c.	0.289	
Cobalt	c.p.h.	0.251	0.409
Copper	f.c.c.	0.362	
Gold	f.c.c.	0.408	
Hafnium	c.p.h.	0.320	0.506
Indium	Face-centred tetragonal		
Iridium	f.c.c.	0.384	
Iron	b.c.c.	0.287	
Lanthanum	c.p.h.	0.376	0.606
Lead	f.c.c.	0.495	
Magnesium	c.p.h.	0.321	0.521
Manganese	Cubic	0.891	
Molybdenum	b.c.c.	0.315	
Nickel	f.c.c.	0.352	
Niobium	b.c.c.	0.330	
Palladium	f.c.c.	0.389	
Platinum	f.c.c.	0.392	
Rhodium	f.c.c.	0.380	
Silver	f.c.c.	0.409	
Tantalum	b.c.c.	0.331	
Thallium	c.p.h.	0.346	0.553
Tin	Body-centred tetragonal		
Titanium	c.p.h.	0.295	0.468
Tungsten	b.c.c.	0.317	
Vanadium	b.c.c.	0.303	
Yttrium	c.p.h.	0.365	0.573
Zinc	c.p.h.	0.267	0.495
Zirconium	c.p.h.	0.323	0.515

Some metals have more than one crystal structure. The most important examples are iron and titanium. As Fig. 2.1 shows, iron changes from b.c.c. to f.c.c. at 914°C but goes back to b.c.c. at 1391°C; and titanium changes from c.p.h. to b.c.c. at 882°C. This multiplicity of crystal structures is called *polymorphism*. But it is obviously out of the question to try to control crystal structure simply by changing the temperature (iron is useless as a structural material well below 914°C). Polymorphism can, however, be brought about at room temperature by alloying. Indeed, many stainless steels are f.c.c. rather than b.c.c. and, especially at low temperatures, have much better ductility and toughness than ordinary carbon steels.

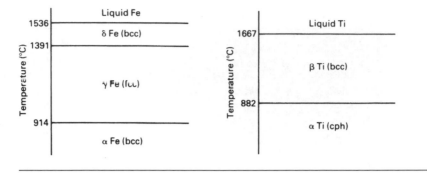

Figure 2.1   Some metals have more than one crystal structure. The most important examples of this *polymorphism* are in iron and titanium.

This is why stainless steel is so good for cryogenic work: the fast fracture of a steel vacuum flask containing liquid nitrogen would be embarrassing, to say the least, but stainless steel is essential for the vacuum jackets needed to cool the latest superconducting magnets down to liquid helium temperatures, or for storing liquid hydrogen or oxygen.

If molten metals (or, more usually, alloys) are cooled very fast – faster than about $10^6\,\mathrm{K\,s^{-1}}$ – there is no time for the randomly arranged atoms in the liquid to switch into the orderly arrangement of a solid crystal. Instead, a *glassy* or *amorphous* solid is produced which has essentially a "frozen-in" liquid structure. This structure – which is termed *dense random packing (drp)* – can be modelled very well by pouring ball-bearings into a glass jar and shaking them down to maximise the packing density. It is interesting to see that, although this structure is disordered, it has well-defined characteristics. For example, the packing density is always 64%, which is why corn was always sold in bushels (1 bushel = 8 UK gallons): provided the corn was always shaken down well in the sack a bushel always gave $0.64 \times 8 = 5.12$ gallons of corn material! It has only recently become practicable to make glassy metals in quantity but, because their structure is so different from that of "normal" metals, they have some very unusual and exciting properties.

Structures of solutions and compounds

As you can see from the tables in Chapter 1, few metals are used in their pure state – they nearly always have other elements added to them which turn them into *alloys* and give them better mechanical properties. The alloying elements will always dissolve in the basic metal to form *solid solution*, although the solubility can vary between <0.01% and 100% depending on the combinations of elements we choose. As examples, the iron in a carbon steel can only dissolve 0.007% carbon at room temperature; the copper in brass can dissolve more than 30% zinc; and the copper–nickel system – the basis of the monels and the cupronickels – has complete solid solubility.

There are two basic classes of solid solution. In the first, small atoms (like carbon, boron and most gases) fit between the larger metal atoms to give *interstitial solid solutions* (Fig. 2.2a). Although this interstitial solubility is usually limited to a few per cent it can have a large effect on properties. Indeed, as we shall see later, interstitial solutions of carbon in iron are largely responsible for the enormous range of strengths that we can get from carbon steels. It is much more common, though, for the dissolved atoms to have a similar size to those of the host metal. Then the dissolved atoms simply replace some of the host atoms to give a *substitutional solid solution* (Fig. 2.2b). Brass and cupronickel are good examples of the large solubilities that this atomic substitution can give.

Solutions normally tend to be *random* so that one cannot predict which of the sites will be occupied by which atoms (Fig. 2.2c). But if A atoms prefer to have A neighbours, or B atoms prefer B neighbours, the solution can *cluster* (Fig. 2.2d); and when A atoms prefer B neighbours the solution can *order* (Fig. 2.2e).

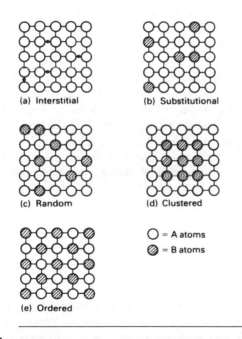

(a) Interstitial

(b) Substitutional

(c) Random

(d) Clustered

(e) Ordered

O = A atoms

⊘ = B atoms

Figure 2.2   Solid-solution structures. In *interstitial* solutions small atoms fit into the spaces between large atoms. In *substitutional* solutions similarly sized atoms replace one another. If A–A, A–B and B–B bonds have the same strength then this replacement is *random*. But unequal bond strengths can give *clustering* or *ordering*.

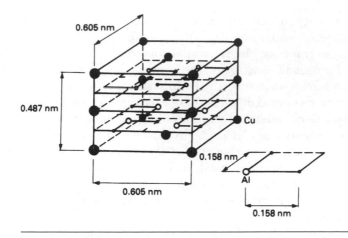

Figure 2.3   The crystal structure of the "intermetallic" compound $CuAl_2$. The structures of compounds are usually more complicated than those of pure metals.

Many alloys contain more of the alloying elements than the host metal can dissolve. Then the surplus must separate out to give regions that have a high concentration of the alloying element. In a few alloys these regions consist of a solid solution based on the *alloying* element. (The lead–tin alloy system, on which most soft solders are based, Table 1.6, is a nice example of this – the lead can only dissolve 2% tin at room temperature and any surplus tin will separate out as regions of *tin* containing 0.3% dissolved lead.) In most alloy systems, however, the surplus atoms of the alloying element separate out as *chemical compounds*. An important example of this is in the aluminium–copper system (the basis of the 2000 series alloys, Table 1.4) where surplus copper separates out as the compound $CuAl_2$. $CuAl_2$ is hard and is not easily cut by dislocations. And when it is finely dispersed throughout the alloy it can give *very* big increases in strength. Other important compounds are Ni_3Al, Ni_3Ti, Mo_2C and TaC (in super-alloys) and Fe_3C (in carbon steels). Figure 2.3 shows the crystal structure of $CuAl_2$. As with most compounds, it is quite complicated.

Phases

The things that we have been talking about so far – metal crystals, amorphous metals, solid solutions, and solid compounds – are all *phases*. A phase is a region of material that has uniform physical and chemical properties. Water is a phase – any one drop of water is the same as the next. Ice is another phase – one splinter of ice is the same as any other. But the mixture of ice and water in your glass at dinner is not a single phase because its properties vary as you move from water to ice. Ice + water is a *two-phase* mixture.

Grain and phase boundaries

A pure metal, or a solid solution, is single-phase. It is certainly possible to make single crystals of metals or alloys but it is difficult and the expense is only worth it for high-technology applications such as single-crystal turbine blades or single-crystal silicon for microchips. Normally, any single-phase metal is *polycrystalline* – it is made up of millions of small crystals, or *grains*, "stuck" together by *grain boundaries* (Fig. 2.4). Because of their unusual structure, grain boundaries have special properties of their own. First, the lower bond density in the boundary is associated with a boundary surface-energy: typically 0.5 Joules per square metre of boundary area ($0.5\,J\,m^{-2}$). Secondly, the more open structure of the boundary can give much faster diffusion in the boundary plane than in the crystal on either side. And finally, the extra space makes it easier for outsized impurity atoms to dissolve in the boundary. These atoms tend to *segregate* to the boundaries, sometimes very strongly. Then an *average* impurity concentration of a few parts per million can give a *local* concentration of 10% in the boundary with very damaging effects on the fracture toughness.*

As we have already seen, when an alloy contains more of the alloying element than the host metal can dissolve, it will split up into *two* phases. The two phases are "stuck" together by *interphase boundaries* which, again, have special properties of their own. We look first at two phases which have different chemical compositions but the same crystal structure (Fig. 2.5a). Provided they are oriented in the right way, the crystals can be made to match up at the boundary. Then, although there is a sharp change in chemical composition, there is no structural change, and the energy of this *coherent* boundary is low (typically $0.05\,J\,m^{-2}$). If the two crystals have slightly different lattice spacings, the boundary is still coherent but has some strain (and more energy)

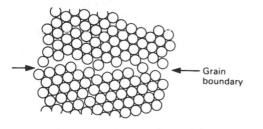

Figure 2.4   The structure of a typical grain boundary. In order to "bridge the gap" between two crystals of different orientation the atoms in the grain boundary have to be packed in a less ordered way. The packing density in the boundary is then as low as 50%.

---

* Henry Bessemer, the great Victorian ironmaster and the first person to mass-produce mild steel, was nearly bankrupted by this. When he changed his suppliers of iron ore, his steel began to crack in service. The new ore contained phosphorus, which we now know segregates badly to grain boundaries. Modern steels must contain less than $\approx 0.05\%$ phosphorus as a result.

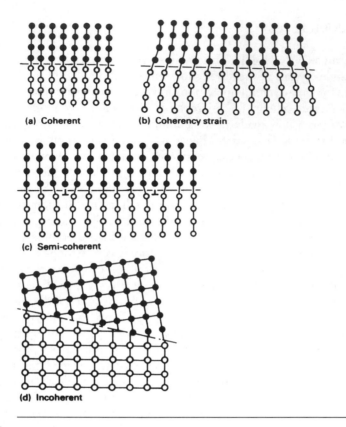

(a) Coherent

(b) Coherency strain

(c) Semi-coherent

(d) Incoherent

Figure 2.5   Structures of interphase boundaries.

associated with it (Fig. 2.5b). The strain obviously gets bigger as the boundary grows sideways: full coherency is usually possible only with small second-phase particles. As the particle grows, the strain builds up until it is relieved by the injection of dislocations to give a *semi-coherent* boundary (Fig. 2.5c). Often the two phases which meet at the boundary are large, and differ in both chemical composition *and* crystal structure. Then the boundary between them is *incoherent*; it is like a grain boundary across which there is also a change in chemical composition (Fig. 2.5d). Such a phase boundary has a high energy – comparable with that of a grain boundary – and around $0.5\,\mathrm{J\,m^{-2}}$.

Shapes of grains and phases

Grains come in all shapes and sizes, and both shape and size can have a big effect on the properties of the polycrystalline metal (a good example is mild steel – its strength can be *doubled* by a ten-times decrease in grain size). Grain shape is strongly affected by the way in which the metal is processed. Rolling oi forging, for instance, can give stretched-out (or "textured") grains; and in casting the solidifying grains are often elongated in the direction of the easiest heat loss. But if there are no external effects like these, then the energy of the grain boundaries is the important thing. This can be illustrated very nicely by looking at a "two-dimensional" array of soap bubbles in a thin glass cell. The soap film minimises its overall energy by straightening out; and at the corners of the bubbles the films meet at angles of 120° to balance the surface tensions (Fig. 2.6a). Of course a polycrystalline metal is three-dimensional, but the same principles apply: the grain boundaries try to form themselves into flat planes, and these planes always try to meet at 120°. A grain shape does indeed exist which not only satisfies these conditions but also packs together to fill space. It has 14 faces, and is therefore called a tetrakaidecahedron (Fig. 2.6b). This shape is remarkable, not only for the properties just given, but because it appears in almost every physical science (the shape of cells in plants, of bubbles in foams, of grains in metals and of Dirichlet cells in solid-state physics).*

If the metal consists of *two* phases then we can get more shapes. The simplest is when a single-crystal particle of one phase forms inside a grain of another

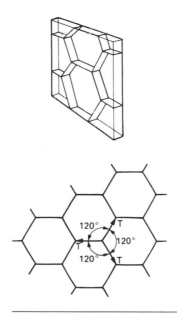

Figure 2.6   (a) The surface energy of a "two-dimensional" array of soap bubbles is minimised if the soap films straighten out. Where films meet the forces of surface tension must balance. This can only happen if films meet in "120° three-somes".

(*Continued*)

* For a long time it was thought that soap foams, grains in metals and so on were icosahedra. It took Lord Kelvin (of the degree K) to get it right.

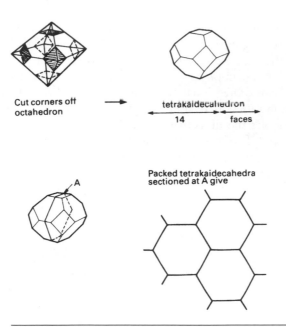

Cut corners off
octahedron   →   tetrakaidecahedron

14   faces

A

Packed tetrakaidecahedra
sectioned at A give

Figure 2.6   (*Continued*) (**b**) In a three-dimensional polycrystal the grain boundary energy is minimised if the boundaries flatten out. These flats must meet in 120° three-somes to balance the grain boundary tensions. If we fill space with equally sized tetrakaidecahedra we will satisfy these conditions. Grains in polycrystals therefore tend to be shaped like tetrakaidecahedra when the grain-boundary energy is the dominating influence.

phase. Then, if the energy of the interphase boundary is *isotropic* (the same for all orientations), the second-phase particle will try to be *spherical* in order to minimise the interphase boundary energy (Fig. 2.7a). Naturally, if coherency is possible along some planes, but not along others, the particle will tend to grow as a *plate*, extensive along the low-energy planes but narrow along the high-energy ones (Fig. 2.7b). Phase shapes get more complicated when interphase boundaries and grain boundaries meet. Figure 2.7(c) shows the shape of a second-phase particle that has formed at a grain boundary. The particle is shaped by two spherical caps which meet the grain boundary at an angle θ. This angle is set by the balance of boundary tensions

$$2\gamma_{\alpha\beta} \cos\theta = \gamma_{gb} \tag{2.1}$$

where $\gamma_{\alpha\beta}$ is the tension (or energy) of the interphase boundary and γ_{gb} is the grain boundary tension (or energy).

In some alloys, $\gamma_{\alpha\beta}$ can be $\leqslant \gamma_{gb}/2$ in which case $\theta = 0$. The second phase will then spread along the boundary as a thin layer of β. This "wetting" of the grain boundary can be a great nuisance – if the phase is brittle then cracks can spread along the grain boundaries until the metal falls apart completely. A favourite scientific party trick is to put some aluminium sheet in a dish of

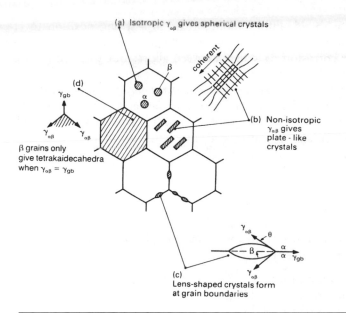

Figure 2.7   Many metals are made up of *two* phases. This figure shows some of the shapes that they can have when boundary energies dominate. To keep things simple we have sectioned the tetrakaidecahedral grains in the way that we did in Fig. 2.6(b). Note that Greek letters are often used to indicate phases. We have called the major phase α and the second phase β. But γ is the symbol for the energy (or tension) of grain boundaries (γ_{gb}) and interphase interfaces ($\gamma_{\alpha\beta}$).

molten gallium and watch the individual grains of aluminium come apart as the gallium whizzes down the boundary.

The second phase can, of course, form complete grains (Fig. 2.7d). But only if $\gamma_{\alpha\beta}$ and γ_{gb} are similar will the phases have tetrakaidecahedral shapes where they come together. In general, $\gamma_{\alpha\beta}$ and γ_{gb} may be quite different and the grains then have more complicated shapes.

Summary: constitution and structure

The structure of a metal is defined by two things. The first is the *constitution*:

(a) The overall composition – the elements (or *components*) that the metal contains and the relative weights of each of them.
(b) The number of phases, and their relative weights.
(c) The composition of each phase.

The second is the geometric information about *shape and size*:

(d) The shape of each phase.
(e) The sizes and spacings of the phases.

Armed with this information, we are in a strong position to re-examine the mechanical properties, and explain the great differences in strength, or toughness, or corrosion resistance between alloys. But where does this information come from? The *constitution* of an alloy is summarised by its phase diagram – the subject of the next chapter. The *shape and size* are more difficult, since they depend on the details of how the alloy was made. But, as we shall see from later chapters, a fascinating range of microscopic processes operates when metals are cast, or worked or heat-treated into finished products; and by understanding these, shape and size can, to a large extent, be predicted.

Examples

2.1 Describe, in a few words, with an example or sketch where appropriate, what is meant by each of the following:

(a) polymorphism;
(b) dense random packing;
(c) an interstitial solid solution;
(d) a substitutional solid solution;
(e) clustering in solid solutions;
(f) ordering in solid solutions;
(g) an intermetallic compound;
(h) a phase in a metal;
(i) a grain boundary;
(j) an interphase boundary;
(k) a coherent interphase boundary;
(l) a semi-coherent interphase boundary;
(m) an incoherent interphase boundary;
(n) the constitution of a metal;
(o) a component in a metal.

2.2 Why do impurity atoms segregate to grain boundaries?

2.3 A large furnace flue operating at 440°C was made from a steel containing 0.10% phosphorus as an impurity. After two years in service, specimens were removed from the flue and tested for fracture toughness. The value of K_c was $30 \, \text{MPa m}^{1/2}$, compared to $100 \, \text{MPa m}^{1/2}$ for new steel. Because of this, the flue had to be scrapped for safety reasons. Explain this dramatic drop in toughness.

2.4 Indicate the shapes that the following adopt *when boundary energies dominate*:

(a) a polycrystalline pure metal (isotropic γ_{gb});
(b) an intermetallic precipitate inside a grain (isotropic $\gamma_{\alpha\beta}$);
(c) an intermetallic precipitate at a grain boundary ($\gamma_{\alpha\beta} > \gamma_{\text{gb}}/2$);
(d) an intermetallic precipitate at a grain boundary ($\gamma_{\alpha\beta} < \gamma_{\text{gb}}/2$).

Chapter 3
Equilibrium constitution and phase diagrams

Introduction

Whenever you have to report on the structure of an alloy – because it is a possible design choice, or because it has mysteriously failed in service – the first thing you should do is reach for its *phase diagram*. It tells you what, at equilibrium, the *constitution* of the alloy should be. The real constitution may not be the equilibrium one, but the equilibrium constitution gives a base line from which other non-equilibrium constitutions can be inferred.

Using phase diagrams is like reading a map. We can explain how they work, but you will not feel confident until you have used them. Hands-on experience is essential. So, although this chapter introduces you to phase diagrams, it is important for you to work through the "Teaching Yourself Phase Diagrams" section at the end of the book. This includes many short examples which give you direct experience of using the diagrams. The whole thing will only take you about four hours and we have tried to make it interesting, even entertaining. But first, a reminder of some essential definitions.

Definitions

An *alloy* is a metal made by taking a pure metal and adding other elements (the "alloying elements") to it. Examples are brass ($Cu + Zn$) and monel ($Ni + Cu$).

The *components* of an alloy are the elements which make it up. In brass, the components are copper and zinc. In monel they are nickel and copper. The components are given the atomic symbols, e.g. Cu, Zn or Ni, Cu.

An *alloy system* is all the alloys you can make with a given set of components: "the Cu–Zn system" describes all the alloys you can make from copper and zinc. A *binary* alloy has two components; a *ternary* alloy has three.

A *phase* is a region of material that has uniform physical and chemical properties. Phases are often given Greek symbols, like α or β. But when a phase consists of a solid solution of an alloying element in a host metal, a clearer symbol can be used. As an example, the phases in the lead–tin system may be symbolised as (Pb) – for the solution of tin in *lead*, and (Sn) – for the solution of lead in *tin*.

The *composition* of an alloy, or of a phase in an alloy, is usually measured in weight %, and is given the symbol W. Thus, in an imaginary A–B alloy system:

$$W_A = \frac{\text{wt of A}}{\text{wt of A} + \text{wt of B}} \times 100\%, \qquad (3.1)$$

$$W_B = \frac{\text{wt of B}}{\text{wt of A} + \text{wt of B}} \times 100\%, \qquad (3.2)$$

and

$$W_A + W_B = 100\%. \qquad (3.3)$$

Sometimes it is helpful to define the atom (or mol)%, given by

$$X_A = \frac{\text{atoms of A}}{\text{atoms of A} + \text{atoms of B}} \times 100\% \qquad (3.4)$$

and so on.

The *constitution* of an alloy is described by

(a) the overall composition;
(b) the number of phases;
(c) the composition of each phase;
(d) the proportion by weight of each phase.

An alloy has its *equilibrium constitution* when there is no further tendency for the constitution to change with time.

The equilibrium diagram or *phase diagram* summarises the equilibrium constitution of the alloy system.

The lead–tin phase diagram

And now for a real phase diagram. We have chosen the lead–tin diagram (Fig. 3.1) as our example because it is pretty straightforward and we already know a bit about it. Indeed, if you have soldered electronic components together or used soldered pipe fittings in your hot-water layout, you will already have had some direct experience of this system.

As in all binary phase diagrams, we have plotted the composition of the alloy on the horizontal scale (in weight %), and the temperature on the vertical scale. The diagram is simply a two-dimensional map (drawn up from experimental data on the lead–tin system) which shows us where the various phases are in composition–temperature space. But how do we use the diagram in practice? As a first example, take an alloy of overall composition 50 wt% lead at 170°C. The *constitution point* (Fig. 3.2a) lies inside a two-phase field. So, *at equilibrium,*

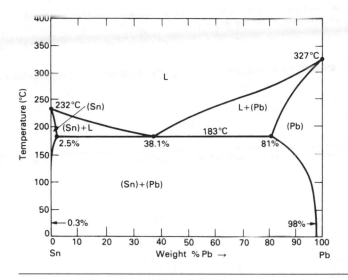

Figure 3.1   The phase diagram for the lead–tin alloy system. There are three phases: L – a liquid solution of lead and tin; (Pb) – a solid solution of tin in lead; and (Sn) – a solid solution of lead in tin. The diagram is divided up into six *fields* – three of them are single-phase, and three are two-phase.

the alloy must be a two-phase mixture: it must consist of "lumps" of (Sn) and (Pb) stuck together. More than this, the diagram tells us (Fig. 3.2b) that the (Sn) phase in our mixture contains 2% lead dissolved in it (it is 98% tin) and the (Pb) phase is 85% lead (it has 15% tin dissolved in it). And finally the diagram tells us (Fig. 3.2c) that the mixture contains 58% by weight of the (Pb) phase. To summarise, then, with the help of the phase diagram, we now know what the equilibrium *constitution* of our alloy is – we know:

(a) the overall composition (50 wt% lead + 50 wt% tin),
(b) the number of phases (two),
(c) the composition of each phase (2 wt% lead, 85 wt% lead),
(d) the proportion of each phase (58 wt% (Pb), 42 wt% (Sn)).

What we *don't* know is how the lumps of (Sn) and (Pb) are sized or shaped. And we can only find *that* out by cutting the alloy open and looking at it with a microscope.*

---

* A whole science, called metallography, is devoted to this. The oldest method is to cut the alloy in half, polish the cut faces, etch them in acid to colour the phases differently, and look at them in the light microscope. But you don't even need a microscope to see some grains. Look at any galvanised steel fire-escape or cast brass door knob and you will see the grains, etched by acid rain or the salts from people's hands.

UXBRIDGE COLLEGE
LEARNING CENTRE

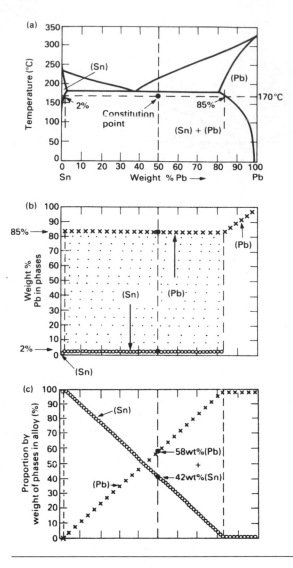

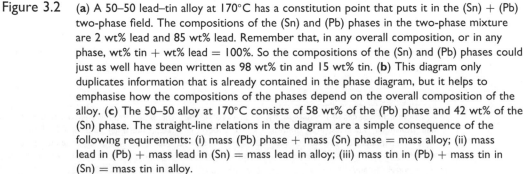

Figure 3.2   (a) A 50–50 lead–tin alloy at 170°C has a constitution point that puts it in the (Sn) + (Pb) two-phase field. The compositions of the (Sn) and (Pb) phases in the two-phase mixture are 2 wt% lead and 85 wt% lead. Remember that, in any overall composition, or in any phase, wt% tin + wt% lead = 100%. So the compositions of the (Sn) and (Pb) phases could just as well have been written as 98 wt% tin and 15 wt% tin. (b) This diagram only duplicates information that is already contained in the phase diagram, but it helps to emphasise how the compositions of the phases depend on the overall composition of the alloy. (c) The 50–50 alloy at 170°C consists of 58 wt% of the (Pb) phase and 42 wt% of the (Sn) phase. The straight-line relations in the diagram are a simple consequence of the following requirements: (i) mass (Pb) phase + mass (Sn) phase = mass alloy; (ii) mass lead in (Pb) + mass lead in (Sn) = mass lead in alloy; (iii) mass tin in (Pb) + mass tin in (Sn) = mass tin in alloy.

Now let's try a few other alloy compositions at 170°C. Using Figs 3.2(b) and 3.2(c) you should be able to convince yourself that the following equilibrium constitutions are consistent.

(a) 25 wt% lead + 75 wt% tin,
(b) two phases,
(c) 2 wt% lead, 85 wt% lead,
(d) 30 wt% (Pb), 70 wt% (Sn).

(a) 75 wt% lead + 25 wt% tin,
(b) two phases,
(c) 2 wt% lead, 85 wt% lead,
(d) 87 wt% (Pb), 13 wt% (Sn).

(a) 85 wt% lead + 15 wt% tin,
(b) one phase (just),
(c) 85 wt% lead,
(d) 100 wt% (Pb).

(a) 95 wt% lead + 5 wt% tin,
(b) one phase,
(c) 95 wt% lead,
(d) 100 wt% (Pb).

(a) 2 wt% lead + 98 wt% tin,
(b) one phase (just),
(c) 2 wt% lead,
(d) 100 wt% (Sn).

(a) 1 wt% lead + 99 wt% tin,
(b) one phase,
(c) 1 wt% lead,
(d) 100 wt% (Sn).

For our second example, we look at alloys at 200°C. We can use exactly the same method, as Fig. 3.3 shows. A typical constitution at 200°C would be:

(a) 50 wt% lead + 50 wt% tin,
(b) two phases,
(c) 45 wt% lead, 82 wt% lead,
(d) 87 wt% (L), 13 wt% (Pb),

and you should have no problem in writing down many others.

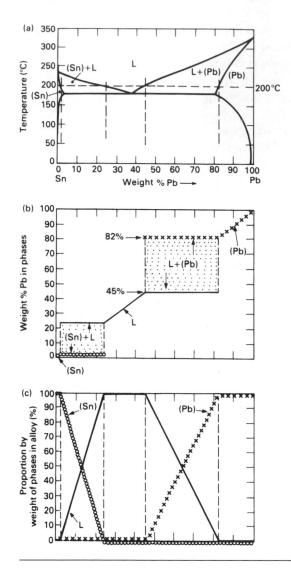

Figure 3.3   Diagrams showing how you can find the equilibrium constitution of any lead–tin alloy at 200°C. Once you have had a little practice you will be able to write down constitutions directly from the phase diagram without bothering about diagrams like (**b**) or (**c**).

Incompletely defined constitutions

There *are* places in the phase diagram where we *can't* write out the full constitution. To start with, let's look at pure tin. At 233°C we have single-phase liquid tin (Fig. 3.4). At 231°C we have single-phase solid tin. At 232°C, the melting point of pure tin, we can either have solid tin about to melt, or

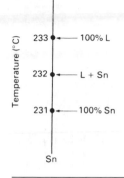

Figure 3.4   At 232°C, the melting point of pure tin, we have a L + Sn two-phase mixture. But, without more information, we can't say what the relative weights of L and Sn *are*.

liquid tin about to solidify, or a mixture of both. If we started with solid tin about to melt we could, of course, supply latent heat of melting at 232°C and get some liquid tin as a result. But the phase diagram knows nothing about external factors like this. Quite simply, the constitution of pure tin at 232°C is incompletely defined because we cannot write down the relative weights of the phases. And the same is, of course, true for pure *lead* at 327°C.

The other place where the constitution is not fully defined is where there is a horizontal line on the phase diagram. The lead–tin diagram has one line like this – it runs across the diagram at 183°C and connects (Sn) of 2.5 wt% lead, L of 38.1% lead and (Pb) of 81% lead. Just above 183°C an alloy of tin + 38.1% lead is single-phase liquid (Fig. 3.5). Just below 183°C it is two-phase, (Sn) + (Pb). At 183°C we have a *three-phase mixture* of L + (Sn) + (Pb) but we can't of course say from the phase diagram what the relative weights of the three phases are.

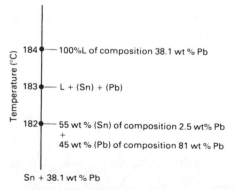

Figure 3.5   At 183°C we have a *three-phase mixture* of L + (Sn) + (Pb). Their relative weights can't be found from the phase diagram.

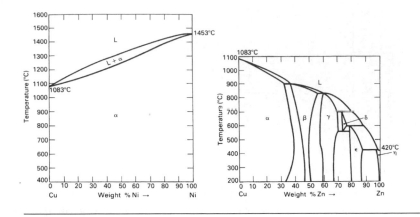

Figure 3.6   (**a**) The copper–nickel diagram is a good deal simpler than the lead–tin one, largely because copper and nickel are completely soluble in one another in the solid state. (**b**) The copper–zinc diagram is much more involved than the lead–tin one, largely because there are extra (*intermediate*) phases in between the end (*terminal*) phases. However, it is still an assembly of single-phase and two-phase fields.

Other phase diagrams

Phase diagrams have been measured for almost any alloy system you are likely to meet: copper–nickel, copper–zinc, gold–platinum, or even water–antifreeze. Some diagrams, like copper–nickel (Fig. 3.6a), are simple; others, like copper–zinc (Fig. 3.6b) are fairly involved; and a few are positively horrendous! But, as Table 3.1 shows, the only real difference between one phase diagram and the next is one of degree.

Table 3.1

Feature	Cu–Ni (Fig. 3.6a)	Pb–Sn (Fig. 3.1)	Cu–Zn (Fig. 3.6b)
Melting points	Two: Cu, Ni	Two: Pb, Sn	Two: Cu, Zn
Three-phase horizontals	None	One: L + (Sn) + (Pb)	Six: $\alpha + \beta + L$ $\beta + \gamma + L$ $\gamma + \delta + L$ $\gamma + \delta + \varepsilon$ $\delta + \varepsilon + L$ $\varepsilon + \eta + L$
Single-phase fields	Two: L, α	Three: L, (Pb), (Sn)	Seven: L, $\alpha, \beta, \gamma, \delta, \varepsilon, \eta$
Two-phase fields	One: L + α	Three: $\begin{cases} L + (Pb) \\ L + (Sn) \\ (Sn) + (Pb) \end{cases}$	Twelve: $\begin{cases} L+\alpha, L+\beta, L+\gamma, \\ L+\delta, L+\varepsilon, L+\eta, \\ \alpha+\beta, \beta+\gamma, \gamma+\delta, \\ \gamma+\varepsilon, \delta+\varepsilon, \varepsilon+\eta \end{cases}$

So that you know where to find the phase diagrams you need we have listed published sources of phase diagrams in Appendix 3. The determination of a typical phase diagram used to provide enough work to keep a doctoral student busy for several years. And yet the most comprehensive of the references runs to over a thousand different phase diagrams!

Examples

Do not attempt examples 3.1 to 3.5 until you have worked through the Teaching Yourself Phase Diagrams section at the end of the book.

3.1 Explain briefly what is meant by the following terms:

(a) a eutectic reaction;
(b) a eutectoid reaction.

3.2 The phase diagram for the copper-antimony system is shown below. The phase diagram contains the intermetallic compound marked "X" on the diagram. Determine the chemical formula of this compound. The atomic weights of copper and antimony are 63.54 and 121.75 respectively.

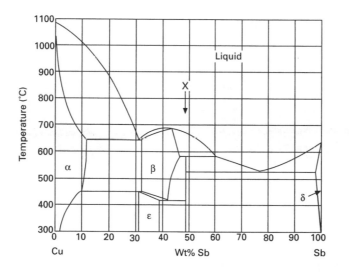

Answer

Cu_2Sb.

3.3   The copper-antimony phase diagram contains two eutectic reactions and one eutectoid reaction. For each reaction:

(a) identify the phases involved;
(b) give the compositions of the phases;
(c) give the temperature of the reaction.

Answers

eutectic at 650°C: L(31% Sb) = α(12% Sb) + β(32% Sb).
eutectic at 520°C: L(77% Sb) = Cu_2Sb + δ(98% Sb).
eutectoid at 420°C: β(42% Sb) = ε(38% Sb) + Cu_2Sb.

3.4   A copper-antimony alloy containing 95 weight% antimony is allowed to cool from 650°C to room temperature. Describe the different phase changes which take place as the alloy is cooled and make labelled sketches of the microstructure to illustrate your answer.

3.5   Sketch a graph of temperature against time for a copper-antimony alloy containing 95 weight% antimony over the range 650°C to 500°C and account for the shape of your plot.

Chapter 4
Case studies in phase diagrams

Introduction

Now for some practical examples of how phase diagrams are used. In the first, a typical design problem, we find out how solders are chosen for different uses. In the second we look at the high-technology area of microchip fabrication and study the production, by zone refining, of ultra-pure silicon. And lastly, for some light-hearted relief, we find out how bubble-free ice is made for up-market cocktails.

Choosing soft solders

Most soft solders are based on our old friend, the lead–tin system (Table 4.1). Soft solders are called "soft" because they are just that – soft mechanically, and soft in the sense that they melt easily. Even so, thin soldered joints can be very strong in shear. And the "thermal" softness of the solder can be a positive advantage. A good example of this is in the soldering of electronic components – like transistors, resistors or integrated circuits – into printed circuit boards. Because the components will be damaged if they get too hot, we want a solder with a low melting point. The phase diagram (Fig. 4.1) shows at a glance that the alloy we want is tin +38% lead. Now that you are familiar with the Teach Yourself course you will know that the particular sort of three-phase horizontal where

$$L \rightleftharpoons \alpha + \beta$$

or, in our case,

$$L \rightleftharpoons (Sn) + (Pb)$$

is called a *eutectic*, from the ancient Greek for "easy melting". And electronic solders are, appropriately enough, marketed under trade names like "Eutectic". But eutectic solders have another property. They become completely molten as soon as they are heated up past 183°C; and they flow nicely into the joints, leaving small tidy fillets of solder around each connection. You can imagine the number of faults that you could get on a crowded circuit board if the solder did *not* flow easily and had to be "put on with a trowel".

There are times, however, when soft solder *is* – almost literally – put on with a trowel. Lead was widely used by the Romans for piping water; and it is only

Table 4.1   Properties of common solders

Type	Composition (wt%)	Melting range (°C)	Typical uses
Soft; eutectic (free-flowing)	62 Sn + 38 Pb	183	Electronic assemblies.
Soft; general-purpose (moderately pasty)	50 Sn + 50 Pb	183–212	Joints in copper water systems; sheet metal work.
Soft; plumbers' (pasty)	35 Sn + 65 Pb	183–244	Wiped joints; car body filling.
Soft; high-melting (free flowing)	5 Sn + 1.5 Ag + 93.5 Pb	296–301	Higher temperatures.
Silver; eutectic (free-flowing)	42 Ag + 19 Cu + 16 Zn + 25 Cd	610–620	High-strength; high-temperature.
Silver; general-purpose (pasty)	38 Ag + 20 Cu + 22 Zn + 20 Cd	605–650	High-strength; high-temperature.

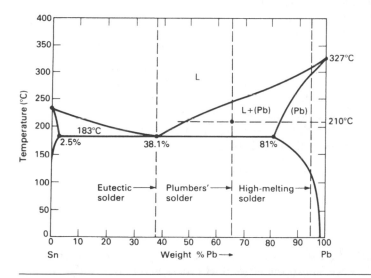

Figure 4.1   The lead–tin phase diagram showing the compositions of the three main soft solders. The diagram tells us that, as soon as *eutectic solder* reaches 183°C, it melts completely and flows easily into joints. *Plumbers' solder*, on the other hand, has a *melting range* – although it starts to melt at 183°C it doesn't become completely molten until it gets up to 244°C. At a middling temperature like 210°C it is a half-solid–half-liquid paste. *High-melting point soft solders* are nearly pure lead. They have a small melting range and, like eutectic solders, flow easily into joints.

within the past fifty years that copper – and more recently polymers – have replaced lead as the major plumbing material. A vital part of the plumber's craft was joining lengths of lead pipe together. This was done by bringing the ends of the pipes up to one another and gradually building a deposit of solder around the joint. The joint was "wiped" by hand with a moleskin pad as fresh solder was added, producing the very fine wiped joints shown in Fig. 4.2. Now, eutectic solder would be useless for this purpose. It would either be fully molten and run all over the place, or it would go solid and stick on the job in unsightly lumps. What is wanted is a *pasty* solder which can be gradually moulded to the shape of the joint. Now, if we look at an alloy of tin +65% lead on the phase diagram we can see that, at around 210°C, the alloy will be a half-molten, half-solid slurry. This will work very nicely to shape without the temperature being too critical. And there is also little chance of melting the lead pipes by accident. So this is the sort of composition that plumbers'

Figure 4.2   When we heat plumbers' solder to about 210°C, and make it "pasty", we can "wipe" it into elegantly curved shapes. Lead pipes have been joined in this way for centuries.

solders have (Table 4.1). Actually, lead–tin alloys with a freezing range like this (called "mushy freezers" in the foundry world) are still used quite widely for things like filling-in dents in car body panels, and soldering car radiators together. But in many cases they are used not so much because they are mushy freezers but because they contain less tin than eutectic solders and are therefore cheaper.

One of the interesting things about the lead–tin diagram (Fig. 4.1) is that all alloys containing between 2.5% and 81% of added lead *start* to melt at 183°C even though some – as we have just seen – don't become completely molten until they are quite a bit hotter. This is obviously a problem when we want to use soft solders in hot surroundings, because they will rapidly lose their strength as they get near to 183°C.

There is a nice illustration of this problem in the model traction engine that we looked at in the first case study in Chapter 1. If you look back you will see that the most critical component in the engine is the copper boiler (Fig. 1.1). The main parts of the boiler are soldered together with a "silver solder" which melts between 610 and 620°C (Table 4.1). This is an extremely strong and ductile alloy which loses little of its structural strength at boiler operating temperatures. But the screwed stays that tie the flat firebox sides together (Figs 1.3, 4.3) present a problem. To make them pressure-tight they need to be sealed with solder. Now, it is easy enough to silver solder the stays where they come out through the outside of the firebox; but it is difficult to do this *inside* the firebox because the gas blow torches that we normally use for silver soldering copper will not burn properly in a confined space (an oxy-acetylene torch will, but it has a very localised flame that can easily burn a hole in thin

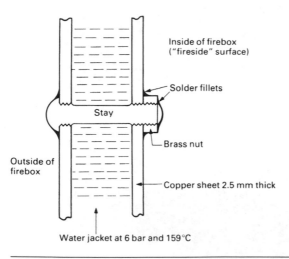

Figure 4.3   The flat plates that make up the firebox of a model steam boiler are tied together with screwed *stays*. To make the threads pressure-tight they must be sealed with solder.

copper). Will soft solder – which of course melts much more easily than silver solder – be adequate for sealing stays inside the firebox?

Model boilers typically work at a gauge pressure of 6 bar. At this pressure the temperature of the saturated water and steam in the boiler is 159°C, only 24°C below the temperature at which lead–tin solders start to melt. Worse, the inner surface of the firebox is hotter even than the water because it is next to the fire. In fact, ordinary soft solders have been used successfully for many years inside model fireboxes, probably because the copper is very good at conducting away the heat of the fire. But there is little room for variation – if you habitually run the boiler on hard scale-forming water, or accidentally let the water level down too far when steaming hard, or increase the boiler pressure and thus temperature to get better performance, you are likely to blow leaks big enough to put the fire out. The solution is to use the high-melting point soft solder in Table 4.1 which is very close to being pure lead, and this is recommended by the leading designers of miniature steam engines.

Pure silicon for microchips

The semiconductor industry would have been impossible had not the process of *zone refining* been invented first. It is the standard way of producing ultrapure materials, both for research and for making silicon and germanium-based devices.

Zone refining works like this. We start with a bar of silicon containing a small uniform concentration C_0 of impurity (Fig. 4.4a). A small tubular electric furnace is put over the left-hand end of the bar, and this is used to melt a short length (Fig. 4.4b). Obviously, the concentration of impurity in this molten section must still be C_0 as no impurity has either left it or come into it. But before we can go any further we must look at the phase diagram (Fig. 4.5). To have equilibrium between the new liquid of composition C_0 and the existing solid in the bar, the solid must be of composition kC_0. And yet the solid already has a composition of C_0, $(1/k)$ times too big. In fact, the situation is rescued because a *local* equilibrium forms between liquid and solid at the interface where they touch. The solid next to this interface loses a small amount of impurity into the liquid by diffusion, and this lowers the composition of the solid to kC_0. Naturally, a gradient is produced in the composition of the solid (Fig. 4.4b), but because solid-state diffusion is relatively sluggish we can neglect the atomic flux that the gradient causes in the present situation.

The next stage in the zone-refining process is to move the furnace slowly and steadily to the right. The left-hand end of the bar will then cool and refreeze but with the equilibrium composition kC_0 (Fig. 4.4c). As the furnace continues to move to the right the freezing solid, because it contains much less impurity than the liquid, rejects the surplus impurity *into* the liquid zone. This has the effect of *increasing* the impurity concentration in the zone, which in turn then increases the impurity concentration in the next layer of freshly frozen solid,

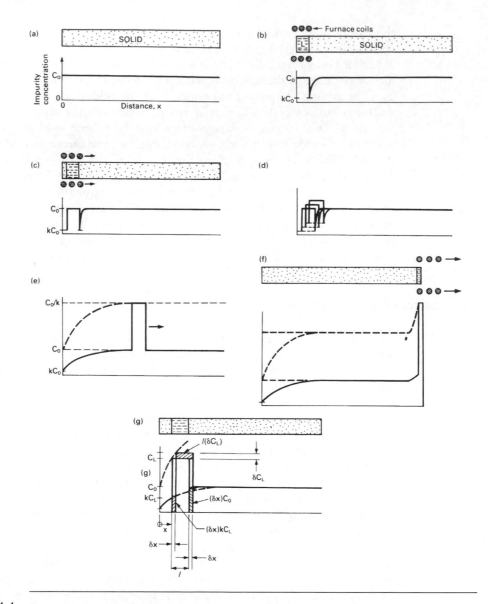

Figure 4.4   Stages in zone refining a bar of impure silicon: (**a**) We start with a bar that has a uniform concentration of impurity, C_0. (**b**) The left-hand end of the bar is melted by a small electric tube furnace, making a liquid *zone*. The bar is encapsulated in a ceramic tube to stop the liquid running away. (**c**) The furnace is moved off to the right, pulling the zone with it. (**d**) As the zone moves, it takes in more impurity from the melted solid on the right than it leaves behind in the freshly frozen solid on the left. The surplus pushes up the concentration of impurity in the zone, which in turn pushes up the concentration of impurity in the next layer of solid frozen from it. (**e**) Eventually we reach steady state. (**f**) When the zone gets to the end of the bar the concentrations in both solid and liquid increase rapidly. (**g**) How we set up eqn. (4.1).

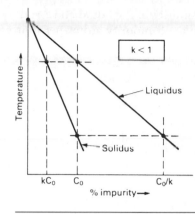

Figure 4.5   Schematic of top left corner of the "silicon-impurity" phase diagram. To make things simple, we assume that the liquidus and solidus lines are straight. The impurity concentration in the solid is then always less than that in the liquid by the factor k (called the *distribution coefficient*).

and so on (Fig. 4.4d). Eventually the concentrations ramp themselves up to the situation shown in Fig. 4.4(e). Here, the solid ahead of the zone has exactly the same composition as the newly frozen solid behind the zone. This means that we have a *steady state* where as much impurity is removed *from* the zone in the form of freshly frozen solid as is taken *into* the zone by melting old solid ahead of it; the composition of the zone itself therefore stays constant.

The final stage takes place when the furnace runs up to the end of the bar (Fig. 4.4f). The freshly frozen solid behind the zone continues to pump surplus impurity into an ever-shortening zone, and the compositions of the liquid and the solid frozen from it ramp themselves up again, eventually reaching in theory (but not in practice) infinite values.

Figure 4.4(f) shows what moving the molten zone along the bar has done to it: we have removed impurity from the left-hand end of the bar and dumped it at the right-hand end; that is, we have *zone refined* the left-hand part of the bar.

Because we need to know how long the refined section of the bar is, it is important to describe the ramping up of the compositions in a quantitative way. We can do this by writing a differential equation which describes what happens as the zone moves from some general position x to a new position $x + \delta x$ (Fig. 4.4g). For a bar of unit cross-section we can write the mass conservation equation

$$(\delta x)\,C_0 \quad - \quad (\delta x)\,k C_L \quad = \quad l(\delta C_L). \qquad (4.1)$$

impurity taken into zone by melting slice of bar of thickness δx	impurity lost from zone in slice of freshly frozen solid of thickness δx	surplus impurity left in fixed-volume zone which increases concentration of impurity in liquid by δC_L

Here C_L is the concentration of impurity in the liquid and l is the zone length. All concentrations are given in units of impurity atoms per unit volume.

Developing eqn. (4.1) we get

$$l(\mathrm{d}C_L) = (C_0 - kC_L)\mathrm{d}x, \tag{4.2}$$

and

$$\frac{l\,\mathrm{d}C_L}{(C_0 - kC_L)} = \mathrm{d}x, \tag{4.3}$$

which we can integrate to give

$$-\frac{l}{k}\ln(C_0 - kC_L) = x + \text{constant}. \tag{4.4}$$

Now the boundary condition is that, when $x = 0$, $C_L = C_0$. Substituting this into eqn. (4.4) allows us to work out what the constant of integration is, and gives the solution

$$-\frac{l}{k}\ln\left(\frac{C_0 - kC_L}{C_0 - kC_0}\right) = x. \tag{4.5}$$

What we really want is the impurity concentration in the solid, C_S. To get this we manipulate eqn. (4.5) as follows.

$$\ln\left(\frac{C_0 - kC_L}{C_0 - kC_0}\right) = -\frac{kx}{l}, \tag{4.6}$$

which we can invert to give

$$C_0 - kC_L = (C_0 - kC_0)\exp\left\{-\frac{kx}{l}\right\}. \tag{4.7}$$

Substituting $C_S = kC_L$ in eqn. (4.7) produces

$$C_0 - C_S = C_0(1 - k)\exp\left\{-\frac{kx}{l}\right\}. \tag{4.8}$$

This gives, at last,

$$C_S = C_0 = \left\{1 - (1 - k)\exp\left\{-\frac{kx}{l}\right\}\right\}, \tag{4.9}$$

which we have plotted schematically in Fig. 4.6.

Figure 4.6 is interesting because it shows that for the best refining performance we need both a long zone and an impurity that is relatively insoluble

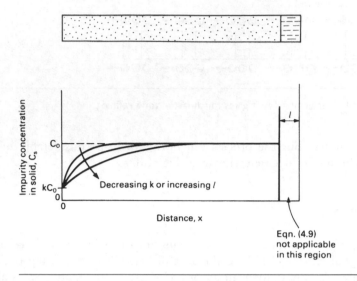

Figure 4.6   Schematic plot of eqn. (4.9) showing how small k and long l give the best zone refining performance.

in the solid (low k). Unfortunately long liquid zones can be destabilised by convection, and impurities with a low k do not come to order! Commercial zone refining processes may therefore involve a large number of passes done one after the other (Fig. 4.7). This obviously adds a lot to the cost of the

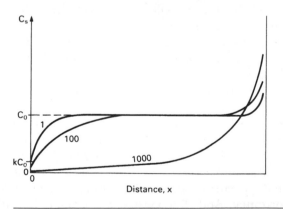

Figure 4.7   If the bar is repeatedly zone refined from left to right then more and more of the impurity will be swept to the right-hand end of the bar. A large number of zone-refining passes may be needed to make the left-hand half of the bar as pure as we need. The right-hand half is cut off and recycled. Note that eqn. (4.9) can only be used to calculate the impurity distribution produced by the *first* pass. A computer program has to be written to handle each subsequent pass.

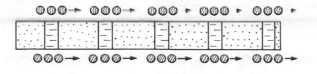

Figure 4.8   A multi-heater arrangement gives much faster zone refining.

pure material, but the process can be speeded up considerably by using the multi-heater arrangement shown in Fig. 4.8.

Making bubble-free ice

People who go in for expensive cocktails like to cool them with ice cubes which are crystal clear – it adds to the aura of bejewelled, refreshing purity. Unfortunately, ice cubes grown in an ordinary fridge are cloudy. So establishments which cater for up-market clients install special machines to make clear ice.

The cloudiness of ordinary ice cubes is caused by thousands of tiny air bubbles. Air dissolves in water, and tap water at 10°C can – and usually *does* – contain 0.0030 wt% of air. In order to follow what this air does when we make an ice cube, we need to look at the phase diagram for the H_2O–air system (Fig. 4.9). As we cool our liquid solution of water + air the first change takes place at about −0.002°C when the composition line hits the liquidus line. At this temperature ice crystals will begin to form and, as the temperature is lowered still further, they will grow. By the time we reach the eutectic three-phase horizontal at −0.0024°C we will have 20 wt% ice (called *primary* ice) in our two-phase mixture, leaving 80 wt% liquid (Fig. 4.9). This liquid will contain the maximum possible amount of dissolved air (0.0038 wt%). As latent heat of freezing is removed at −0.0024°C the three-phase eutectic reaction of

Eutectic liquid $(H_2O + 0.0038$ wt% air$) \rightarrow$ eutectic mixture of (ice) + air

will take place. Finally, when all the eutectic liquid has frozen to ice, we will be left with a two-phase mixture of eutectic [(ice) + air] which we can then cool down below the eutectic temperature of −0.0024°C. In fact, because the solubility of air in ice is so small (Fig. 4.9) almost all the air that *was* dissolved in the eutectic liquid separates out into bubbles of air as soon as we get below the three-phase horizontal. And although this air accounts for only 0.0038% by *weight* of the eutectic mixture, it takes up 2.92% by *volume*. This is why ordinary ice looks cloudy – it contains a network of air bubbles which scatter light very effectively.

Having understood why ordinary ice cubes are cloudy, could we devise a way of making bubble-free ice? Since air bubbles can only form from liquid having the eutectic composition, one obvious way would be to make sure that

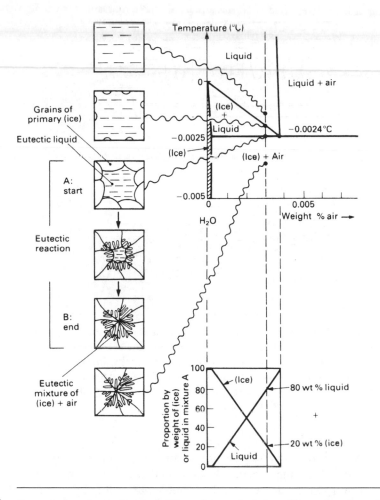

Figure 4.9   Stages in freezing an "ordinary" ice cube can be inferred from the phase diagram for the H_2O + air system.

the liquid into which the ice grains are growing never reaches the eutectic composition. This is in fact the approach used in the two standard ice-making machines (Fig. 4.10). Here, tap water continually flows across the growing ice grains so that the composition of the liquid is always kept at about 0.0030 wt% air. This is safely below the eutectic composition of 0.0038 wt%, and the bubble-free ice thus made is, apparently, to the entire satisfaction of the customers!

Of course, because all liquids dissolve gases, bubbles tend to form not just in ice but in any frozen solid. And this *casting porosity* is a major worry to foundry staff. The approach used in the ice-cube machines is useless for dealing with an enclosed casting, and we have to find other ways. When making high-strength castings the metal is usually melted in a vacuum chamber so that the dissolved

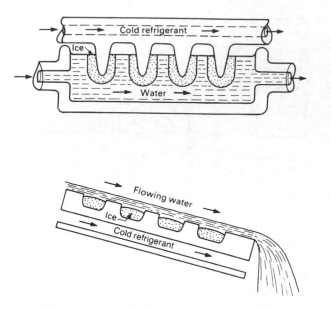

Figure 4.10   The two main types of clear-ice machine. In the top one, the ice grows on the outside of a set of aluminium cold fingers. In the bottom one, the cubes grow in aluminium trays. Cubes are removed when they are big enough by stopping the flow of refrigerant and heating the aluminium electrically. This melts the surface of the ice so that it can fall away from the metal (remember that the easiest way to take the wrapper off an ice lolly is to warm the outside with your hands for a minute!)

gas will diffuse out of the liquid and be pumped away. And for ordinary castings "degassing" chemicals are added which react with the dissolved gases to form gaseous compounds that bubble out of the liquid, or solid compounds that remain in the metal as harmless inclusions.

Examples

4.1   What composition of lead-tin solder is the best choice for joining electronic components? Why is this composition chosen?

4.2   A single-pass zone-refining operation is to be carried out on a uniform bar 2000 mm long. The zone is 2 mm long. Setting the initial impurity concentration $C_0 = 1$ unit of concentration, plot C_S as a function of x for $0 < x < 1000$ mm (a) when $k = 0.01$, (b) when $k = 0.1$.
    [Hint: use eqn. (4.9).]

4.3   A single-pass zone refining operation is to be carried out on a long uniform bar of aluminium containing an even concentration C_0 of copper as a dissolved impurity. The left-hand end of the bar is first melted to produce a short liquid zone of length l and concentration C_L. The zone is then moved along the bar so that fresh solid deposits at the left of the zone and existing solid at the right of the zone melts. The length of the zone remains unchanged. Show that the concentration C_S of the fresh solid is related to the concentration C_L of the liquid from which it forms by the relation

$$C_S = 0.15 C_L.$$

At the end of the zone-refining operation the zone reaches the right-hand end of the bar. The liquid at the left of the zone then begins to solidify so that in time the length of the zone decreases to zero. Derive expressions for the variations of both C_S and C_L with distance x in this final stage. Explain whether or not these expressions are likely to remain valid as the zone length tends to zero.

The aluminium-copper phase diagram is shown below.

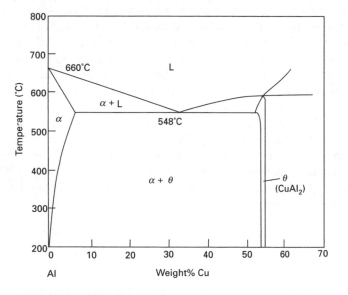

Answer

$$C_L = \frac{C_0}{0.15}\left(\frac{l}{l-x}\right)^{0.85} ; \; C_S = C_0\left(\frac{l}{l-x}\right)^{0.85}.$$

4.4 Up-market radiators for central-heating systems are built up from modules having an internal water space and external heat-transfer fins (typically 2 mm thick and 40 mm deep). They are cast to shape from A1–12 wt% Si alloy (see Fig. A1.31 for the phase diagram). What makes this alloy a good choice for casting such thin sections? Why would A1–6 wt% Si not be such a good choice?

Chapter 5
The driving force for structural change

Introduction

When the structure of a metal changes, it is because there is a *driving force* for the change. When iron goes from b.c.c. to f.c.c. as it is heated, or when a boron dopant diffuses into a silicon semiconductor, or when a powdered superalloy sinters together, it is because each process is pushed along by a driving force.

Now, the mere fact of having a driving force does not guarantee that a change *will* occur. There must also be a *route* that the process can follow. For example, even though boron will *want* to mix in with silicon it can only do this if the route for the process – atomic diffusion – is fast enough. At high temperature, with plenty of thermal energy for diffusion, the doping process *will* be fast; but at low temperature it will be immeasurably slow. The rate at which a structural change actually takes place is then a function of both the *driving force* and the speed, or *kinetics* of the route; and both must have finite values if we are to get a change.

We will be looking at kinetics in Chapter 6. But before we can do this we need to know what we mean by driving forces and how we calculate them. In this chapter we show that driving forces can be expressed in terms of simple thermodynamic quantities, and we illustrate this by calculating driving forces for some typical processes like solidification, changes in crystal structure, and precipitate coarsening.

Driving forces

A familiar example of a change is what takes place when an automobile is allowed to move off down a hill (Fig. 5.1). As the car moves downhill it can be made to do work – perhaps by raising a weight (Fig. 5.1), or driving a machine. This work is called the *free work*, W_f. It is the free work that drives the change of the car going downhill and provides what we term the "driving force" for the change. (The traditional term driving force is rather unfortunate because we don't mean "force", with units of N, but *work*, with units of J).

How can we calculate the free work? The simplest case is when the free work is produced by the decrease of *potential energy*, with

$$W_f = mgh. \tag{5.1}$$

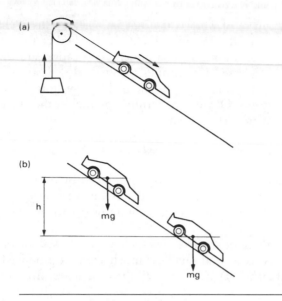

Figure 5.1 (a) An automobile moving downhill can do work. It is this *free work* that drives the process. (b) In the simplest situation the free work can be calculated from the change in potential energy, *mgh*, that takes place during the process.

This equation does, of course, assume that all the potential energy is converted into useful work. This is impossible in practice because some work will be done against friction – in wheel bearings, tyres and air resistance – and the free work must really be written as

$$W_f \le mgh. \tag{5.2}$$

What do we do when there are other ways of doing free work? As an example, if our car were initially moving downhill with velocity v but ended up stationary at the bottom of the hill, we would have

$$W_f \le mgh + \frac{1}{2}mv^2 \tag{5.3}$$

instead. And we could get even more free work by putting a giant magnet at the bottom of the hill! In order to cover all these possibilities we usually write

$$W_f \le -\Delta N, \tag{5.4}$$

where ΔN is the change in the *external energy*. The minus sign comes in because a decrease in external energy (e.g. a decrease in potential energy) gives us a positive output of work W. *External* energy simply means all sources of work that are due solely to directed (i.e. non-random) movements (as in mgh, $\frac{1}{2}mv^2$ and so on).

A quite different source of work is the *internal energy*. This is characteristic of the intrinsic nature of the materials themselves, whether they are moving non-randomly or not. Examples in our present illustration are the chemical energy that could be released by burning the fuel, the elastic strain energy stored in the suspension springs, and the thermal energy stored in the random vibrations of all the atoms. Obviously, burning the fuel in the engine will give us an extra amount of free work given by

$$W_f \leq -\Delta U_b, \tag{5.5}$$

where ΔU_b is the change in internal energy produced by burning the fuel.

Finally, *heat* can be turned into work. If our car were steam-powered, for example, we could produce work by exchanging heat with the boiler and the condenser.

The first law of thermodynamics – which is just a statement of energy conservation – allows us to find out how much work is produced by *all* the changes in N, all the changes in U, and all the heat flows, from the equation

$$W = Q - \Delta U - \Delta N. \tag{5.6}$$

The nice thing about this result is that the inequalities have all vanished. This is because any energy lost in one way (e.g. potential energy lost in friction) must appear somewhere else (e.g. as heat flowing out of the bearings). But eqn. (5.6) gives us the *total* work produced by Q, ΔU and ΔN; and this is not necessarily the *free* work available to drive the change.

In order to see why, we need to look at our car in a bit more detail (Fig. 5.2). We start by assuming that it is surrounded by a large and thermally insulated environment kept at constant thermodynamic temperature T_0 and absolute pressure p_0 (assumptions that are valid for most structural changes in the earth's atmosphere). We define our *system* as: (the automobile + the air needed for burning the fuel + the exhaust gases given out at the back). The system starts off with internal energy U_1, external energy N_1, and volume V_1. As the car travels to the right U, N and V change until, at the end of the change, they end up as U_2, N_2 and V_2. Obviously the total work produced will be

$$W = Q - (U_2 - U_1) - (N_2 - N_1). \tag{5.7}$$

However, the volume of gas put out through the exhaust pipe will be greater than the volume of air drawn in through the air filter and V_2 will be greater than V_1. We thus have to do work W_e in pushing back the environment, given by

$$W_e = p_0(V_2 - V_1). \tag{5.8}$$

The free work, W_f, is thus given by $W_f = W - W_e$, or

$$W_f = Q - (U_2 - U_1) - p_0(V_2 - V_1) - (N_2 - N_1). \tag{5.9}$$

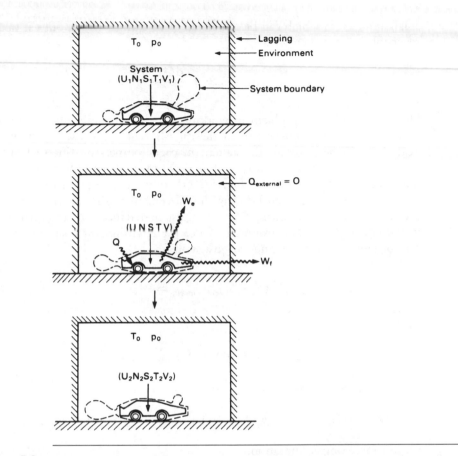

Figure 5.2 Changes that take place when an automobile moves in a thermally insulated environment at constant temperature T_0 and pressure p_0. The environment is taken to be large enough that the change in system volume $V_2 - V_1$ does not increase p_0; and the flow of heat Q across the system boundary does not affect T_0.

Reversibility

A thermodynamic change can take place in two ways – either *reversibly*, or *irreversibly*. In a reversible change, all the processes take place as efficiently as the second law of thermodynamics will allow them to. In this case the second law tells us that

$$\mathrm{d}S = \mathrm{d}Q/T. \tag{5.10}$$

This means that, if we put a small amount of heat $\mathrm{d}Q$ into the system when it is at thermodynamic temperature T we will increase the system entropy by a small amount dS which can be calculated from eqn. (5.10). If our car operates reversibly we can then write

$$S_2 - S_1 = \int_Q \frac{dQ(T)}{T}. \qquad (5.11)$$

However, we have a problem in working out this integral: unless we continuously monitor the movements of the car, we will not know just how much heat dQ will be put into the system in each temperature interval of T to $T + dT$ over the range T_1 to T_2. The way out of the problem lies in seeing that, because $Q_{external} = 0$ (see Fig. 5.2), there is no change in the entropy of the (system + environment) during the movement of the car. In other words, the increase of system entropy $S_2 - S_1$ must be balanced by an equal *decrease* in the entropy of the environment. Since the environment is always at T_0 we do not have to integrate, and can just write

$$(S_2 - S_1)_{environment} = \frac{-Q}{T_0} \qquad (5.12)$$

so that

$$(S_2 - S_1) = -(S_2 - S_1)_{environment} = \frac{Q}{T_0}. \qquad (5.13)$$

This can then be substituted into eqn. (5.9) to give us

$$W_f = -(U_2 - U_1) - p_0(V_2 - V_1) + T_0(S_2 - S_1) - (N_2 - N_1), \qquad (5.14)$$

or, in more compact notation,

$$W_f = -\Delta U - p_0 \Delta V + T_0 \Delta S - \Delta N. \qquad (5.15)$$

To summarise, eqn. (5.15) allows us to find how much *free work* is available for driving a reversible process as a function of the thermodynamic properties of the system (U, V, S, N) and its surroundings (p_0, T_0).

Equation (5.15) was originally derived so that engineers could find out how much work they could get from machines like steam generators or petrol engines. Changes in external energy cannot give continuous outputs of work, and engineers therefore distinguish between ΔN and $-\Delta U - p_0 V + T_0 \Delta S$. They define a function A, called the *availability*, as

$$A \equiv U + p_0 V - T_0 S. \qquad (5.16)$$

The free, or available, work can then be expressed in terms of changes in *availability* and *external energy* using the final result

$$W_f = -\Delta A - \Delta N. \qquad (5.17)$$

Of course, real changes can never be ideally efficient, and some work will be lost in *irreversibilities* (e.g. friction). Equation (5.17) then gives us an over-estimate of W_f. But it is very difficult to calculate irreversible effects in materials processes. We will therefore stick to eqn. (5.17) as the best we can do!

Stability, instability and metastability

The stability of a static mechanical system can, as we know, be tested very easily by looking at how the potential energy is affected by any changes in the orientation or position of the system (Fig. 5.3). The stability of more complex systems can be tested in exactly the same sort of way using W_f (Fig. 5.4).

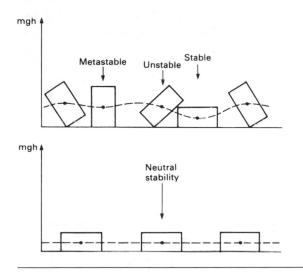

Figure 5.3 Changes in the potential energy of a static mechanical system tell us whether it is in a stable, unstable or metastable state.

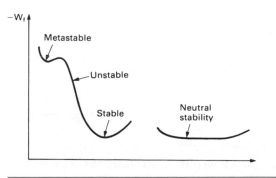

Figure 5.4 The stability of complex systems is determined by changes in the free work W_f. Note the minus sign – systems try to move so that they *produce* the *maximum* work.

The driving force for solidification

How do we actually use eqn. (5.17) to calculate driving forces in materials processes? A good example to begin with is solidification – most metals are melted or solidified during manufacture, and we have already looked at two case studies involving solidification (zone refining, and making bubble-free ice). Let us therefore look at the thermodynamics involved when water solidifies to ice.

We assume (Fig. 5.5) that all parts of the system and of the environment are at the same constant temperature T and pressure p. Let's start with a mixture of ice and water at the melting point T_m (if $p = 1$ atm then $T_m = 273$ K of course). At the melting point, the ice–water system is in a state of *neutral equilibrium*: no free work can be extracted if some of the remaining water is frozen to ice, or if some of the ice is melted to water. If we neglect changes in external energy (freezing ponds don't get up and walk away!) then eqn. (5.17) tells us that $\Delta A = 0$, or

$$(U + pV - T_m S)_{\text{ice}} = (U + pV - T_m S)_{\text{water}}. \tag{5.18}$$

We know from thermodynamics that the enthalpy H is defined by $H \equiv U + pV$, so eqn. (5.18) becomes

$$(H - T_m S)_{\text{ice}} = (H - T_m S)_{\text{water}}. \tag{5.19}$$

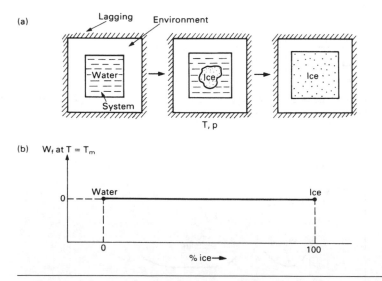

Figure 5.5   (a) Stages in the freezing of ice. All parts of the system and of the environment are at the same constant temperature T and pressure p. (b) An ice–water system at the melting point T_m is in neutral equilibrium.

Thus, for the ice–water change, $\Delta H = T_m \Delta S$, or

$$\Delta S = \frac{\Delta H}{T_m}. \tag{5.20}$$

This is of exactly the same form as eqn. (5.10) and ΔH is simply the "latent heat of melting" that generations of schoolchildren have measured in school physics labs.*

We now take some water at a temperature $T < T_m$. We know that this will have a definite tendency to freeze, so W_f is positive. To *calculate* W_f we have $W_f = -\Delta A$, and $H \equiv U + pV$ to give us

$$W_f = -[(H - TS)_{\text{ice}} - (H - TS)_{\text{water}}], \tag{5.21}$$

or

$$W_f = -\Delta H + T\Delta S. \tag{5.22}$$

If we assume that neither ΔH nor ΔS change much with temperature (which is reasonable for small $T_m - T$) then substituting eqn. (5.20) in eqn. (5.22) gives us

$$W_f(T) = -\Delta H + T\left(\frac{\Delta H}{T_m}\right), \tag{5.23}$$

or

$$W_f(T) = \frac{-\Delta H}{T_m}(T_m - T). \tag{5.24}$$

We can now put some numbers into the equation. Calorimetry experiments tell us that $\Delta H = -334 \,\text{kJ}\,\text{kg}^{-1}$. For water at 272 K, with $T_m - T = 1\,\text{K}$, we find that $W_f = 1.22 \,\text{kJ}\,\text{kg}^{-1}$ (or $22 \,\text{J}\,\text{mol}^{-1}$). 1 kg of water at 272 K thus has 1.22 kJ of free work available to make it turn into ice. The reverse is true at 274 K, of course, where each kg of ice has 1.22 kJ of free work available to make it *melt*.

For large departures from T_m we have to fall back on eqn. (5.21) in order to work out W_f. Thermodynamics people soon got fed up with writing $H - TS$ all the time and invented a new term, the Gibbs function G, defined by

$$G \equiv H - TS. \tag{5.25}$$

---

* To melt ice we have to put heat into the system. This increases the system entropy via eqn. (5.20). Physically, entropy represents *disorder*; and eqn. (5.20) tells us that water is more disordered than ice. We would expect this anyway because the atoms in a liquid are arranged much more chaotically than they are in a crystalline solid. When water freezes, of course, heat *leaves* the system and the entropy *decreases*.

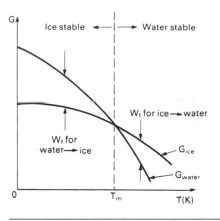

Figure 5.6   Plot of the Gibbs functions for ice and water as functions of temperature. Below the melting point T_m, $G_{water} > G_{ice}$ and ice is the stable state of H_2O; above T_m, $G_{ice} > G_{water}$ and water is the stable state.

Then, for any reversible structural change at constant uniform temperature and pressure

$$W_f = -(G_2 - G_1) = -\Delta G. \qquad (5.26)$$

We have plotted G_{ice} and G_{water} in Fig. 5.6 as a function of temperature in a way that clearly shows how the regions of stability of ice and water are determined by the "driving force", $-\Delta G$.

Solid-state phase changes

We can use exactly the same approach for phase changes in the solid state, like the $\alpha-\gamma$ transformation in iron or the $\alpha-\beta$ transformation in titanium. And, in line with eqn. (5.24), we can write

$$W_f(T) = -\frac{\Delta H}{T_e}(T_e - T), \qquad (5.27)$$

where ΔH is now the latent heat of the *phase transformation* and T_e is the temperature at which the two *solid phases* are in equilibrium. For example, the α and β phases in titanium are in equilibrium at 882°C, or 1155 K. ΔH for the $\alpha-\beta$ reaction is $-3.48\,\text{kJ}\,\text{mol}^{-1}$, so that a departure of 1 K from T_e gives us a W_f of $3.0\,\text{J}\,\text{mol}^{-1}$.

Driving forces for solid-state phase transformations are about one-third of those for solidification. This is just what we would expect: the difference in order between two crystalline phases will be less than the difference in order between a *liquid* and a crystal; the entropy change in the solid-state transformation will be less than in solidification; and $\Delta H/T_t$ will be less than $\Delta H/T_m$.

Precipitate coarsening

Many metals – like nickel-based superalloys, or age-hardened aluminium alloys – depend for their strength on a dispersion of fine second-phase particles. But if the alloys get too hot during manufacture or in service the particles can *coarsen*, and the strength will fall off badly. During coarsening small precipitates shrink, and eventually vanish altogether, whilst large precipitates grow at their expense. Matter is transferred between the precipitates by solid-state diffusion. Figure 5.7 summarises the process. But how do we work out the driving force?

As before, we start with our basic static-system equation

$$W_f = -\Delta A. \tag{5.28}$$

Now the only way in which the system can do free work is by reducing the total energy of α–β interface. Thus

$$\Delta A = 4\pi r_3^2 \gamma - 4\pi r_1^2 \gamma - 4\pi r_2^2 \gamma \tag{5.29}$$

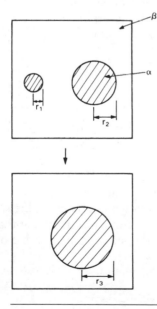

Figure 5.7   Schematic of precipitate coarsening. The small precipitate is shrinking, and the large precipitate is growing at its expense. Material travels between the two by solid-state diffusion.

where γ is the energy of the α–β interface per unit area. Conservation of volume gives

$$\frac{4}{3}\pi r_3^3 = \frac{4}{3}\pi r_1^3 + \frac{4}{3}\pi r_2^3. \tag{5.30}$$

Combining eqns (5.29) and (5.30) gives

$$\Delta A = 4\pi\gamma[(r_1^3 + r_2^3)^{2/3} - (r_1^2 + r_2^2)]. \tag{5.31}$$

For r_1/r_2 in the range 0 to 1 this result is negative. $W_f = -\Delta A$ is therefore positive, and this is what drives the coarsening process.

How large is the driving force for a typical coarsening process? If we put $r_1 = r_2/2$ in eqn. (5.31) we get $\Delta A = -4\pi\gamma(-0.17r_2^2)$. If $\gamma = 0.5\,\text{J}\,\text{m}^{-2}$ and $r_2 = 10^{-7}\,\text{m}$ our two precipitates give us a free work of $10^{-14}\,\text{J}$, or about $7\,\text{J}\,\text{mol}^{-1}$. And this is large enough to make coarsening quite a problem. One way of getting over this is to choose alloying elements that give us *coherent* precipitates. γ is then only about $0.05\,\text{J}\,\text{m}^{-2}$ (see Chapter 2) and this brings W_f down to only $0.7\,\text{J}\,\text{mol}^{-1}$.

Grain growth

The grain boundary energy tied up in a polycrystalline metal works in the same sort of way to give us a driving force for *grain* coarsening. As we shall see in Chapter 13, grain coarsening can cause us big problems when we try to weld high-strength steels together. A typical $\gamma_{\text{gb}}(0.5\,\text{J}\,\text{m}^{-2})$ and grain size $(100\,\mu\text{m})$ give us a W_f of about $2 \times 10^{-2}\,\text{J}\,\text{mol}^{-1}$.

Recrystallisation

When metals are deformed plastically at room temperature the dislocation density goes up enormously (to $\approx 10^{15}\,\text{m}^{-2}$). Each dislocation has a strain energy of about $Gb^2/2$ per unit length and the total dislocation strain energy in a cubic metre of deformed metal is about $2\,\text{MJ}$, equivalent to $15\,\text{J}\,\text{mol}^{-1}$. When cold worked metals are heated to about $0.6T_m$, new strain-free grains nucleate and grow to consume all the cold-worked metal. This is called – for obvious reasons – *recrystallisation*. Metals are much softer when they have been recrystallised (or "annealed"). And provided metals are annealed often enough they can be deformed almost indefinitely.

Sizes of driving forces

In Table 5.1 we have listed typical driving forces for structural changes. These range from $10^6\,\text{J}\,\text{mol}^{-1}$ for oxidation to $2 \times 10^{-2}\,\text{J}\,\text{mol}^{-1}$ for grain growth.

Table 5.1   Driving forces for structural change

Change	$-\Delta G$ (J mol^{-1})
Chemical reaction – oxidation	0 to 10^6
Chemical reaction – formation of intermetallic compounds	300 to 5×10^4
Diffusion in solid solutions (dilute ideal solutions: between solute concentrations $2c$ and c at 1000 K)	6×10^3
Solidification or melting (1°C departure from T_m)	8 to 22
Polymorphic transformations (1°C departure from T_e)	1 to 8
Recrystallisation (caused by cold working)	≈ 15
Precipitate coarsening	0.7 to 7
Grain growth	2×10^{-2}

With such a huge range of driving force we would expect structural changes in materials to take place over a very wide range of timescales. However, as we shall see in the next three chapters, kinetic effects are just as important as driving forces in deciding how fast a structural change will go.

Examples

5.1   Calculate the free work available to drive the following processes *per kg of material.*

(a) Solidification of molten copper at 1080°C. (For copper, $T_m = 1083°C$, $\Delta H = 13.02\,\text{kJ mol}^{-1}$, atomic weight = 63.54).

(b) Transformation from β–Ti to α–Ti at 800°C. (For titanium, $T_e = 882°C$, $\Delta H = 3.48\,\text{kJ mol}^{-1}$, atomic weight = 47.90).

(c) Recrystallisation of cold-worked aluminium with a dislocation density of $10^{15}\,\text{m}^{-2}$. (For aluminium, $G = 26\,\text{GPa}$, $b = 0.286\,\text{nm}$, density = $2700\,\text{kg m}^{-3}$).

Hint – write the units out in all the steps of your working.

Answers

(a) 453 J; (b) 5157 J; (c) 393 J.

5.2   The microstructure of normalised carbon steels contains colonies of "pearlite". Pearlite consists of thin, alternating parallel plates of α-Fe and iron carbide, Fe_3C (see Figs A1.40 and A1.41). When carbon steels containing pearlite are heated at about 700°C for several hours, it is observed that the plates of Fe_3C start to change shape, and eventually become "spheroidised" (each plate turns into a large number of small spheres of Fe_3C). What provides the driving force for this shape change?

5.3 When manufacturing components by the process of *powder metallurgy*, the metal is first converted into a fine powder (by atomising liquid metal and solidifying the droplets). Powder is then compacted into the shape of the finished component, and heated at a high temperature. After several hours, the particles of the powder fuse ("sinter") together to form a polycrystalline solid with good mechanical strength. What provides the driving force for the sintering process?

5.4 In a die-casting operation 0.1 kg charges of molten zinc alloy at 410°C are injected at a rate of 120 charges per hour. The liquid metal has a specific heat of 500 J kg^{-1} K^{-1} and a freezing temperature of 400°C. The latent heat of solidification of the alloy is 2000 J kg^{-1}. The casting is ejected from the die as soon as it is solid. The insulated die is water cooled at a rate $C(T_d - T_0)$, where T_d is the die temperature, $T_0 = 20°C$ is the ambient temperature and C has the value 1 J s^{-1} K^{-1}. Estimate the average die temperature after a long period of steady operation.

Answer

43°C.

Chapter 6
Kinetics of structural change:
I – diffusive transformations

Introduction

The speed of a structural change is important. Some changes occur in only fractions of a second; others are so slow that they become a problem to the engineer only when a component is held at a high temperature for some years. To a geologist the timescale is even wider: during volcanic eruptions, phase changes (such as the formation of glasses) may occur in milliseconds; but deep in the Earth's crust other changes (such as the formation of mineral deposits or the growth of large natural diamonds) occur at rates which can be measured only in terms of millennia.

Predicting the speed of a structural change is rather like predicting the speed of an automobile. The driving force alone tells us nothing about the speed – it is like knowing the energy content of the petrol. To get at the speed we need to understand the details of how the petrol is converted into movement by the engine, transmission and road gear. In other words, we need to know about the *mechanism* of the change.

Structural changes have two types of mechanism: *diffusive* and *displacive*. Diffusive changes require the diffusion of atoms (or molecules) through the material. Displacive changes, on the other hand, involve only the minor "shuffling" of atoms about their original positions and are limited by the propagation of shear waves through the solid at the speed of sound. Most structural changes occur by a diffusive mechanism. But one displacive change is important: the quench hardening of carbon steels is only possible because a displacive transformation occurs during the quench. This chapter and the next concentrate on diffusive transformations; we will look at displacive transformations in Chapter 8.

Solidification

Most metals are melted or solidified at some stage during their manufacture and solidification provides an important as well as an interesting example of a diffusive change. We saw in Chapter 5 that the driving force for solidification was given by

$$W_f = -\Delta G. \qquad (6.1)$$

For small $(T_m - T)$, ΔG was found from the relation

$$\Delta G \approx \frac{\Delta H}{T_m}(T_m - T). \tag{6.2}$$

In order to predict the speed of the process we must find out how quickly individual atoms or molecules diffuse under the influence of this driving force.

We begin by examining the solidification behaviour of a rather unlikely material – phenyl salicylate, commonly called "salol". Although organic compounds like salol are of more interest to chemical engineers than materials people, they provide excellent laboratory demonstrations of the processes which underlie solidification. Salol is a colourless, transparent material which melts at about 43°C. Its solidification behaviour can be followed very easily in the following way. First, a thin glass cell is made up by gluing two microscope slides together as shown in Fig. 6.1. Salol crystals are put into a shallow glass dish which is heated to about 60°C on a hotplate. At the same time the cell is warmed up on the hotplate, and is filled with molten salol by putting it into the dish. (Trapped air can be released from the cell by lifting the open end with a pair of tweezers.) The filled cell is taken out of the dish and the contents are frozen by holding the cell under the cold-water tap. The cell is then put on to a temperature-gradient microscope stage (see Fig. 6.2). The salol above the cold block stays solid, but the solid above the hot block melts. A stationary solid-liquid interface soon forms at the position where $T = T_m$, and can be seen in the microscope.

To get a driving force the cell is pushed towards the cold block, which cools the interface below T_m. The solid then starts to grow into the liquid and the growth speed can be measured against a calibrated scale in the microscope eyepiece. When the interface is cooled to 35°C the speed is about 0.6 mm min^{-1}. At 30°C the speed is 2.3 mm min^{-1}. And the maximum growth speed, of 3.7 mm min^{-1}, is obtained at an interface temperature of 24°C (see Fig. 6.3). At still lower temperatures the speed *decreases*. Indeed, if the interface is cooled to −30°C, there is hardly any growth at all.

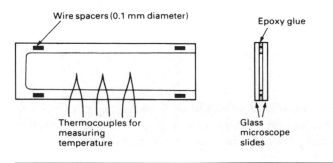

Figure 6.1   A glass cell for solidification experiments.

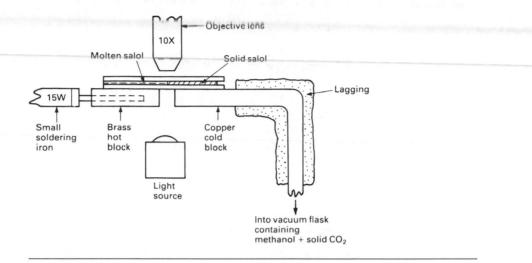

Figure 6.2 The solidification of salol can be followed very easily on a temperature-gradient microscope stage. This can be made up from standard laboratory equipment and is mounted on an ordinary transmission light microscope.

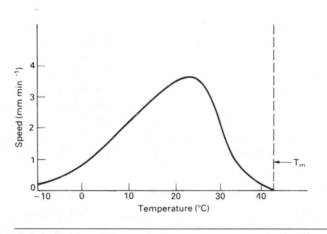

Figure 6.3 The solidification speed of salol at different temperatures.

Equation (6.2) shows that the driving force increases almost linearly with decreasing temperature; and we might well expect the growth speed to do the same. The decrease in growth rate below 24°C is therefore quite unexpected; but it can be accounted for perfectly well in terms of the movements of molecules at the solid–liquid interface. We begin by looking at solid and liquid salol in equilibrium at T_m. Then $\Delta G = 0$ from eqn. (6.2). In other words, if a molecule is taken from the liquid and added to the solid then the change in Gibbs free energy, ΔG, is zero (see Fig. 6.4). However, in order to move from positions in the liquid to positions in the solid, each molecule must first free

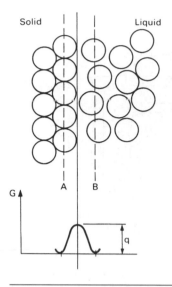

Figure 6.4   Solid and liquid in equilibrium at T_m.

itself from the attractions of the neighbouring liquid molecules: specifically, it must be capable of overcoming the energy barrier q in Fig. 6.4. Due to thermal agitation the molecules vibrate, oscillating about their mean positions with a frequency v (typically about $10^{13}\ \mathrm{s}^{-1}$). The average thermal energy of each molecule is $3kT_m$, where k is Boltzmann's constant. But as the molecules vibrate they collide and energy is continually transferred from one molecule to another. Thus, at any instant, there is a certain probability that a particular molecule has more or less than the average energy $3kT_m$. Statistical mechanics then shows that the probability, p, that a molecule will, at any instant, have an energy $\geqslant q$ is

$$p = \mathrm{e}^{-q/kT_m}. \tag{6.3}$$

We now apply this result to the layer of liquid molecules immediately next to the solid-liquid interface (layer B in Fig. 6.4). The number of liquid molecules that have enough energy to climb over the energy barrier at any instant is

$$n_{\mathrm{B}}p = n_{\mathrm{B}}\mathrm{e}^{-q/kT_m}. \tag{6.4}$$

In order for these molecules to jump from liquid positions to solid positions they must be moving in the correct direction. The number of times each liquid molecule oscillates towards the solid is $v/6$ per second (there are six possible directions in which a molecule can move in three dimensions, only one of which is from liquid to solid). Thus the number of molecules that jump from liquid to solid per second is

$$\frac{v}{6}n_{\mathrm{B}}\mathrm{e}^{-q/kT_m}. \tag{6.5}$$

In the same way, the number of molecules that jump in the reverse direction from solid to liquid per second is

$$\frac{v}{6}n_A e^{-q/kT_m}.$$

(6.6)

The net number of molecules jumping from liquid to solid per second is therefore

$$n_{net} = \frac{v}{6}(n_B - n_A)e^{-q/kT_m}.$$

(6.7)

In fact, because $n_B \approx n_A$, n_{net} is zero at T_m, and the solid–liquid interface is in a state of *dynamic equilibrium*.

Let us now cool the interface down to a temperature $T(<T_m)$, producing a driving force for solidification. This will bias the energies of the A and B molecules in the way shown in Fig. 6.5. Then the number of molecules jumping from liquid to solid per second is

$$\frac{v}{6}n_B e^{-\{q-(1/2)\Delta G\}/kT}$$

(6.8)

and the number jumping from solid to liquid is

$$\frac{v}{6}n_A e^{-\{q+(1/2)\Delta G\}/kT}.$$

(6.9)

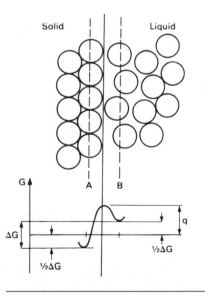

Figure 6.5 A solid–liquid interface at temperature $T(<T_m)$. ΔG is the free work done when one atom or molecule moves from B to A.

The net number jumping from liquid to solid is therefore

$$n_{\text{net}} = \frac{v}{6}n_{\text{B}}e^{-\{q-(1/2)\Delta G\}/kT} - \frac{v}{6}n_{\text{A}}e^{-\{q+(1/2)\Delta G\}/kT}. \tag{6.10}$$

Taking $n_{\text{A}} = n_{\text{B}} = n$ gives

$$n_{\text{net}} = \frac{v}{6}ne^{-q/kT}(e^{\Delta G/2kT} - e^{-\Delta G/2kT}). \tag{6.11}$$

Now ΔG is usually much less than $2kT$, so we can use the approximation $e^x \approx 1 + x$ for small x. Equation (6.11) then becomes

$$n_{\text{net}} = \frac{v}{6}ne^{-q/kT}\frac{\Delta G}{kT}. \tag{6.12}$$

Finally, we can replace ΔG by $\Delta H(T_m - T)/T_m$; and the theory of atomic vibrations tells us that $v \approx kT/h$, where h is Planck's constant. The equation for n_{net} thus reduces to

$$n_{\text{net}} = \frac{n}{6h}e^{-q/kT}\frac{\Delta H(T_m - T)}{T_m}. \tag{6.13}$$

The distance moved by the solid-liquid interface in 1 second is given by

$$v \approx d\frac{n_{\text{net}}}{n}, \tag{6.14}$$

where d is the molecular diameter. So the solidification rate is given by

$$v \approx \frac{d}{6h}e^{-q/kT}\frac{\Delta H(T_m - T)}{T_m}. \tag{6.15}$$

This function is plotted out schematically in Fig. 6.6. Its shape corresponds well with that of the experimental plot in Fig. 6.3. Physically, the solidification rate increases below T_m because the driving force increases as $T_m - T$. But at a low enough temperature the $e^{-q/kT}$ term starts to become important: there is less thermal energy available to help molecules jump from liquid to solid and the rate begins to decrease. At absolute zero there is no thermal energy at all; and even though the driving force is enormous the interface is quite unable to move in response to it.

Heat-flow effects

When crystals grow they give out latent heat. If this is not removed from the interface then the interface will warm up to T_m and solidification will stop. In

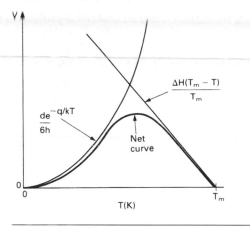

Figure 6.6   How the solidification rate should vary with temperature.

practice, latent heat will be removed from the interface by conduction through the solid and convection in the liquid; and the extent to which the interface warms up will depend on how fast heat is generated there, and how fast that heat is removed.

In chemicals like salol the molecules are elongated (non-spherical) and a lot of energy is needed to rotate the randomly arranged liquid molecules into the specific orientations that they take up in the crystalline solid. Then q is large, $e^{-q/kT}$ is small, and the interface is very sluggish. There is plenty of time for latent heat to flow away from the interface, and its temperature is hardly affected. The solidification of salol is therefore *interface controlled*: the process is governed almost entirely by the kinetics of molecular diffusion at the interface.

In metals the situation is quite the opposite. The spherical atoms move easily from liquid to solid and the interface moves quickly in response to very small undercoolings. Latent heat is generated rapidly and the interface is warmed up almost to T_m. The solidification of metals therefore tends to be *heat-flow controlled* rather than interface controlled.

The effects of heat flow can be illustrated nicely by using sulphur as a demonstration material. A thin glass cell (as in Fig. 6.1, but without any thermocouples) is filled with melted flowers of sulphur. The cell is transferred to the glass plate of an overhead projector and allowed to cool. Sulphur crystals soon form, and grow rapidly into the liquid. The edges of the growing crystals can be seen clearly on the projection screen. As growth progresses, the rate of solidification decreases noticeably – a direct result of the build-up of latent heat at the solid–liquid interface.

In spite of this dominance of heat flow, the solidification speed of pure metals still obeys eqn. (6.15), and depends on temperature as shown in Fig. 6.6. But measurements of $v(T)$ are almost impossible for metals. When the undercooling at the interface is big enough to measure easily ($T_m - T \approx 1°C$) then the velocity of the interface is so large (as much as $1\,\mathrm{m\,s^{-1}}$) that one does not have enough time to measure its temperature. However, as we shall see in a later case study, the kinetics of eqn. (6.15) have allowed the development of a whole new range of *glassy metals* with new and exciting properties.

Solid-state phase changes

We can use the same sort of approach to look at phase changes in solids, like the $\alpha-\gamma$ transformation in iron. Then, as we saw in Chapter 5, the driving force is given by

$$\Delta G \approx \frac{\Delta H}{T_e}(T_e - T).$$

(6.16)

And the speed with which the $\alpha-\gamma$ interface moves is given by

$$v \approx \frac{d}{6h}e^{-q/kT}\frac{\Delta H(T_e - T)}{T_e}$$

(6.17)

where q is the energy barrier at the $\alpha-\gamma$ interface and ΔH is the latent heat of the $\alpha-\gamma$ phase change.

Diffusion-controlled kinetics

Most metals in commercial use contain quite large quantities of impurity (e.g. as alloying elements, or in contaminated scrap). Solid-state transformations in impure metals are usually limited by the diffusion of these impurities through the bulk of the material.

We can find a good example of this *diffusion-controlled* growth in plain carbon steels. As we saw in the "Teaching Yourself Phase Diagrams" course, when steel is cooled below 723°C there is a driving force for the eutectoid reaction of

γ(f.c.c. iron + 0.80 wt% dissolved carbon) $\rightarrow$ α(b.c.c. iron + 0.035 wt% dissolved carbon) + Fe_3C (6.67 wt% carbon).

Provided the driving force is not too large, the α and Fe_3C grow alongside one another to give the layered structure called "pearlite" (see Fig. 6.7). Because the α contains 0.765 wt% carbon less than the γ it must reject this excess carbon into the γ as it grows. The rejected carbon is then transferred to the $Fe_3C-\gamma$ interface, providing the extra 5.87 wt% carbon that the Fe_3C needs to grow. The transformation is controlled by the rate at which the carbon atoms can diffuse through the γ from α to Fe_3C; and the driving force must be

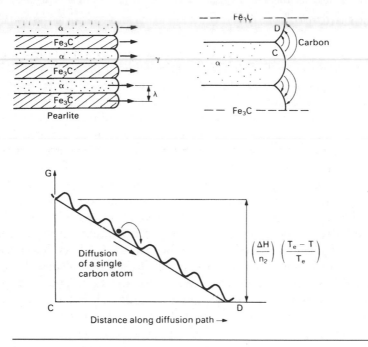

Figure 6.7 How pearlite grows from undercooled Y during the eutectoid reaction. The transformation is limited by diffusion of carbon in the Y, and driving force must be shared between all the diffusional energy barriers. Note that ΔH is in units of J kg^{-1}; n_2 is the number of carbon atoms that diffuse from α to Fe$_3$C when I kg of Y is transformed. $(\Delta H/n_2)([T_e - T]/T_e)$ is therefore the free work done when a *single* carbon atom goes from α to Fe$_3$C.

shared between all the energy barriers crossed by the diffusing carbon atoms (see Fig. 6.7).

The driving force applied across any one energy barrier is thus

$$\left(\frac{1}{n_1}\right)\left(\frac{\Delta H}{n_2}\right)\left(\frac{T_e - T)}{T_e}\right)$$

where n_1 is the number of barriers crossed by a typical carbon atom when it diffuses from α to Fe$_3$C. The speed at which the pearlite grows is then given by

$$v \alpha e^{-q/kT} \frac{\Delta H(T_e - T)}{n_1 n_2 T_e}, \tag{6.18}$$

which has the same form as eqn. (6.15).

Shapes of grains and phases

We saw in Chapter 2 that, when boundary energies were the dominant factor, we could easily predict the shapes of the grains or phases in a material. Isotropic energies gave tetrakaidecahedral grains and spherical (or lens-shaped) second-phase particles. Now boundary energies can only dominate when the material has been allowed to come quite close to equilibrium. If a structural change is taking place the material will *not* be close to equilibrium and the mechanism of the transformation will affect the shapes of the phases produced.

The eutectoid reaction in steel is a good example of this. If we look at the layered structure of pearlite we can see that the flat Fe_3C-α interfaces contain a large amount of boundary energy. The total boundary energy in the steel would be much less if the Fe_3C were accommodated as spheres of Fe_3C rather than extended plates (a sphere gives the minimum surface-to-volume ratio). The high energy of pearlite is "paid for" because it allows the eutectoid reaction to go much more quickly than it would if spherical phases were involved. Specifically, the "co-operative" growth of the Fe_3C and α plates shown in Fig. 6.7 gives a small diffusion distance CD; and, because the transformation is diffusion-limited, this gives a high growth speed. Even more fascinating is the fact that, the bigger the driving force, the finer is the structure of the pearlite (i.e. the smaller is the *interlamellar spacing* λ – see Fig. 6.7). The smaller diffusion distance allows the speed of the reaction to keep pace with the bigger driving force; and this pays for the still higher boundary energy of the structure (as $\lambda \to 0$ the total boundary energy $\to \infty$).

We can see very similar effects during solidification. You may have noticed long rod-shaped crystals of ice growing on the surface of a puddle of water during the winter. These crystals often have side branches as well, and are therefore called "dendrites" after the Greek word meaning "tree". Nearly all metals solidify with a dendritic structure (Fig. 6.8), as do some organic compounds. Although a dendritic shape gives a large surface-to-volume ratio it also encourages latent heat to flow away from the solid–liquid interface (for the same sort of reason that your hands lose heat much more rapidly if you wear gloves than if you wear mittens). And the faster solidification that we get as a consequence "pays for" the high boundary energy. Large driving forces produce fine dendrites – which explains why one can hardly see the dendrites in an iced lollipop grown in a freezer ($-10°C$); but they are obvious on a freezing pond ($-1°C$).

To summarise, the shapes of the grains and phases produced during transformations reflect a balance between the need to minimise the total boundary energy and the need to maximise the speed of transformation. Close to equilibrium, when the driving force for the transformation is small, the grains and phases are primarily shaped by the boundary energies. Far from equilibrium, when the driving force for the transformation is large, the structure depends strongly on the *mechanism* of the transformation. Further,

Figure 6.8   Most metals solidify with a dendritic structure. It is hard to see dendrites growing in metals but they can be seen very easily in transparent organic compounds like camphene which — because they have spherical molecules — solidify just like metals.

even the *scale* of the structure depends on the driving force — the larger the driving force the finer the structure.

Examples

6.1   The solidification speed of salol is about $2.3\,\mathrm{mm\,min^{-1}}$ at $10°C$. Using eqn. (6.15) estimate the energy barrier q that must be crossed by molecules moving from liquid sites to solid sites. The melting point of salol is $43°C$ and its latent heat of fusion is $3.2 \times 10^{-20}\,\mathrm{J\,molecule^{-1}}$. Assume that the molecular diameter is about 1 nm.

Answer

$6.61 \times 10^{-20}\,\mathrm{J}$, equivalent to $39.8\,\mathrm{kJ\,mol^{-1}}$.

6.2   Glass ceramics are a new class of high-technology crystalline ceramic. They are made by taking complex amorphous glasses (like SiO_2–Al_2O_3–Li_2O) and making them devitrify (crystallise). For a particular glass it is found that: (a) no devitrification occurs above a temperature of $1000°C$; (b) the rate of devitrification is a maximum at $950°C$; (c) the rate of devitrification is negligible below $700°C$. Give reasons for this behaviour.

6.3 Samples cut from a length of work-hardened mild steel bar were annealed for various times at three different temperatures. The samples were then cooled to room temperature and tested for hardness. The results are given below.

Annealing temperature (°C)	Vickers hardness	Time at annealing temperature (minutes)
600	180	0
	160	10
	135	20
	115	30
	115	60
620	180	0
	160	4
	135	9
	115	13
	115	26
645	180	0
	160	1.5
	135	3.5
	115	5
	115	10

Estimate the time that it takes for recrystallisation to be completed at each of the three temperatures.

Estimate the time that it would take for recrystallisation to be completed at an annealing temperature of 700°C. Because the new strain-free grains grow by diffusion, you may assume that the rate of recrystallisation follows Arrhenius' law, i.e. the *time* for recrystallisation, t_r, is given by

$$t_r = Ae^{Q/RT},$$

where A is a constant, Q is the activation energy for self-diffusion in the ferrite lattice, R is the gas constant and T is the temperature in kelvin.

Answers

30, 13, 5 minutes respectively; 0.8 minutes.

6.4 Explain the following:

(a) metals usually solidify by the growth of dendrites into the liquid;

(b) the dimensions and spacing of the dendrites decrease as the growth rate increases.

6.5 Explain the following:

    (a) the eutectoid reaction in steel produces pearlite, even though this increases the interfacial energy per unit volume;

    (b) the dimensions and spacing of the plates in pearlite decrease as the growth rate increases.

Chapter 7
Kinetics of structural change: II – nucleation

Introduction

We saw in Chapter 6 that diffusive transformations (like the growth of metal crystals from the liquid during solidification, or the growth of one solid phase at the expense of another during a polymorphic change) involve a mechanism in which atoms are attached to the surfaces of the growing crystals. This means that diffusive transformations can only take place if crystals of the new phase are already present. But how do these crystals – or *nuclei* – form in the first place?

Nucleation in liquids

We begin by looking at how crystals nucleate in liquids. Because of thermal agitation the atoms in the liquid are in a state of continual movement. From time to time a small group of atoms will, purely by chance, come together to form a tiny crystal. If the liquid is above T_m the crystal will, after a very short time, shake itself apart again. But if the liquid is below T_m there is a chance that the crystal will be thermodynamically stable and will continue to grow. How do we calculate the probability of finding stable nuclei below T_m?

There are two work terms to consider when a nucleus forms from the liquid. Equations (6.1) and (6.2) show that work of the type $|\Delta H|(T_m-T)/T_m$ is available to help the nucleus form. If ΔH is expressed as the latent heat given out when *unit volume* of the solid forms, then the total available energy is $(4/3)\pi r^3 |\Delta H|(T_m-T)/T_m$. But this is offset by the work $4\pi r^2 \gamma_{SL}$ needed to create the solid–liquid interface around the crystal. The net work needed to form the crystal is then

$$W_f = 4\pi r^2 \gamma_{SL} - \frac{4}{3}\pi r^3 |\Delta H|\frac{(T_m - T)}{T_m}. \qquad (7.1)$$

This result has been plotted out in Fig. 7.1. It shows that there is a maximum value for W_f corresponding to a critical radius r^*. For $r < r^* (\mathrm{d}W_f/\mathrm{d}r)$ is positive, whereas for $r > r^*$ it is negative. This means that if a random fluctuation produces a nucleus of size $r < r^*$ it will be unstable: the system can do free work if the nucleus loses atoms and r decreases. The opposite is true

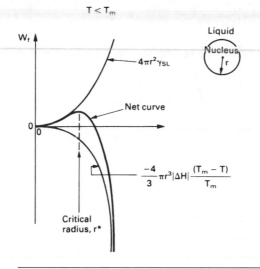

$T < T_m$

W_f

$4\pi r^2 \gamma_{SL}$

Liquid

Nucleus

r

Net curve

0

$\dfrac{-4}{3}\pi r^3 |\Delta H| \dfrac{(T_m - T)}{T_m}$

Critical
radius, r^*

Figure 7.1   The work needed to make a spherical nucleus.

when a fluctuation gives a nucleus with $r > r^*$. Then, free work is done when the nucleus *gains* atoms, and it will tend to grow. To summarise, if random fluctuations in the liquid give crystals with $r > r^*$ then stable nuclei will form, and solidification can begin.

To calculate r^* we differentiate eqn. (7.1) to give

$$\frac{dW_f}{dr} = 8\pi r \gamma_{SL} - 4\pi r^2 |\Delta H| \frac{(T_m - T)}{T_m}. \tag{7.2}$$

We can then use the condition that $dW_f/dr = 0$ at $r = r^*$ to give

$$r^* = \frac{2\gamma_{SL} T_m}{|\Delta H|(T_m - T)} \tag{7.3}$$

for the critical radius.

We are now in a position to go back and look at what is happening in the liquid in more detail. As we said earlier, small groups of liquid atoms are continually shaking themselves together to make tiny crystals which, after a short life, shake themselves apart again. There is a high probability of finding small crystals, but a small probability of finding large crystals. And the probability of finding crystals containing more than 10^2 atoms ($r \gtrsim 1$ nm) is negligible. As Fig. 7.2 shows, we can estimate the temperature T_{hom} at which nucleation will first occur by setting $r^* = 1$ nm in eqn. (7.3). For typical values of γ_{SL}, T_m and ΔH we then find that $T_m - T_{\text{hom}} \approx 100$ K, so an enormous undercooling is needed to make nucleation happen.

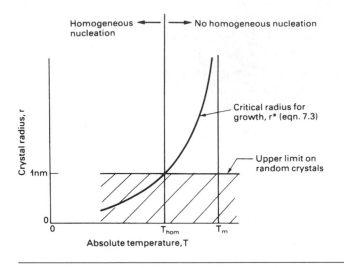

Figure 7.2   Homogeneous nucleation will take place when the random crystals can grow, i.e. when $r > r^*$.

This sort of nucleation – where the only atoms involved are those of the material itself – is called *homogeneous nucleation*. It cannot be the way materials usually solidify because (usually) an undercooling of 1°C or less is all that is needed. Homogeneous nucleation has been observed in ultraclean laboratory samples. But it is the exception, not the rule.

Heterogeneous nucleation

Normally, when a pond of water freezes over, or when a metal casting starts to solidify, nucleation occurs at a temperature only a few degrees below T_m. How do we explain this easy nucleation? Well, liquids like pond water or foundry melts inevitably contain solid particles of dirt. These act as catalysts for nucleation: they give a random crystal a "foothold", so to speak, and allow it to grow more easily. It is this *heterogeneous* nucleation which is responsible for solidification in all practical materials situations.

Heterogeneous nucleation is most likely to occur when there is a strong tendency for the crystal to stick to the surface of the catalyst. This sticking tendency can be described by the angle of contact, θ, shown in Fig. 7.3: the smaller θ, the better the adhesion. Anyone who has tried to get electronic solder to stick to a strip of copper will understand this well. If the copper is tarnished the solder will just roll around as a molten blob with $\theta = 180°$, and will not stick to the surface at all. If the tarnished copper is fluxed then θ may decrease to 90°: the molten solder will stay put on the copper but it will not spread. Only when the copper is both clean and fluxed will θ be zero, allowing the solder to "wet" the copper.

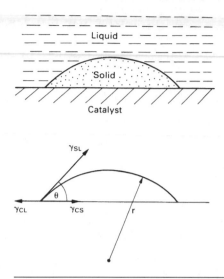

Figure 7.3   Heterogeneous nucleation takes place on the surface of a solid catalyst. For the catalyst to be effective there must be a strong tendency for it to be "wetted" by the crystal, i.e. θ must be small.

If we know the contact angle we can work out r^* quite easily. We assume that the nucleus is a spherical cap of radius r and use standard mathematical formulae for the area of the solid–liquid interface, the area of the catalyst–solid interface and the volume of the nucleus. For $0 \leqslant \theta \leqslant 90°$ these are:

$$\text{solid–liquid area} = 2\pi r^2 (1 - \cos\theta); \tag{7.4}$$

$$\text{catalyst–solid area} = \pi r^2 (1 - \cos^2\theta); \tag{7.5}$$

$$\text{nucleus volume} = \frac{2\pi r^3}{3}\left\{1 - \frac{3}{2}\cos\theta + \frac{1}{2}\cos^3\theta\right\}. \tag{7.6}$$

Then, by analogy with eqn. (7.1) we can write

$$W_f = 2\pi r^2 (1 - \cos\theta)\gamma_{SL} + \pi r^2 (1 - \cos^2\theta)\gamma_{CS} - \pi r^2 (1 - \cos^2\theta)\gamma_{CL}$$

$$- \frac{2\pi r^3}{3}\left\{1 - \frac{3}{2}\cos\theta + \frac{1}{2}\cos^3\theta\right\}|\Delta H|\frac{(T_m - T)}{T_m}. \tag{7.7}$$

Note that this equation has two energy terms that did not appear in eqn. (7.1). The first, $\pi r^2(1 - \cos^2\theta)\gamma_{CS}$, is the energy needed to create the new interface between the catalyst and the solid. The second, $-\pi r^2(1 - \cos^2\theta)\gamma_{CL}$, is the energy released because the area of the catalyst–liquid interface is smaller after nucleation than it was before.

As it stands, eqn. (7.7) contains too many unknowns. But there is one additional piece of information that we can use. The interfacial energies, γ_{SL},

γ_{CS} and γ_{CL} act as surface tensions in just the way that a soap film has both a surface energy and a surface tension. This means that the mechanical equilibrium around the edge of the nucleus can be described by the triangle of forces

$$\gamma_{CL} = \gamma_{CS} + \gamma_{SL} \cos \theta. \qquad (7.8)$$

If we substitute this result into eqn. (7.7) we get the interfacial energy terms to reduce to

$$2\pi r^2 (1 - \cos \theta)\gamma_{SL} + \pi r^2 (1 - \cos^2 \theta)(-\gamma_{SL} \cos \theta),$$

or

$$2\pi r^2 \left\{ 1 - \frac{3}{2}\cos\theta + \frac{1}{2}\cos^3\theta \right\} \gamma_{SL}. \qquad (7.9)$$

The complete result for W_f then becomes

$$W_f = \left\{ 1 - \frac{3}{2}\cos\theta + \frac{1}{2}\cos^3\theta \right\} \left\{ 2\pi r^2 \gamma_{SL} - \frac{2\pi r^3}{3}|\Delta H|\frac{(T_m - T)}{T_m} \right\}. \qquad (7.10)$$

Finally, if we use the condition that $dW_f/dr = 0$ at $r = r^*$ we get

$$r^* = \frac{2\gamma_{SL} T_m}{|\Delta H|(T_m - T)} \qquad (7.11)$$

for the critical radius in heterogeneous nucleation.

If we compare eqns (7.11) and (7.3) we see that the expressions for the critical radius are identical for both homogeneous and heterogeneous nucleation. But the expressions for the *volume* of the critical nucleus are not. For homogeneous nucleation the critical volume is

$$V^*_{hom} = \frac{4}{3}\pi (r^*_{hom})^3 \qquad (7.12)$$

whereas for heterogeneous nucleation it is

$$V^*_{het} = \frac{2}{3}\pi (r^*_{het})^3 \left\{ 1 - \frac{3}{2}\cos\theta + \frac{1}{2}\cos^3\theta \right\}. \qquad (7.13)$$

The maximum statistical fluctuation of 10^2 atoms is the same in both homogeneous and heterogeneous nucleation. If Ω is the volume occupied by one atom in the nucleus then we can easily see that

$$V^*_{hom} = 10^2 \Omega = V^*_{het}. \qquad (7.14)$$

Equating the right hand terms of eqns (7.12) and (7.13) then tells us that

$$r^*_{\text{het}} = \frac{r^*_{\text{hom}}}{\left(\frac{1}{2}\{1 - \frac{3}{2}\cos\theta + \frac{1}{2}\cos^3\theta\}\right)^{1/3}}. \tag{7.15}$$

If the nucleus wets the catalyst well, with $\theta = 10°$, say, then eqn. (7.15) tells us that $r^*_{\text{het}} = 18.1 r^*_{\text{hom}}$. In other words, if we arrange our 10^2 atoms as a spherical cap on a good catalyst we get a much bigger crystal radius than if we arrange them as a sphere. And, as Fig. 7.4 explains, this means that heterogeneous nucleation always "wins" over homogeneous nucleation.

It is easy to estimate the undercooling that we would need to get heterogeneous nucleation with a 10° contact angle. From eqns (7.11) and (7.3) we have

$$\frac{2\gamma_{\text{SL}} T_m}{|\Delta H|(T_m - T_{\text{het}})} = 18.1 \times \frac{2\gamma_{\text{SL}} T_m}{|\Delta H|(T_m - T_{\text{hom}})}, \tag{7.16}$$

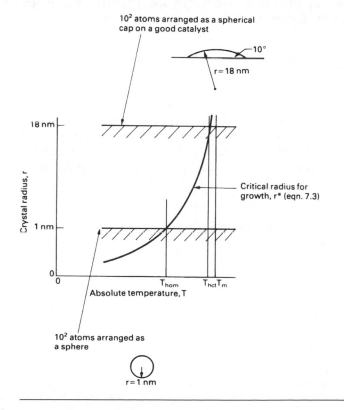

Figure 7.4   Heterogeneous nucleation takes place at higher temperatures because the maximum random fluctuation of 10^2 atoms gives a bigger crystal radius if the atoms are arranged as a spherical cap.

which gives

$$T_m - T_{\text{het}} = \frac{T_m - T_{\text{hom}}}{18.1} \approx \frac{10^2\,\text{K}}{18.1} \approx 5\,\text{K}. \qquad (7.17)$$

And it is nice to see that this result is entirely consistent with the small undercoolings that we usually see in practice.

You can observe heterogeneous nucleation easily in carbonated drinks like "fizzy" lemonade. These contain carbon dioxide which is dissolved in the drink under pressure. When a new bottle is opened the pressure on the liquid immediately drops to that of the atmosphere. The liquid becomes supersaturated with gas, and a driving force exists for the gas to come out of solution in the form of bubbles. The materials used for lemonade bottles – glass or plastic – are poor catalysts for the heterogeneous nucleation of gas bubbles and are usually very clean, so you can swallow the drink before it loses its "fizz". But ordinary blackboard chalk (for example), is an excellent former of bubbles. If you drop such a nucleant into a newly opened bottle of carbonated beverage, spectacular heterogeneous nucleation ensues. Perhaps it is better put another way. Chalk makes lemonade fizz up.

Nucleation in solids

Nucleation in solids is very similar to nucleation in liquids. Because solids usually contain high-energy defects (like dislocations, grain boundaries and surfaces) new phases usually nucleate heterogeneously; homogeneous nucleation, which occurs in defect-free regions, is rare. Figure 7.5 summarises the various ways in which nucleation can take place in a typical polycrystalline solid; and

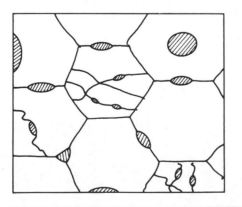

Figure 7.5   Nucleation in solids. Heterogeneous nucleation can take place at defects like dislocations, grain boundaries, interphase interfaces and free surfaces. Homogeneous nucleation, in defect-free regions, is rare.

Examples 7.2 and 7.3 illustrate how nucleation theory can be applied to a solid-state situation.

Summary

In this chapter we have shown that diffusive transformations can only take place if *nuclei* of the new phase can form to begin with. Nuclei form because random atomic vibrations are continually making tiny crystals of the new phase; and if the temperature is low enough these tiny crystals are thermodynamically stable and will grow. In *homogeneous* nucleation the nuclei form as spheres within the bulk of the material. In *heterogeneous* nucleation the nuclei form as spherical caps on defects like solid surfaces, grain boundaries or dislocations. Heterogeneous nucleation occurs much more easily than homogeneous nucleation because the defects give the new crystal a good "foothold". Homogeneous nucleation is rare because materials almost always contain defects.

Postscript

Nucleation – of one sort or another – crops up almost everywhere. During winter plants die and people get frostbitten because ice nucleates heterogeneously inside cells. But many plants have adapted themselves to prevent heterogeneous nucleation; they can survive down to the homogeneous nucleation temperature of −40°C. The "vapour" trails left by jet aircraft consist of tiny droplets of water that have nucleated and grown from the water vapour produced by combustion. Sub-atomic particles can be tracked during high-energy physics experiments by firing them through superheated liquid in a "bubble chamber": the particles trigger the nucleation of gas bubbles which show where the particles have been. And the food industry is plagued by nucleation problems. Sucrose (sugar) has a big molecule and it is difficult to get it to crystallise from aqueous solutions. That is fine if you want to make caramel – this clear, brown, tooth-breaking substance is just amorphous sucrose. But the sugar refiners have big problems making granulated sugar, and will go to great lengths to get adequate nucleation in their sugar solutions. We give examples of how nucleation applies specifically to materials in a set of case studies on phase transformations in Chapter 9.

Examples

7.1 The temperature at which ice nuclei form homogeneously from under-cooled water is −40°C. Find r^* given that $\gamma = 25\,\mathrm{mJ\,m^{-2}}$, $|\Delta H| = 335\,\mathrm{kJ\,kg^{-1}}$, and $T_m = 273\,\mathrm{K}$. Estimate the number of H_2O molecules needed to make a

critical-sized nucleus. Why do ponds freeze over when the temperature falls
below 0°C by only a few degrees?

[The density of ice is $0.92\,\mathrm{Mg\,m^{-3}}$. The atomic weights of hydrogen and
oxygen are 1.01 and 16.00 respectively.]

Answers

r^*, 1.11 nm; 176 molecules.

7.2 An alloy is cooled from a temperature at which it has a single-phase structure
(α) to a temperature at which the equilibrium structure is two-phase ($\alpha + \beta$).
During cooling, small precipitates of the β phase nucleate heterogeneously at
α grain boundaries. The nuclei are lens-shaped as shown below.

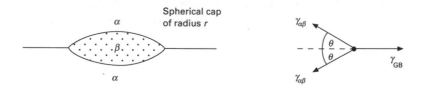

Show that the free work needed to produce a nucleus is given by

$$W_f = \left\{ 1 - \frac{3}{2}\cos\theta + \frac{1}{2}\cos^3\theta \right\}\left\{ 4\pi r^2 \gamma_{\alpha\beta} - \frac{4}{3}\pi r^3 |\Delta G| \right\}$$

where $|\Delta G|$ is the free work produced when unit volume of β forms from α.
You may assume that mechanical equilibrium at the edge of the lens requires
that

$$\gamma_{\mathrm{GB}} = 2\gamma_{\alpha\beta}\cos\theta.$$

Hence, show that the critical radius is given by
$$r^* = 2\gamma_{\alpha\beta}/|\Delta G|.,$$

7.3 Pure titanium is cooled from a temperature at which the b.c.c. phase is stable
to a temperature at which the c.p.h. phase is stable. As a result, lens-shaped
nuclei of the c.p.h. phase form at the grain boundaries. Estimate the number
of atoms needed to make a critical-sized nucleus given the following data:
$\Delta H = 3.48\,\mathrm{kJ\,mol^{-1}}$; atomic weight $= 47.90$; $T_e - T = 30\,\mathrm{K}$; $T_e = 882°\mathrm{C}$;
$\gamma = 0.1\,\mathrm{Jm^{-2}}$; density of the c.p.h. phase $= 4.5\,\mathrm{Mg\,m^{-3}}$; $\theta = 5°$.

Answer

67 atoms.

Chapter 8
Kinetics of structural change: III – displacive transformations

Introduction

So far we have only looked at transformations which take place by diffusion: the so-called *diffusive* transformations. But there is one very important class of transformation – the *displacive* transformation – which can occur without any diffusion at all.

The most important displacive transformation is the one that happens in carbon steels. If you take a piece of 0.8% carbon steel "off the shelf" and measure its mechanical properties you will find, roughly, the values of hardness, tensile strength and ductility given in Table 8.1. But if you test a piece that has been heated to red heat and then quenched into cold water, you will find a dramatic increase in hardness (4 times or more), and a big decrease in ductility (it is practically zero) (Table 8.1).

The two samples have such divergent mechanical properties because they have radically different structures: the structure of the as-received steel is shaped by a diffusive transformation, but the structure of the quenched steel is shaped by a displacive change. But what are displacive changes? And why do they take place?

In order to answer these questions as directly as possible we begin by looking at diffusive and displacive transformations in *pure iron* (once we understand how pure iron transforms we will have no problem in generalising to iron–carbon alloys). Now, as we saw in Chapter 2, iron has different crystal structures at different temperatures. Below 914°C the stable structure is b.c.c., but above 914°C it is f.c.c. If f.c.c. iron is cooled below 914°C the structure becomes thermodynamically unstable, and it tries to change back to b.c.c. This f.c.c. → b.c.c. transformation usually takes place by a diffusive mechanism.

Table 8.1   Mechanical properties of 0.8% carbon steel

Property	As-received	Heated to red heat and water-quenched
H(GPa)	2	9
σ_{TS}(MPa)	600	Limited by brittleness
ε_f%	10	≈ 0

But in exceptional conditions it can occur by a displacive mechanism instead. To understand how iron can transform displacively we must first look at the details of how it transforms by diffusion.

The diffusive f.c.c. → b.c.c. transformation in pure iron

We saw in Chapter 6 that the speed of a diffusive transformation depends strongly on temperature (see Fig. 6.6). The diffusive f.c.c. → b.c.c. transformation in iron shows the same dependence, with a maximum speed at perhaps 700°C (see Fig. 8.1). Now we must be careful not to jump to conclusions about Fig. 8.1. This plots the speed of an individual b.c.c.–f.c.c. interface, measured in metres per second. If we want to know the overall rate of the transformation (the *volume* transformed per second) then we need to know the *area* of the b.c.c.–f.c.c. interface as well.

The total area of b.c.c.–f.c.c. interface is obviously related to the number of b.c.c. nuclei. As Fig. 8.2 shows, fewer nuclei mean a smaller interfacial area and a smaller volume transforming per second. Indeed, if there are no nuclei at all, then the rate of transformation is obviously zero. The overall rate of transformation is thus given approximately by

$$\text{Rate (volume s}^{-1}) \propto \text{No. of nuclei} \times \text{speed of interface.} \qquad (8.1)$$

We know that the interfacial speed varies with temperature; but would we expect the number of nuclei to depend on temperature as well?

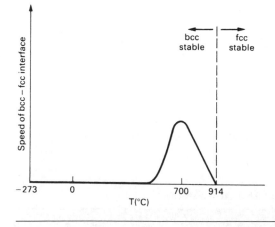

Figure 8.1   The diffusive f.c.c. → b.c.c. transformation in iron. The vertical axis shows the speed of the b.c.c.–f.c.c. interface at different temperatures. Note that the transformation can take place extremely rapidly, making it very difficult to measure the interface speeds. The curve is therefore only semi-schematic.

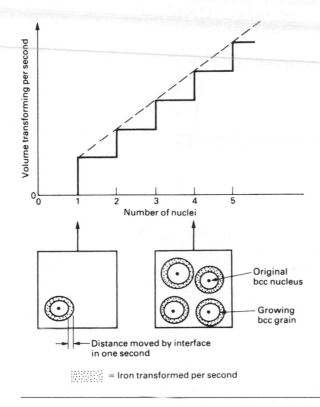

Figure 8.2   In a diffusive transformation the volume transforming per second increases linearly with the number of nuclei.

The nucleation rate is, in fact, critically dependent on temperature, as Fig. 8.3 shows. To see why, let us look at the heterogeneous nucleation of b.c.c. crystals at grain boundaries. We have already looked at grain boundary nucleation in Examples 7.2 and 7.3. Example 7.2 showed that the critical radius for grain boundary nucleation is given by

$$r^* = 2\gamma_{\alpha\beta}/|\Delta G|. \tag{8.2}$$

Since $|\Delta G| = |\Delta H|(T_e - T)/T_e$ (see eqn. 6.16), then

$$r^* = \frac{2\gamma_{\alpha\beta}}{|\Delta H|} \frac{T_e}{(T_e - T)}. \tag{8.3}$$

Grain boundary nucleation will not occur in iron unless it is cooled below perhaps 910°C. At 910°C the critical radius is

$$r^*_{910} = \frac{2\gamma_{\alpha\beta}}{|\Delta H|} \frac{(914 + 273)}{(914 - 910)} = \frac{2\gamma_{\alpha\beta}}{|\Delta H|} \times 297. \tag{8.4}$$

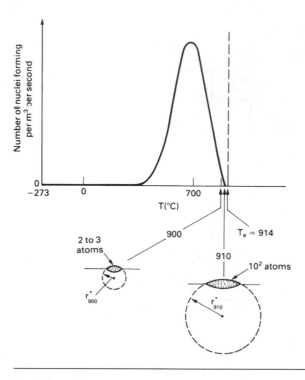

Figure 8.3 The diffusive f.c.c. → b.c.c. transformation in iron: how the number of nuclei depends on temperature (semi-schematic only).

But at 900°C the critical radius is

$$r^*_{900} = \frac{2\gamma_{\alpha\beta}}{|\Delta H|} \frac{(914+273)}{(914-900)} = \frac{2\gamma_{\alpha\beta}}{|\Delta H|} \times 85. \tag{8.5}$$

Thus

$$(r^*_{910}/r^*_{900}) = (297/85) = 3.5. \tag{8.6}$$

As Fig. 8.3 shows, grain boundary nuclei will be geometrically similar at all temperatures. The *volume* V^* of the lens-shaped nucleus will therefore scale as $(r^*)^3$, i.e.

$$(V^*_{910}/V^*_{900}) = 3.5^3 = 43. \tag{8.7}$$

Now, nucleation at 910°C will only take place if we get a random fluctuation of about 10^2 atoms (which is the maximum fluctuation that we can expect in practice). Nucleation at 900°C, however, requires a random fluctuation of

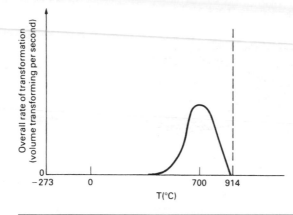

Figure 8.4   The diffusive f.c.c. → b.c.c. transformation in iron: overall rate of transformation as a function of temperature (semi-schematic).

only $(10^2/43)$ atoms.* The chances of assembling this small number of atoms are obviously far greater than the chances of assembling 10^2 atoms, and grain-boundary nucleation is thus much more likely at 900°C than at 910°C. At low temperature, however, the nucleation rate starts to decrease. With less thermal energy it becomes increasingly difficult for atoms to diffuse together to form a nucleus. And at 0 K (where there is no thermal energy at all) the nucleation rate must be zero.

The way in which the overall transformation rate varies with temperature can now be found by multiplying the dependences of Figs 8.1 and 8.3 together. This final result is shown in Fig. 8.4. Below about 910°C there is enough undercooling for nuclei to form at grain boundaries. There is also a finite driving force for the *growth* of nuclei, so the transformation can begin to take place. As the temperature is lowered, the number of nuclei increases, and so does the rate at which they grow: the transformation rate increases. The rate reaches a maximum at perhaps 700°C. Below this temperature diffusion starts to dominate, and the rate decreases to zero at absolute zero.

The time–temperature–transformation diagram

It is standard practice to plot the rates of diffusive transformations in the form of time–temperature–transformation (TTT) diagrams, or "C-curves". Figure 8.5 shows the TTT diagram for the diffusive f.c.c. → b.c.c. transformation in pure

---

* It is really rather meaningless to talk about a nucleus containing only two or three atoms! To define a b.c.c. crystal we would have to assemble at least 20 or 30 atoms. But it will still be far easier to fluctuate 30 atoms into position than to fluctuate 100. Our argument is thus valid qualitatively, if not quantitatively.

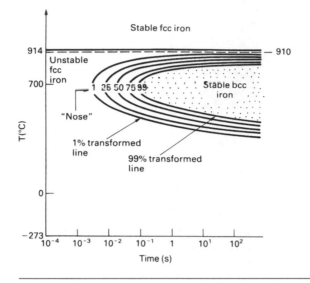

Figure 8.5   The diffusive f.c.c. → b.c.c. transformation in iron: the time–temperature–transformation (TTT) diagram, or "C-curve". The 1% and 99% curves represent, for all practical purposes, the *start* and *end* of the transformation. Semi-schematic only.

iron. The general shape of the C-curves directly reflects the form of Fig. 8.4. In order to see why, let us start with the "1% transformed" curve on the diagram. This gives the time required for 1% of the f.c.c. to transform to b.c.c. at various temperatures. Because the transformation *rate* is zero at both 910°C and −273°C (Fig. 8.4) the *time* required to give 1% transformation must be infinite at these temperatures. This is why the 1% curve tends to infinity as it approaches both 910°C and −273°C. And because the transformation rate is a maximum at say 700°C (Fig. 8.4) the time for 1% transformation must be a minimum at 700°C, which is why the 1% curve has a "nose" there. The same arguments apply, of course, to the 25%, 50%, 75% and 99% curves.

The displacive f.c.c. → b.c.c. transformation

In order to get the iron to transform *displacively* we proceed as follows. We start with f.c.c. iron at 914°C which we then cool to room temperature at a rate of about $10^{5}°C\,s^{-1}$. As Fig. 8.6 shows, we will miss the nose of the 1% curve, and we would expect to end up with f.c.c. iron at room temperature. F.c.c. iron at room temperature would be undercooled by nearly 900°C, and there would be a huge driving force for the f.c.c. → b.c.c. transformation. Even so, the TTT diagram tells us that we might expect f.c.c. iron to survive for years at room temperature before the diffusive transformation could get under way.

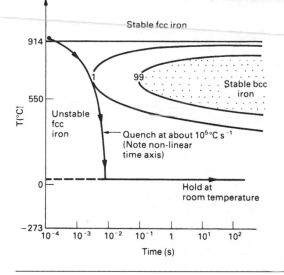

Figure 8.6 If we quench f.c.c. iron from 914°C to room temperature at a rate of about $10^{5}°C\,s^{-1}$ we expect to prevent the diffusive f.c.c. → b.c.c. transformation from taking place. In reality, below 550°C the iron will transform to b.c.c. by a *displacive* transformation instead.

In reality, below 550°C the driving force becomes so large that it cannot be contained; and the iron transforms from f.c.c. to b.c.c. by the *displacive* mechanism. Small lens-shaped grains of b.c.c. nucleate at f.c.c. grain boundaries and move across the grains *at speeds approaching the speed of sound in iron* (Fig. 8.7). In the "switch zone" atomic bonds are broken and remade in such a way that the structure "switches" from f.c.c. to b.c.c.. This is very similar to the breaking and remaking of bonds that goes on when a *dislocation* moves through a crystal. In fact there are strong parallels between *displacive transformations* and *plastic deformation*. Both happen almost instantaneously at speeds that are limited by the propagation of lattice vibrations through the crystal. Both happen at low as well as at high temperatures. And both happen by the precisely sequenced switching of one atom after another. As Table 8.2 shows, most characteristics of displacive transformations are quite different from those of diffusive transformations.

Details of martensite formation

As Fig. 8.8 shows, the martensite lenses are coherent with the parent lattice. Figure 8.9 shows how the b.c.c. lattice is produced by atomic movements of the f.c.c. atoms in the "switch zone". As we have already said, at ≈550°C martensite lenses form and grow almost instantaneously. As the lenses grow the lattice planes distort (see Fig. 8.8) and some of the driving force for the f.c.c. → b.c.c.

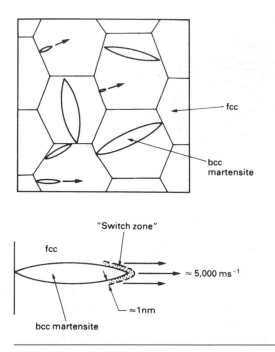

Figure 8.7   The *displacive* f.c.c. → b.c.c. transformation in iron. B.c.c. lenses nucleate at f.c.c. grain boundaries and grow almost instantaneously. The lenses stop growing when they hit the next grain boundary. Note that, when a new phase in *any* material is produced by a displacive transformation it is always referred to as "martensite". Displacive transformations are often called "martensitic" transformations as a result.

transformation is removed as strain energy. Fewer lenses nucleate and grow, and eventually the transformation stops. In other words, provided we keep the temperature constant, the displacive transformation is self-stabilising (see Fig. 8.10). To get more martensite we must cool the iron down to a lower temperature (which gives more driving force). Even at this lower temperature, the displacive transformation will stop when the extra driving force has been used up in straining the lattice. In fact, to get 100% martensite, we have to cool the iron down to $\approx 350°C$ (Fig. 8.10).

The martensite transformation in steels

To make martensite in pure iron it has to be cooled very fast: at about $10^{5}°C\,s^{-1}$. Metals can only be cooled at such large rates if they are in the form of thin foils. How, then, can martensite be made in sizeable pieces of 0.8% carbon steel? As we saw in the "Teaching Yourself Phase Diagrams" course, a 0.8% carbon steel is a "eutectoid" steel: when it is cooled relatively slowly it transforms by diffusion into pearlite (the eutectoid mixture of $\alpha + Fe_3C$).

Table 8.2   Characteristics of transformations

Displacive (also called diffusionless, shear, or martensitic)	Diffusive
Atoms move over distances $\lesssim$ interatomic spacing.	Atoms move over distances of 1 to 10^6 interatomic spacings.
Atoms move by making and breaking interatomic bonds and by minor "shuffling".	Atoms move by thermally activated diffusion from site to site.
Atoms move one after another in precise sequence ("military" transformation).	Atoms hop randomly from site to site (although more hop "forwards" than "backwards") ("civilian" transformation).
Speed of transformation $\approx$ velocity of lattice vibrations through crystal (essentially independent of temperature); transformation can occur at temperatures as low as 4 K.	Speed of transformation depends strongly on temperature; transformation does not occur below $0.3\,T_m$ to $0.4\,T_m$.
Extent of transformation (volume transformed) depends on temperature only.	Extent of transformation depends on time as well as temperature.
Composition cannot change (because atoms have no time to diffuse, they stay where they are).	Diffusion allows compositions of individual phases to change in alloyed systems.
Always specific crystallographic relationship between martensite and parent lattice.	Sometimes have crystallographic relationships between phases.

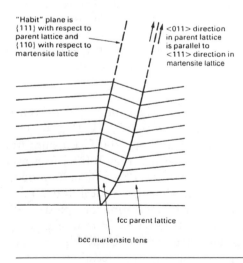

"Habit" plane is {111} with respect to parent lattice and {110} with respect to martensite lattice

<011> direction in parent lattice is parallel to <111> direction in martensite lattice

fcc parent lattice

bcc martensite lens

Figure 8.8   Martensites are always coherent with the parent lattice. They grow as thin lenses on preferred planes and in preferred directions in order to cause the least distortion of the lattice. The crystallographic relationships shown here are for pure iron.

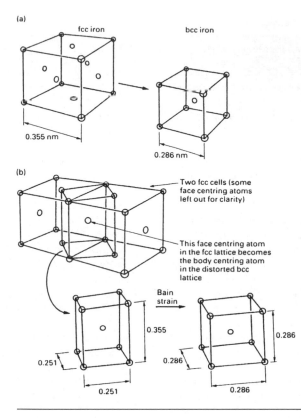

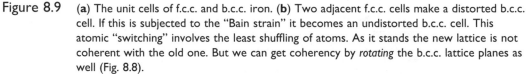

Figure 8.9   (a) The unit cells of f.c.c. and b.c.c. iron. (b) Two adjacent f.c.c. cells make a distorted b.c.c. cell. If this is subjected to the "Bain strain" it becomes an undistorted b.c.c. cell. This atomic "switching" involves the least shuffling of atoms. As it stands the new lattice is not coherent with the old one. But we can get coherency by *rotating* the b.c.c. lattice planes as well (Fig. 8.8).

The eutectoid reaction can only start when the steel has been cooled below 723°C. The nose of the C-curve occurs at ≈ 525°C (Fig. 8.11), about 175°C lower than the nose temperature of perhaps 700°C for pure iron (Fig. 8.5). Diffusion is much slower at 525°C than it is at 700°C. As a result, a cooling rate of ≈ 200°C s^{-1} misses the nose of the 1% curve and produces martensite.

Pure iron martensite has a lattice which is identical to that of ordinary b.c.c. iron. But the displacive and diffusive transformations produce different *large-scale* structures: myriad tiny lenses of martensite instead of large equiaxed grains of b.c.c. iron. Now, fine-grained materials are harder than coarse-grained ones because grain boundaries get in the way of dislocations (see Chapter 2). For this reason pure iron martensite is about twice as hard as ordinary b.c.c. iron. The grain size argument cannot, however, be applied to the 0.8% carbon steel because pearlite not only has a very fine grain size but also contains a large

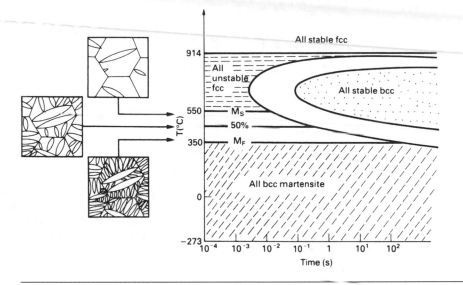

Figure 8.10   The displacive f.c.c. → b.c.c. transformation in iron: the volume of martensite produced is a function of temperature only, and does not depend on time. Note that the temperature at which martensite starts to form is labelled M_s (martensite start); the temperature at which the martensite transformation finishes is labelled M_F (martensite finish).

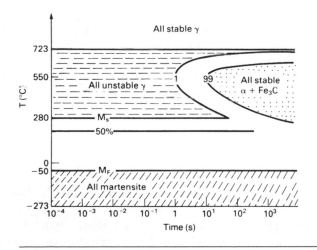

Figure 8.11   The TTT diagram for a 0.8% carbon (eutectoid) steel. We will miss the nose of the 1% curve if we quench the steel at $\approx 200°C\,s^{-1}$. Note that if the steel is quenched into cold water not all the γ will transform to martensite. The steel will contain some "retained" γ which can only be turned into martensite if the steel is cooled below the M_F temperature of $-50°C$.

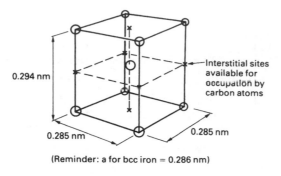

0.294 nm

Interstitial sites available for occupation by carbon atoms

0.285 nm

0.285 nm

(Reminder: a for bcc iron = 0.286 nm)

Figure 8.12   The structure of 0.8% carbon martensite. During the transformation, the carbon atoms put themselves into the interstitial sites shown. To make room for them the lattice stretches along one cube direction (and contracts slightly along the other two). This produces what is called a *face-centred tetragonal* unit cell. Note that only a small proportion of the labelled sites actually contain a carbon atom.

volume fraction of the hard iron carbide phase. Yet 0.8% carbon martensite is *five* times harder than pearlite. The explanation lies with the 0.8% carbon. Above 723°C the carbon dissolves in the f.c.c. iron to form a random solid solution. The carbon atoms are about 40% smaller in diameter than the iron atoms, and they are able to squeeze into the space between the iron atoms to form an interstitial solution. When the steel is quenched, the iron atoms transform displacively to martensite. It all happens so fast that the carbon atoms are frozen in place and remain in their original positions. Under normal conditions b.c.c. iron can only dissolve 0.035% carbon.* The martensite is thus grossly oversaturated with carbon and something must give. Figure 8.12 shows what happens. The carbon atoms make room for themselves by stretching the lattice along one of the cube directions to make a *body-centred tetragonal* unit cell. Dislocations find it very difficult to move through such a highly strained structure, and the martensite is very hard as a result.

A martensite miscellany

Martensite transformations are not limited just to metals. Some ceramics, like zirconia, have them; and even the obscure system of (argon + 40 atom% nitrogen) forms martensite when it is cooled below 30 K. Helical protein crystals in some bacteria undergo a martensitic transformation and the shape change helps the bacteria to burrow into the skins of animals and people!

---

* This may seem a strange result – after all, only 68% of the volume of the b.c.c. unit cell is taken up by atoms, whereas the figure is 74% for f.c.c. Even so, the largest holes in b.c.c. (diameter 0.0722 nm) are smaller than those in f.c.c. (diameter 0.104 nm).

The martensite transformation in steel is associated with a volume change which can be made visible by a simple demonstration. Take a 100 mm length of fine piano wire and run it horizontally between two supports. Hang a light weight in the middle and allow a small amount of slack so that the string is not quite straight. Then connect the ends of the string to a variable low-voltage d.c. source. Rack up the voltage until the wire glows bright red. The wire will sag quite a bit as it expands from room temperature to 800°C. Then cut the power. The wire will cool rapidly and as the γ contracts the wire will move upwards. Below M_s the γ will transform almost instantaneously to martensite. The wire will move sharply downwards (because the martensite occupies a bigger volume than the γ). It can be seen to "shiver" as the shear waves run through the wire, and if you listen very hard you can hear a faint "pinging" sound as well. Down at room temperature you should be able to snap the wire easily, showing that you have indeed made brittle martensite.

One of the unique features of martensite transformations is that they are *structurally reversible*. This means that if we *displacively* transform martensite back to the high-temperature phase each atom in the martensite will go back to its original position in the high-temperature lattice (see Fig. 8.13). In fact, small heat engines have been made to work using this principle – as the alloy is cycled between high and low temperatures, the structure cycles between the high-temperature phase and martensite and the alternating shape change makes the mechanism move back and forth. A range of so-called "memory" alloys has been developed which apply this principle to devices ranging from

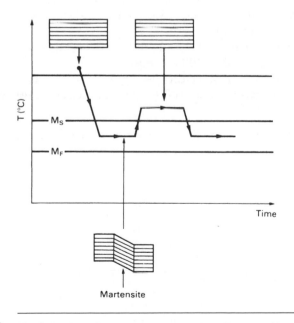

Figure 8.13 Displacive transformations are geometrically reversible.

self-closing rivets to self-erecting antennae for spacecraft. This allows of strange psychokinetic jokes: a bent (but straightened out) tea spoon made of memory-steel, inserted into a cup of hot tea, spontaneously distorts (remembering its earlier bent shape), thereby embarrassing guests and affording its owner the kind of pleasure that only practical jokers can comprehend.

Examples

8.1 Compare and contrast the main features of

(a) diffusive transformations,
(b) displacive transformations.

8.2 Describe the structure of 0.8% carbon martensite

(a) at the nm-scale level,
(b) at the μm-scale level.

Why is 0.8% carbon martensite approximately five times harder than pearlite?

8.3 Sketch the time-temperature-transformation (TTT) diagram for a plain carbon steel of eutectoid composition which exhibits the following features:

(i) At 650°C, transformation of the austenite (fcc phase) is 1% complete after 10 seconds and is 99% complete after 100 seconds.
(ii) The fastest rate of transformation occurs at 550°C. At this temperature, transformation of the austenite is 1% complete after 1 second and is 99% complete after 10 seconds.
(iii) At 360°C, transformation of the austenite is 1% complete after 10 seconds and is 99% complete after 300 seconds.
(iv) The martensite start temperature is 240°C. 90% of the austenite has transformed to martensite at 100°C. The martensite finish temperature is 50°C.

Explain briefly the shape of the lines drawn on the TTT diagram.

Chapter 9
Case studies in phase transformations

Introduction

We now apply the thermodynamic and kinetic theory of Chapters 5–8 to four problems: making rain; getting fine-grained castings; growing crystals for semiconductors; and making amorphous metals.

Making rain

Often, during periods of drought, there is enough moisture in the atmosphere to give clouds, but rain does not fall. The frustration of seeing the clouds but not getting any rain has stimulated all manner of black magic: successful rainmakers have been highly honoured members of society since society existed. But it is only recently that rain-making has acquired some scientific basis. The problem is one of heterogeneous nucleation.

A cloud is cloudy because it is a suspension of vast numbers of minute, spherical water droplets. The droplets are too small and light to fall under gravity; and they are stable, that is to say they do not coarsen and become large enough to fall.*

Rain falls when (surprisingly) the droplets freeze to ice. If a droplet freezes it becomes a stable ice nucleus which then grows by attracting water vapour from the surrounding droplets. The ice particle quickly grows to a size at which gravity can pull it down, and it falls. On cold days it falls as hail or snow; but in warmer weather it remelts in the warmer air near the ground and falls as rain. The difficult stage in making rain is getting the ice to nucleate in the first place. If the water droplets are clean then they will not contain any heterogeneous nucleation catalysts. Ice can then only form if the cloud is cooled to the homogeneous nucleation temperature, and that is a very low temperature (−40°C; see Example 7.1). Clouds rarely get this cold. So the ice must nucleate on something, heterogeneously. Industrial pollution will do: the smoke of a steel works contains so much dirt that the rainfall downwind of it is significantly increased (Fig. 9.1). But building a steelworks to produce rain is unnecessarily extravagant. There are cheaper ways.

---

* We saw in Chapter 5 that there is a driving force tending to make dispersions of precipitates in alloys *coarsen*; and we would expect a dispersion of droplets in water vapour to do the same. Water droplets in clouds, however, carry electrostatic charges; and this gives a different result for the driving force.

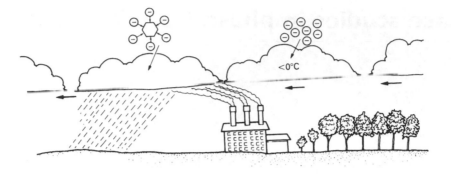

Figure 9.1   Rain falls when the water droplets in clouds turn to ice. This can only happen if the clouds are below 0°C to begin with. If the droplets are clean, ice can form only in the unlikely event that the clouds cool down to the homogeneous nucleation temperature of −40°C. When dust particles are present they can catalyse nucleation at temperatures quite close to 0°C. This is why there is often heavy rainfall downwind of factory chimneys.

The crystal structure of ice is hexagonal, with lattice constants of $a = 0.452$ nm and $c = 0.736$ nm. The inorganic compound silver iodide also has a hexagonal structure, with lattice constants ($a = 0.458$ nm, $c = 0.749$ nm) that are almost identical to those of ice. So if you put a crystal of silver iodide into supercooled water, it is almost as good as putting in a crystal of ice: more ice can grow on it easily, at a low undercooling (Fig. 9.2).

This is an example of heterogeneous nucleation. The good matching between ice and silver iodide means that the interface between them has a low energy: the contact angle is very small and the undercooling needed to nucleate ice decreases from 40°C to 4°C. In artificial rainmaking silver iodide, in the form of a very fine powder of crystals, is either dusted into the cloud from a plane flying above it, or is shot into it with a rocket from below. The powder "seeds" ice crystals which grow, and start to fall, taking the silver iodide with them. But if the ice, as it grows, takes on snow-flake forms, and the tips of the snow flakes break off as they fall, then the process (once started) is self-catalysing: each old generation of falling ice crystals leaves behind a new generation of tiny ice fragments to seed the next lot of crystals, and so on.

There are even better catalysts for ice nucleation than silver iodide. The most celebrated ice nucleating catalyst, produced by the microorganism *Pseudomonas syringae*, is capable of forming nuclei at undetectably small undercoolings. The organism is commonly found on plant leaves and, in this situation, it is a great nuisance: the slightest frost can cause the leaves to freeze and die. A mutant of the organism has been produced which lacks the ability to nucleate ice (the so-called "ice-minus" mutant). American bio-engineers have proposed that the ice-minus organism should be released into the wild, in the hope that it will displace the natural organism and solve the frost-damage problem; but environmentalists have threatened law suits if this goes ahead.

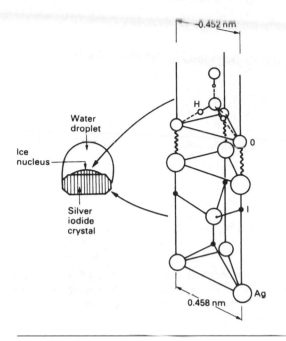

Figure 9.2   The excellent crystallographic matching between silver iodide and ice makes silver iodide a very potent nucleating agent for ice crystals. When clouds at sub-zero temperatures are seeded with AgI dust, spectacular rainfall occurs.

Interestingly, ice nucleation in organisms is not always a bad thing. Take the example of the alpine plant *Lobelia teleki*, which grows on the slopes of Mount Kenya. The ambient temperature fluctuates daily over the range $-10°C$ to $+10°C$, and subjects the plant to considerable physiological stress. It has developed a cunning response to cope with these temperature changes. The plant manufactures a potent biogenic nucleating catalyst: when the outside temperature falls through $0°C$ some of the water in the plant freezes and the latent heat evolved stops the plant cooling any further. When the outside temperature goes back up through $0°C$, of course, some ice melts back to water; and the latent heat absorbed now helps keep the plant cool. By removing the barrier to nucleation, the plant has developed a thermal buffering mechanism which keeps it at an even temperature in spite of quite large variations in the temperature of the environment.

Fine-grained castings

Many engineering components – from cast-iron drain covers to aluminium alloy cylinder heads – are *castings*, made by pouring molten metal into a mould of the right shape, and allowing it to go solid. The casting process can be

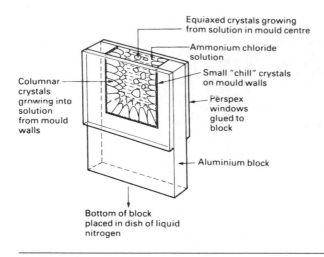

Equiaxed crystals growing from solution in mould centre

Ammonium chloride solution

Columnar crystals growing into solution from mould walls

Small "chill" crystals on mould walls

Perspex windows glued to block

Aluminium block

Bottom of block placed in dish of liquid nitrogen

Figure 9.3   A simple laboratory set-up for observing the casting process directly. The mould volume measures about $50 \times 50 \times 6\,mm$. The walls are cooled by putting the bottom of the block into a dish of liquid nitrogen. The windows are kept free of frost by squirting them with alcohol from a wash bottle every 5 minutes.

modelled using the set-up shown in Fig. 9.3. The mould is made from aluminium but has Perspex side windows to allow the solidification behaviour to be watched. The casting "material" used is ammonium chloride solution, made up by heating water to 50°C and adding ammonium chloride crystals until the solution just becomes saturated. The solution is then warmed up to 75°C and poured into the cold mould. When the solution touches the cold metal it cools very rapidly and becomes highly supersaturated. Ammonium chloride nuclei form heterogeneously on the aluminum and a thin layer of tiny *chill* crystals forms all over the mould walls. The chill crystals grow competitively until they give way to the much bigger *columnar* crystals (Figs 9.3 and 9.4). After a while the top surface of the solution cools below the saturation temperature of 50°C and crystal nuclei form heterogeneously on floating particles of dirt. The nuclei grow to give *equiaxed* (spherical) crystals which settle down into the bulk of the solution. When the casting is completely solid it will have the grain structure shown in Fig. 9.5. This is the classic casting structure, found in any cast-metal ingot.

This structure is far from ideal. The first problem is one of *segregation*: as the long columnar grains grow they push impurities ahead of them.* If, as is usually the case, we are casting *alloys*, this segregation can give big differences in

---

* This is, of course, just what happens in zone refining (Chapter 4). But segregation in zone refining is much more complete than it is in casting. In casting, some of the rejected impurities are trapped between the dendrites so that only a proportion of the impurities are pushed into the liquid ahead of the growth front. Zone refining, on the other hand, is done under such carefully controlled conditions that dendrites do not form. The solid–liquid interface is then totally flat, and impurity trapping cannot occur.

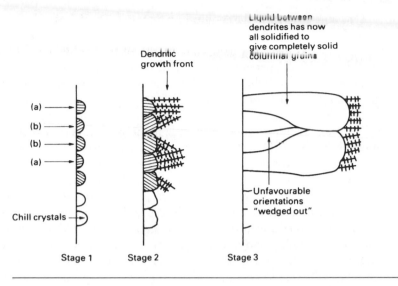

Figure 9.4   Chill crystals nucleate with random crystal orientations. They grow in the form of *dendrites*. Dendrites always lie along specific crystallographic directions. Crystals oriented like (a) will grow further into the liquid in a given time than crystals oriented like (b); (b)-type crystals will get "wedged out" and (a)-type crystals will dominate, eventually becoming columnar grains.

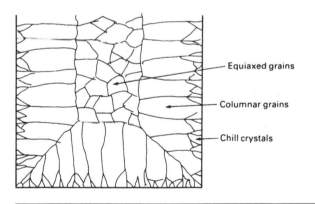

Figure 9.5   The grain structure of the solid casting.

composition – and therefore in properties – between the outside and the inside of the casting. The second problem is one of *grain size*. As we mentioned in Chapter 8, fine-grained materials are harder than coarse-grained ones. Indeed, the yield strength of steel can be doubled by a ten-times decrease in grain size. Obviously, the big columnar grains in a typical casting are a source of weakness. But how do we get rid of them?

One cure is to cast *at* the equilibrium temperature. If, instead of using undersaturated ammonium chloride solution, we pour *saturated* solution into the mould, we get what is called "big-bang" nucleation. As the freshly poured solution swirls past the cold walls, heterogeneous nuclei form in large numbers. These nuclei are then swept back into the bulk of the solution where they act as growth centres for equiaxed grains. The final structure is then almost entirely equiaxed, with only a small columnar region. For some alloys this technique (or a modification of it called "rheocasting") works well. But for most it is found that, if the molten metal is not superheated to begin with, then parts of the casting will freeze prematurely, and this may prevent metal reaching all parts of the mould.

The traditional cure is to use *inoculants*. Small catalyst particles are added to the melt just before pouring (or even poured into the mould *with* the melt) in order to nucleate as many crystals as possible. This gets rid of the columnar region altogether and produces a fine-grained equiaxed structure throughout the casting. This important application of heterogeneous nucleation sounds straightforward, but a great deal of trial and error is needed to find effective catalysts. The choice of AgI for seeding ice crystals was an unusually simple one; finding successful inoculants for metals is still nearer black magic than science. Factors other than straightforward crystallographic matching are important: surface defects, for instance, can be crucial in attracting atoms to the catalyst; and even the smallest quantities of impurity can be adsorbed on the surface to give monolayers which may poison the catalyst. A notorious example of erratic surface nucleation is in the field of electroplating: electroplaters often have difficulty in getting their platings to "take" properly. It is well known (among experienced electroplaters) that pouring condensed milk into the plating bath can help.

Single crystals for semiconductors

Materials for semiconductors have to satisfy formidable standards. Their electrical properties are badly affected by the scattering of carriers which occurs at impurity atoms, or at dislocations, grain boundaries and free surfaces. We have already seen (in Chapter 4) how zone refining is used to produce the ultrapure starting materials. The next stage in semiconductor processing is to grow large single crystals under carefully controlled conditions: grain boundaries are eliminated and a very low dislocation density is achieved.

Figure 9.6 shows part of a typical integrated circuit. It is built on a single-crystal wafer of silicon, usually about 300μm thick. The wafer is doped with an impurity such as boron, which turns it into a p-type semiconductor (bulk doping is usually done after the initial zone refining stage in a process known as zone levelling). The localized n-type regions are formed by firing pentavalent impurities (e.g. phosphorus) into the surface using an ion gun. The circuit is completed by the vapour-phase deposition of silica insulators and aluminium interconnections.

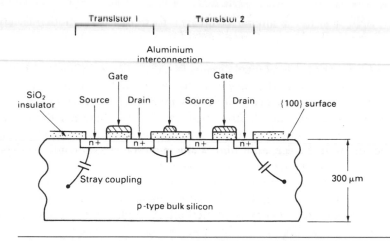

Figure 9.6   A typical integrated circuit. The silicon wafer is cut from a large single crystal using a chemical saw – mechanical sawing would introduce too many dislocations.

Growing single crystals is the very opposite of pouring fine-grained castings. In castings we want to undercool as much of the liquid as possible so that nuclei can form everywhere. In crystal growing we need to start with a single seed crystal of the right orientation and the last thing that we want is for stray nuclei to form. Single crystals are grown using the arrangement shown in Fig. 9.7. The seed crystal fits into the bottom of a crucible containing the molten silicon. The crucible is lowered slowly out of the furnace and the crystal grows into the liquid. The only region where the liquid silicon is undercooled

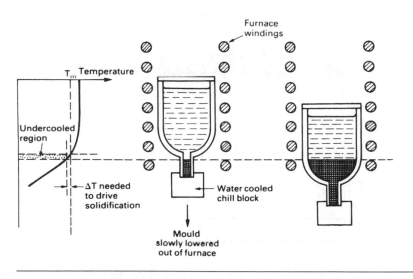

Figure 9.7   Growing single crystals for semiconductor devices.

is right next to the interface, and even there the undercooling is very small. So there is little chance of stray nuclei forming and nearly all runs produce single crystals.

Conventional integrated circuits like that shown in Fig. 9.6 have two major draw-backs. First, the *device density* is limited: silicon is not a very good insulator, so leakage occurs if devices are placed too close together. And second, device speed is limited: stray capacitance exists between the devices and the substrate which imposes a time constant on switching. These problems would be removed if a very thin film of single-crystal silicon could be deposited on a highly insulating oxide such as silica (Fig. 9.8).

Single-crystal technology has recently been adapted to do this, and has opened up the possibility of a new generation of ultra-compact high-speed devices. Figure 9.9 shows the method. A single-crystal wafer of silicon is first

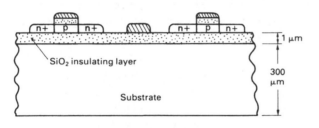

Figure 9.8   A silicon-on-insulator integrated circuit.

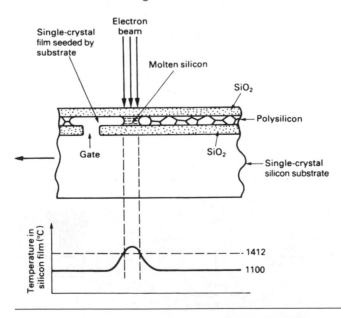

Figure 9.9   How single-crystal films are grown from polysilicon. The electron beam is line-scanned in a direction at right angles to the plane of the drawing.

coated with a thin insulating layer of SiO_2 with a slot, or "gate", to expose the underlying silicon. Then, polycrystalline silicon ("polysilicon") is vapour deposited onto the oxide, to give a film a few microns thick. Finally, a capping layer of oxide is deposited on the polysilicon to protect it and act as a mould.

The sandwich is then heated to 1100°C by scanning it from below with an electron beam (this temperature is only 312°C below the melting point of silicon). The polysilicon at the gate can then be melted by line scanning an electron beam across the top of the sandwich. Once this is done the sandwich is moved slowly to the left under the line scan: the molten silicon at the gate undercools, is seeded by the silicon below, and grows to the right as an oriented single crystal. When the single-crystal film is complete the overlay of silica is dissolved away to expose oriented silicon that can be etched and ion implanted to produce completely isolated components.

Amorphous metals

In Chapter 8 we saw that, when carbon steels were quenched from the austenite region to room temperature, the austenite could not transform to the equilibrium low-temperature phases of ferrite and iron carbide. There was no time for diffusion, and the austenite could only transform by a diffusionless (shear) transformation to give the metastable martensite phase. The martensite transformation can give enormously altered mechanical properties and is largely responsible for the great versatility of carbon and low-alloy steels. Unfortunately, few alloys undergo such useful shear transformations. But are there other ways in which we could change the properties of alloys by quenching?

An idea of the possibilities is given by the old high-school chemistry experiment with sulphur crystals ("flowers of sulphur"). A 10 ml beaker is warmed up on a hot plate and some sulphur is added to it. As soon as the sulphur has melted the beaker is removed from the heater and allowed to cool slowly on the bench. The sulphur will solidify to give a disc of polycrystalline sulphur which breaks easily if pressed or bent. Polycrystalline sulphur is obviously very brittle.

Now take another batch of sulphur flowers, but this time heat it well past its melting point. The liquid sulphur gets darker in colour and becomes more and more viscous. Just before the liquid becomes completely unpourable it is decanted into a dish of cold water, quenching it. When we test the properties of this *quenched* sulphur we find that we have produced a tough and rubbery substance. We have, in fact, produced an *amorphous* form of sulphur with radically altered properties.

This principle has been used for thousands of years to make glasses. When silicates are cooled from the molten state they often end up being amorphous, and many polymers are amorphous too. What makes it easy to produce amorphous sulphur, glasses and polymers is that their high viscosity stops crystallisation taking place. Liquid sulphur becomes unpourable at 180°C because the sulphur polymerises into long cross-linked chains of sulphur atoms. When

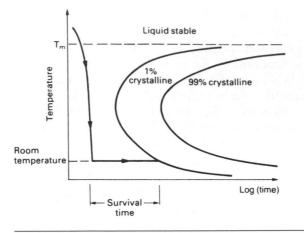

Figure 9.10   Sulphur, glasses and polymers turn into viscous liquids at high temperature. The atoms in the liquid are arranged in long polymerised chains. The liquids are viscous because it is difficult to get these bulky chains to slide over one another. It is also hard to get the atoms to regroup themselves into crystals, and the kinetics of crystallisation are very slow. The liquid can easily be cooled past the nose of the C-curve to give a metastable supercooled liquid which can survive for long times at room temperature.

this polymerised liquid is cooled below the solidification temperature it is very difficult to get the atoms to regroup themselves into crystals. The C-curve for the liquid-to-crystal transformation (Fig. 9.10) lies well to the right, and it is easy to cool the melt past the nose of the C-curve to give a supercooled liquid at room temperature.

There are formidable problems in applying these techniques to metals. Liquid metals do not polymerise and it is very hard to stop them crystallising when they are undercooled. In fact, cooling rates in excess of $10^{10}°C\,s^{-1}$ are needed to make pure metals amorphous. But current rapid-quenching technology has made it possible to make amorphous *alloys*, though their compositions are a bit daunting ($Fe_{40}Ni_{40}P_{14}B_6$ for instance). This is so heavily alloyed that it crystallises to give compounds; and in order for these compounds to grow the atoms must add on from the liquid in a particular sequence. This slows down the crystallisation process, and it is possible to make amorphous $Fe_{40}Ni_{40}P_{14}B_6$ using cooling rates of only $10^{5}°C\,s^{-1}$.

Amorphous alloys have been made commercially for the past 20 years by the process known as melt spinning (Fig. 9.11). They have some remarkable and attractive properties. Many of the iron-based alloys are ferromagnetic. Because they are amorphous, and literally without structure, they are excellent soft magnets: there is nothing to pin the magnetic domain walls, which move easily at low fields and give a very small coercive force. These alloys are now being used for the cores of small transformers and relays. Amorphous alloys have no dislocations (you can only have dislocations in *crystals*) and they are

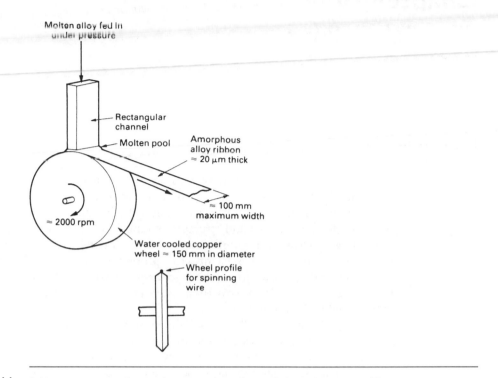

Figure 9.11 Ribbons or wires of amorphous metal can be made by melt spinning. There is an upper limit on the thickness of the ribbon: if it is too thick it will not cool quickly enough and the liquid will crystallise.

therefore very hard. But, exceptionally, they are ductile too; ductile enough to be cut using a pair of scissors. Finally, recent alloy developments have allowed us to make amorphous metals in sections up to 5 mm thick. The absence of dislocations makes for very low mechanical damping, so amorphous alloys are now being used for the striking faces of high-tech. golf clubs!

Examples

9.1 Why is it undesirable to have a columnar grain structure in castings? Why is a fine equiaxed grain structure the most desirable option? What factors determine the extent to which the grain structure is columnar or equiaxed?

9.2 Why is it easy to produce amorphous polymers and glasses, but difficult to produce amorphous metals?

9.3 A cast ingot of pure aluminium has a structure which consists mainly of large columnar grains, whereas a cast ingot of aluminium containing Al–Ti inoculant powder consists of small equiaxed grains. Explain this difference in structure.

Chapter 10
The light alloys

Introduction

No fewer than 14 pure metals have densities $\leqslant 4.5\,\mathrm{Mg\,m^{-3}}$ (see Table 10.1). Of these, titanium, aluminium and magnesium are in common use as structural materials. Beryllium is difficult to work and is toxic, but it is used in moderate quantities for heat shields and structural members in rockets. Lithium is used as an alloying element in aluminium to lower its density and save weight on airframes. Yttrium has an excellent set of properties and, although scarce, may eventually find applications in the nuclear-powered aircraft project. But the majority are unsuitable for structural use because they are chemically reactive or have low melting points.[*]

Table 10.2 shows that alloys based on aluminium, magnesium and titanium may have better stiffness/weight and strength/weight ratios than steel. Not only that; they are also corrosion resistant (with titanium exceptionally so); they are non-toxic; and titanium has good creep properties. So although the light alloys were originally developed for use in the aerospace industry, they are now much more widely used. The dominant use of aluminium alloys is in building and construction: panels, roofs, and frames. The second-largest consumer is the container and packaging industry; after that come transportation systems (the fastest-growing sector, with aluminium replacing steel and cast iron in cars and mass-transit systems); and the use of aluminium as an electrical conductor. Magnesium is lighter but more expensive. Titanium alloys are mostly used in aerospace applications where the temperatures are too high for aluminium or magnesium; but its extreme corrosion resistance makes it attractive in chemical engineering, food processing and bio-engineering. The growth in the use of these alloys is rapid: nearly 7% per year, higher than any other metals, and surpassed only by polymers.

The light alloys derive their strength from *solid solution hardening*, *age* (or *precipitation*) *hardening*, and *work hardening*. We now examine the principles behind each hardening mechanism, and illustrate them by drawing examples from our range of generic alloys.

---

[*] There are, however, many *non-structural* applications for the light metals. Liquid sodium is used in large quantities for cooling nuclear reactors and in small amounts for cooling the valves of high-performance i.c. engines (it conducts heat 143 times better than water but is less dense, boils at 883°C, and is safe as long as it is kept in a sealed system.) Beryllium is used in windows for X-ray tubes. Magnesium is a catalyst for organic reactions. And the reactivity of calcium, caesium and lithium makes them useful as residual gas scavengers in vacuum systems.

Table 10.1 The light metals

Metal	Density (Mg m^{-3})	T_m(°C)	Comments
Titanium	4.50	1667	High T_m – excellent creep resistance.
Yttrium	4.47	1510	Good strength and ductility; scarce.
Barium	3.50	729	
Scandium	2.99	1538	Scarce.
Aluminium	2.70	660	
Strontium	2.60	770	Reactive in air/water.
Caesium	1.87	28.5	Creeps/melts; very reactive in air/water.
Beryllium	1.85	1287	Difficult to process; very toxic.
Magnesium	1.74	649	
Calcium	1.54	839	Reactive in air/water.
Rubidium	1.53	39 ⎫	
Sodium	0.97	98 ⎬	Creep/melt; very reactive in air/water.
Potassium	0.86	63	
Lithium	0.53	181 ⎭	

Table 10.2 Mechanical properties of structural light alloys

Alloy	Density ρ (Mg m^{-3})	Young's modulus E (GPa)	Yield strength σ_y (MPa)	E/ρ^*	$E^{1/2}/\rho^*$	$E^{1/3}/\rho^*$	σ_y/ρ^*	Creep temperature (°C)
Al alloys	2.7	71	25–600	26	3.1	1.5	9–220	150–250
Mg alloys	1.7	45	70–270	25	4.0	2.1	41–160	150–250
Ti alloys	4.5	120	170–1280	27	2.4	1.1	38–280	400–600
(Steels)	(7.9)	(210)	(220–1600)	27	1.8	0.75	28–200	(400–600)

* See Chapter 25 and Fig. 25.7 for more information about these groupings.

Solid solution hardening

When other elements dissolve in a metal to form a solid solution they make the metal harder. The solute atoms differ in size, stiffness and charge from the solvent atoms. Because of this the randomly distributed solute atoms interact with dislocations and make it harder for them to move. The theory of solution hardening is rather complicated, but it predicts the following result for the yield strength

$$\sigma_y \propto \varepsilon_s^{3/2} C^{1/2}, \tag{10.1}$$

where C is the solute concentration. ε_s is a term which represents the "mismatch" between solute and solvent atoms. The form of this result is just what we would expect: badly matched atoms will make it harder for dislocations to move than well-matched atoms; and a large population of solute atoms will obstruct dislocations more than a sparse population.

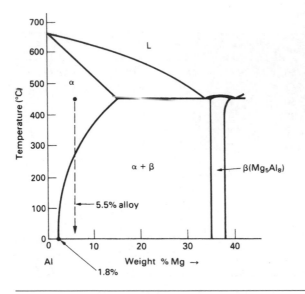

Figure 10.1   The aluminium end of the Al–Mg phase diagram.

Of the generic aluminium alloys (see Chapter 1, Table 1.4), the 5000 series derives most of its strength from solution hardening. The Al–Mg phase diagram (Fig. 10.1) shows why: at room temperature aluminium can dissolve up to 1.8 wt% magnesium at equilibrium. In practice, Al–Mg alloys can contain as much as 5.5 wt% Mg in solid solution at room temperature – a supersaturation of $5.5 - 1.8 = 3.7$ wt%. In order to get this supersaturation the alloy is given the following schedule of heat treatments.

(a) Hold at 450°C ("solution heat treat")

This puts the 5.5% alloy into the single phase (α) field and all the Mg will dissolve in the Al to give a random substitutional solid solution.

(b) Cool moderately quickly to room temperature

The phase diagram tells us that, below 275°C, the 5.5% alloy has an *equilibrium* structure that is two-phase, $\alpha + Mg_5Al_8$. If, then, we cool the alloy *slowly* below 275°C, Al and Mg atoms will diffuse together to form precipitates of the intermetallic compound Mg_5Al_8. However, below 275°C, diffusion is slow and the C-curve for the precipitation reaction is well over to the right (Fig. 10.2). So if we cool the 5.5% alloy moderately quickly we will miss the nose of the C-curve. None of the Mg will be taken out of solution as Mg_5Al_8,

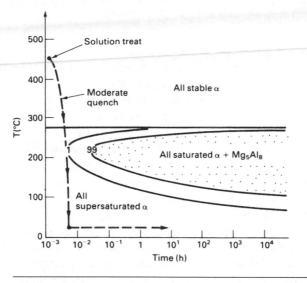

Figure 10.2 Semi-schematic TTT diagram for the precipitation of Mg_5Al_8 from the Al–5.5 wt% Mg solid solution.

and we will end up with a supersaturated solid solution at room temperature. As Table 10.3 shows, this supersaturated Mg gives a substantial increase in yield strength.

Solution hardening is not confined to 5000 series aluminium alloys. The other alloy series all have elements dissolved in solid solution; and they are all solution strengthened to some degree. But most aluminium alloys owe their strength to fine precipitates of intermetallic compounds, and solution strengthening is not dominant as it is in the 5000 series. Turning to the other light alloys, the most widely used titanium alloy (Ti–6 Al 4V) is dominated by solution hardening (Ti effectively dissolves about 7 wt% Al, and has complete solubility for V). Finally, magnesium alloys can be solution strengthened with Li, Al, Ag and Zn, which dissolve in Mg by between 2 and 5 wt%.

Table 10.3 Yield strengths of 5000 series (Al–Mg) alloys

Alloy	(wt% Mg)	σ_y (MPa) (annealed condition)	
5005	0.8	40	
5050	1.5	55	
5052	2.5	90	
5454	2.7	120	supersaturated
5083	4.5	145	
5456	5.1	160	

Age (precipitation) hardening

When the phase diagram for an alloy has the shape shown in Fig. 10.3 (a solid solubility that decreases markedly as the temperature falls), then the potential for *age* (or *precipitation*) *hardening* exists. The classic example is the Duralumins, or 2000 series aluminium alloys, which contain about 4% copper.

The Al–Cu phase diagram tells us that, between 500°C and 580°C, the 4% Cu alloy is single phase: the Cu dissolves in the Al to give the random substitutional solid solution α. Below 500°C the alloy enters the two-phase field of $\alpha + CuAl_2$. As the temperature decreases the amount of $CuAl_2$ increases, and at room temperature the equilibrium mixture is 93 wt% $\alpha + 7$ wt% $CuAl_2$. Figure 10.4(a) shows the microstructure that we would get by cooling an Al–4 wt% Cu alloy *slowly* from 550°C to room temperature. In slow cooling the driving force for the precipitation of $CuAl_2$ is small and the nucleation rate is low (see Fig. 8.3). In order to accommodate the equilibrium amount of $CuAl_2$ the few nuclei that do form grow into large precipitates of $CuAl_2$ spaced well apart. Moving dislocations find it easy to avoid the precipitates and the alloy is rather soft. If, on the other hand, we cool the alloy rather *quickly*, we produce a much finer structure (Fig. 10.4b). Because the driving force is large the nucleation rate is high (see Fig. 8.3). The precipitates, although small, are closely spaced: they get in the way of moving dislocations and make the alloy harder.

There are limits to the precipitation hardening that can be produced by direct cooling: if the cooling rate is too high we will miss the nose of the C-curve for the precipitation reaction and will not get any precipitates at all! But large increases in yield strength *are* possible if we *age harden* the alloy.

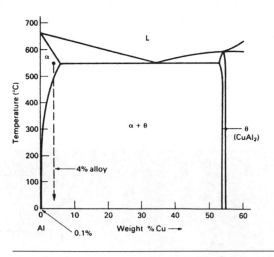

Figure 10.3   The aluminium end of the Al–Cu phase diagram.

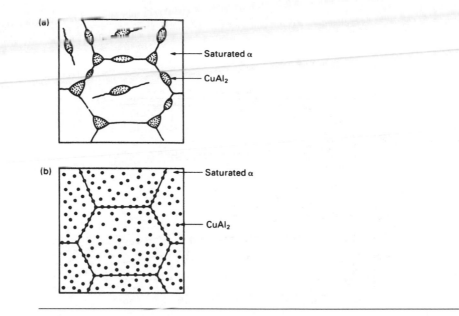

(a)

— Saturated α

— CuAl₂

(b)

— Saturated α

— CuAl₂

Figure 10.4 Room temperature microstructures in the Al + 4 wt% Cu alloy. **(a)** Produced by slow cooling from 550°C. **(b)** Produced by moderately fast cooling from 550°C. The precipitates in **(a)** are large and far apart. The precipitates in **(b)** are small and close together.

To age harden our Al–4 wt% Cu alloy we use the following schedule of heat treatments.

(a) Solution heat treat at 550°C. This gets all the Cu into solid solution.
(b) Cool rapidly to room temperature by quenching into water or oil ("quench").* We will miss the nose of the C-curve and will end up with a highly supersaturated solid solution at room temperature (Fig. 10.5).
(c) Hold at 150°C for 100 hours ("age"). As Fig. 10.5 shows, the supersaturated α will transform to the equilibrium mixture of saturated $\alpha + CuAl_2$. But it will do so under a very high driving force and will give a very fine (and very strong) structure.

Figure 10.5, as we have drawn it, is oversimplified. Because the transformation is taking place at a low temperature, where the atoms are not very mobile, it is not easy for the $CuAl_2$ to separate out in one go. Instead, the transformation takes place in four distinct stages. These are shown in Figs 10.6(a)–(e). The progression may appear rather involved but it is a good illustration of

---

* The C-curve nose is ≈150°C higher for Al–4 Cu than for Al–5.5 Mg (compare Figs 10.5 and 10.2). Diffusion is faster, and a more rapid quench is needed to miss the nose.

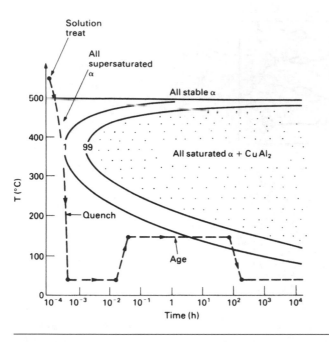

Figure 10.5   TTT diagram for the precipitation of CuAl$_2$ from the Al + 4 wt% Cu solid solution. Note that the *equilibrium* solubility of Cu in Al at room temperature is only 0.1 wt% (see Fig. 10.3). The quenched solution is therefore carrying 4/0.1 = 40 times as much Cu as it wants to.

much of the material in the earlier chapters. More importantly, each stage of the transformation has a direct effect on the yield strength.

Four separate hardening mechanisms are at work during the ageing process:

(a) Solid solution hardening

At the start of ageing the alloy is mostly strengthened by the 4 wt% of copper that is trapped in the supersaturated α. But when the GP zones form, almost all of the Cu is removed from solution and the solution strengthening virtually disappears (Fig. 10.7).

(b) Coherency stress hardening

The coherency strains around the GP zones and θ'' precipitates generate stresses that help prevent dislocation movement. The GP zones give the larger hardening effect (Fig. 10.7).

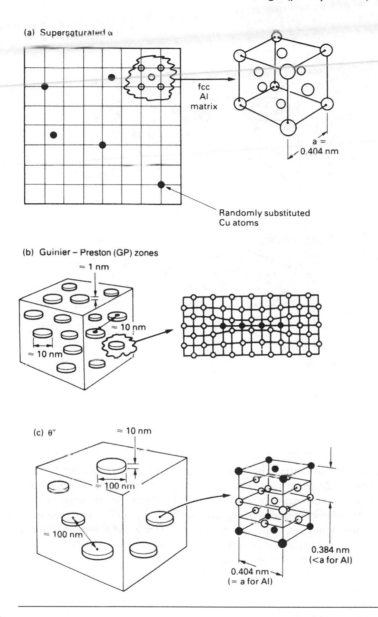

(a) Supersaturated α

fcc
Al
matrix

a =
0.404 nm

Randomly substituted
Cu atoms

(b) Guinier – Preston (GP) zones

≈ 1 nm

≈ 10 nm

≈ 10 nm

(c) θ″

≈ 10 nm

≈ 100 nm

≈ 100 nm

0.384 nm
(<a for Al)

0.404 nm
(= a for Al)

Figure 10.6   Stages in the precipitation of $CuAl_2$. Disc-shaped GP zones (**b**) nucleate homogeneously from supersaturated solid solution (**a**). The disc faces are perfectly coherent with the matrix. The disc edges are also coherent, but with a large *coherency strain*. (**c**) Some of the GP zones grow to form precipitates called θ''. (The remaining GP zones dissolve and transfer Cu to the growing θ'' by diffusion through the matrix.) Disc faces are perfectly coherent. Disc edges are coherent, but the mismatch of lattice parameters between the θ'' and the Al matrix generates coherency strain.

(*Continued*)

(d) θ′

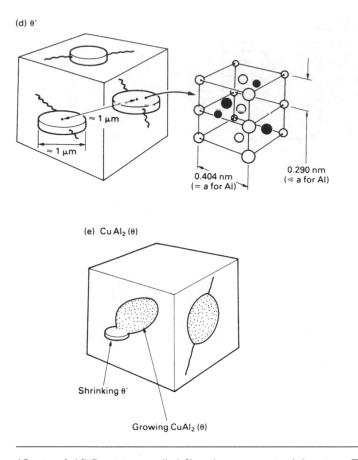

(e) Cu Al₂ (θ)

Shrinking θ′

Growing CuAl₂ (θ)

Figure 10.6   (*Continued*) (**d**) Precipitates called θ′ nucleate at matrix dislocations. The θ″ precipitates all dissolve and transfer Cu to the growing θ′. Disc faces are still perfectly coherent with the matrix. But disc edges are now *incoherent*. Neither faces nor edges show coherency strain, but for different reasons. (**e**) Equilibrium CuAl₂ (θ) nucleates at grain boundaries and at θ′–matrix interfaces. The θ′ precipitates all dissolve and transfer Cu to the growing θ. The CuAl₂ is completely *incoherent* with the matrix (see structure in Fig. 2.3). Because of this it grows as *rounded* rather than disc-shaped particles.

(c) Precipitation hardening

The precipitates can obstruct the dislocations directly. But their effectiveness is limited by two things: dislocations can either *cut through* the precipitates, or they can *bow around* them (Fig. 10.8).

Resistance to cutting depends on a number of factors, of which the shearing resistance of the precipitate lattice is only one. In fact the cutting stress *increases* with ageing time (Fig. 10.7).

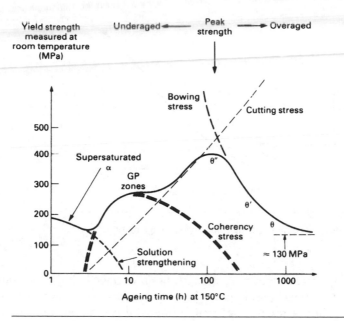

Figure 10.7   The yield strength of quenched Al–4 wt% Cu changes dramatically during ageing at 150°C.

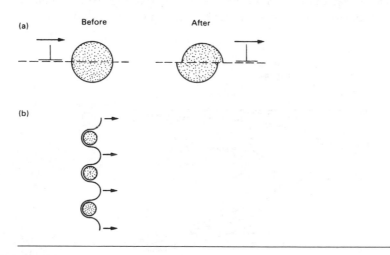

Figure 10.8   Dislocations can get past precipitates by (a) cutting or (b) bowing.

Bowing is easier when the precipitates are far apart. During ageing the precipitate spacing increases from 10 nm to 1 μm and beyond (Fig. 10.9). The bowing stress therefore decreases with ageing time (Fig. 10.7).

The four hardening mechanisms add up to give the overall variation of yield strength shown in Fig. 10.7. *Peak strength is reached if the transformation is stopped at θ″.* If the alloy is aged some more the strength will *decrease*; and the

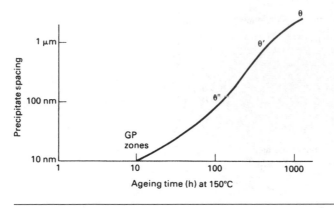

Figure 10.9   The gradual increase of particle spacing with ageing time.

only way of recovering the strength of an overaged alloy is to solution-treat it at 550°C, quench, and start again! If the alloy is not aged for long enough, then it will not reach peak strength; but this can be put right by more ageing.

Although we have chosen to age our alloy at 150°C, we could, in fact, have aged it at any temperature below 180°C (see Fig. 10.10). The lower the ageing temperature, the longer the time required to get peak hardness. In practice, the

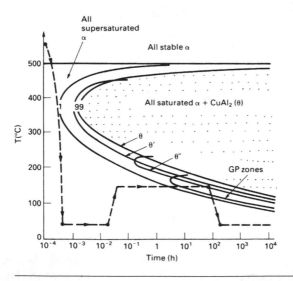

Figure 10.10   Detailed TTT diagram for the Al–4 wt% Cu alloy. We get peak strength by ageing to give θ''. The lower the ageing temperature, the longer the ageing time. Note that GP zones do not form above 180°C: if we age above this temperature we will fail to get the peak value of yield strength.

Table 10.4   Yield strengths of heat treatable alloys

Alloy series	Typical composition (wt%)	σ_y (MPa)	
		Slowly cooled	Quenched and aged
2000	Al + 4 Cu + Mg, Si, Mn	130	465
6000	Al + 0.5 Mg 0.5 Si	85	210
7000	Al + 6 Zn + Mg, Cu, Mn	300	570

ageing time should be long enough to give good control of the heat treatment operation without being too long (and expensive).

Finally, Table 10.4 shows that copper is not the only alloying element that can age-harden aluminium. Magnesium and titanium can be age hardened too, but not as much as aluminium.

Work hardening

Commercially pure aluminium (1000 series) and the non-heat-treatable aluminium alloys (3000 and 5000 series) are usually work hardened. The work hardening superimposes on any solution hardening, to give considerable extra strength (Table 10.5).

Work hardening is achieved by cold rolling. The yield strength increases with strain (reduction in thickness) according to

$$\sigma_y = A\varepsilon^n, \tag{10.2}$$

where A and n are constants. For aluminium alloys, n lies between 1/6 and 1/3.

Thermal stability

Aluminium and magnesium melt at just over 900 K. Room temperature is $0.3\,T_m$, and 100°C is $0.4\,T_m$. Substantial diffusion can take place in these alloys if they are used for long periods at temperatures approaching 80–100°C.

Table 10.5   Yield strengths of work-hardened aluminium alloys

Alloy number	σ_y (MPa)		
	Annealed	"Half hard"	"Hard"
1100	35	115	145
3005	65	140	185
5456	140	300	370

Several processes can occur to reduce the yield strength: loss of solutes from supersaturated solid solution, overageing of precipitates and recrystallisation of cold-worked microstructures.

This lack of *thermal stability* has some interesting consequences. During supersonic flight frictional heating can warm the skin of an aircraft to 150°C. Because of this, Rolls-Royce had to develop a special age-hardened aluminium alloy (RR58) which would not over-age during the lifetime of the Concorde supersonic airliner. When aluminium cables are fastened to copper busbars in power circuits contact resistance heating at the junction leads to interdiffusion of Cu and Al. Massive, brittle plates of $CuAl_2$ form, which can lead to joint failures; and when light alloys are welded, the properties of the heat-affected zone are usually well below those of the parent metal.

Examples

10.1 An alloy of Al–4 weight% Cu was heated to 550°C for a few minutes and was then quenched into water. Samples of the quenched alloy were aged at 150°C for various times before being quenched again. Hardness measurements taken from the re-quenched samples gave the following data:

Ageing time (h)	0	10	100	200	1000
Hardness (MPa)	650	950	1200	1150	1000

Account briefly for this behaviour.

Peak hardness is obtained after 100 h at 150°C. Estimate how long it would take to get peak hardness at (a) 130°C, (b) 170°C.

[Hint: use Fig. 10.10.]

Answers

(a) 10^3 h; (b) 10 h.

10.2 A batch of 7000 series aluminium alloy rivets for an aircraft wing was inadvertently over-aged. What steps can be taken to reclaim this batch of rivets?

10.3 Two pieces of work-hardened 5000 series aluminium alloy plate were butt welded together by arc welding. After the weld had cooled to room temperature, a series of hardness measurements was made on the surface of the fabrication. Sketch the variation in hardness as the position of the hardness indenter passes across the weld from one plate to the other. Account for the form of the hardness profile, and indicate its practical consequences.

10.4 One of the major uses of aluminium is for making beverage cans. The body is cold-drawn from a single slug of 3000 series non-heat treatable alloy because this has the large ductility required for the drawing operation. However, the top of the can must have a much lower ductility in order to allow the ring-pull to work (the top must tear easily). Which alloy would you select for the top from Table 10.5? Explain the reasoning behind your choice. Why are non-heat treatable alloys used for can manufacture?

10.5   A sample of Al–4 wt% Cu was cooled slowly from 550°C to room temperature. The yield strength of the slowly cooled sample was 130 MPa. A second sample of the alloy was quenched into cold water from 550°C and was then aged at 150°C for 100 hours. The yield strength of the quenched-and-aged sample was 450 MPa. Explain the difference in yield strength. [Both yield strengths were measured at 20°C.]

Chapter 11
Steels: 1 – carbon steels

Introduction

Iron is one of the oldest known metals. Methods of extracting* and working it have been practised for thousands of years, although the large-scale production of carbon steels is a development of the ninetenth century. From these carbon steels (which still account for 90% of all steel production) a range of alloy steels has evolved: the low alloy steels (containing up to 6% of chromium, nickel, etc.); the stainless steels (containing, typically, 18% chromium and 8% nickel) and the tool steels (heavily alloyed with chromium, molybdenum, tungsten, vanadium and cobalt).

We already know quite a bit about the transformations that take place in steels and the microstructures that they produce. In this chapter we draw these features together and go on to show how they are instrumental in determining the mechanical properties of steels. We restrict ourselves to carbon steels; alloy steels are covered in Chapter 12.

Carbon is the cheapest and most effective alloying element for hardening iron. We have already seen in Chapter 1 (Table 1.1) that carbon is added to iron in quantities ranging from 0.04 to 4 wt% to make low, medium and high carbon steels, and cast iron. The mechanical properties are strongly dependent on both the carbon content and on the type of heat treatment. Steels and cast iron can therefore be used in a very wide range of applications (see Table 1.1).

Microstructures produced by slow cooling ("normalising")

Carbon steels as received "off the shelf" have been worked at high temperature (usually by rolling) and have then been cooled slowly to room temperature ("normalised"). The room-temperature microstructure should then be close to equilibrium and can be inferred from the Fe—C phase diagram (Fig. 11.1) which we have already come across in the Phase Diagrams course (XXX). Table 11.1 lists the *phases* in the Fe—Fe$_3$C system and Table 11.2 gives details of the composite eutectoid and eutectic structures that occur during slow cooling.

---

* People have sometimes been able to avoid the tedious business of extracting iron from its natural ore. When Commander Peary was exploring Greenland in 1894 he was taken by an Eskimo to a place near Cape York to see a huge, half-buried meteorite. This had provided metal for Eskimo tools and weapons for over a hundred years. Meteorites usually contain iron plus about 10% nickel: a direct delivery of low-alloy iron from the heavens.

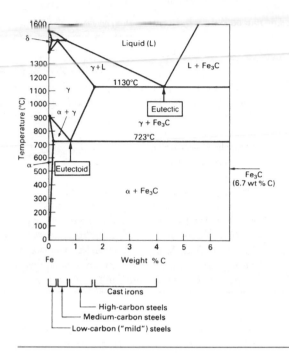

Figure 11.1 The left-hand part of the iron–carbon phase diagram. There are five phases in the Fe–Fe$_3$C system: L, δ, γ, α and Fe$_3$C (see Table 11.1).

Table 11.1   Phases in the Fe–Fe$_3$C system

Phase	Atomic packing	Description and comments
Liquid	d.r.p.	Liquid solution of C in Fe.
δ	b.c.c.	Random interstitial solid solution of C in b.c.c. Fe. Maximum solubility of 0.08 wt% C occurs at 1492°C. Pure δ Fe is the stable polymorph between 1391°C and 1536°C (see Fig. 2.1).
γ (also called "austenite")	f.c.c.	Random interstitial solid solution of C in f.c.c. Fe. Maximum solubility of 1.7 wt% C occurs at 1130°C. Pure γ Fe is the stable polymorph between 914°C and 1391°C (see Fig. 2.1).
α (also called "ferrite")	b.c.c.	Random interstitial solid solution of C in b.c.c. Fe. Maximum solubility of 0.035 wt% C occurs at 723°C. Pure α Fe is the stable polymorph below 914°C (see Fig. 2.1).
Fe$_3$C (also called "iron carbide" or "cementite")	Complex	A hard and brittle chemical compound of Fe and C containing 25 atomic % (6.7 wt%) C.

Table 11.2   Composite structures produced during the slow cooling of Fe–C alloys

Name of structure	Description and comments
Pearlite	The composite eutectoid structure of alternating plates of α and Fe$_3$C produced when γ containing 0.80 wt% C is cooled below 723°C (see Fig. 6.7 and Phase Diagrams p. 406). Pearlite nucleates at γ grain boundaries. It occurs in low, medium and high carbon steels. It is sometimes, quite wrongly, called a phase. It is not a phase but is a *mixture* of the two separate phases α and Fe$_3$C in the proportions of 88.5% by weight of α to 11.5% by weight of Fe$_3$C. Because grains are single crystals it is *wrong* to say that Pearlite forms in grains: we say instead that it forms in *nodules*.
Ledeburite	The composite eutectic structure of alternating plates of γ and Fe$_3$C produced when liquid containing 4.3 wt% C is cooled below 1130°C. Again, *not* a phase! Ledeburite only occurs during the solidification of cast irons, and even then the γ in ledeburite will transform to $\alpha +$ Fe$_3$C at 723°C.

Figures 11.2–11.6 show how the room temperature microstructure of carbon steels depends on the carbon content. The limiting case of pure iron (Fig. 11.2) is straight-forward: when γ iron cools below 914°C α grains nucleate at γ grain boundaries and the microstructure transforms to α. If we cool a steel of

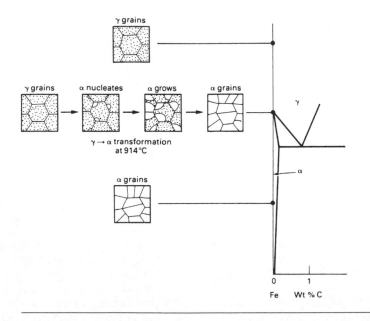

Figure 11.2   Microstructures during the slow cooling of pure iron from the hot working temperature.

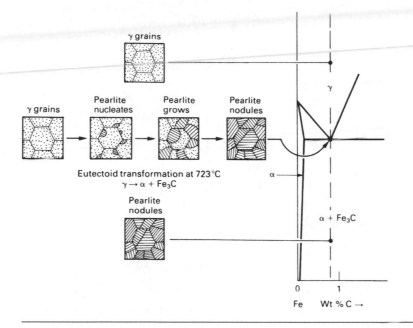

Figure 11.3   Microstructures during the slow cooling of a eutectoid steel from the hot working temperature. As a point of detail, when pearlite is cooled to room temperature, the concentration of carbon in the α decreases slightly, following the $\alpha/\alpha + Fe_3C$ boundary. The excess carbon reacts with iron at the α–Fe_3C interfaces to form more Fe_3C. This "plates out" on the surfaces of the existing Fe_3C plates which become very slightly thicker. The composition of Fe_3C is independent of temperature, of course.

eutectoid composition (0.80 wt% C) below 723°C pearlite nodules nucleate at grain boundaries (Fig. 11.3) and the microstructure transforms to pearlite. If the steel contains less than 0.80% C (a *hypoeutectoid* steel) then the γ starts to transform as soon as the alloy enters the $\alpha + \gamma$ field (Fig. 11.4). "Primary" α nucleates at γ grain boundaries and grows as the steel is cooled from A_3 to A_1. At A_1 the remaining γ (which is now of eutectoid composition) transforms to pearlite as usual. The room temperature microstructure is then made up of primary α + pearlite. If the steel contains more than 0.80% C (a *hypereutectoid* steel) then we get a room-temperature microstructure of primary Fe_3C plus pearlite instead (Fig. 11.5). These structural differences are summarised in Fig. 11.6.

Mechanical properties of normalised carbon steels

Figure 11.7 shows how the mechanical properties of normalised carbon steels change with carbon content. Both the yield strength and tensile strength increase linearly with carbon content. This is what we would expect: the Fe_3C

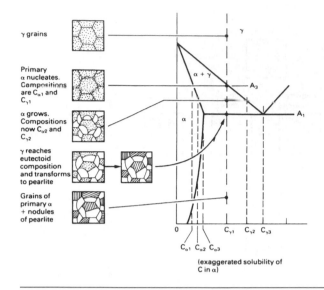

Figure 11.4   Microstructures during the slow cooling of a hypoeutectoid steel from the hot working temperature. A_3 is the standard labelling for the temperature at which α first appears, and A_1 is standard for the eutectoid temperature. *Hypo*eutectoid means that the carbon content is *below* that of a eutectoid steel (in the same sense that hypodermic means "under the skin"!).

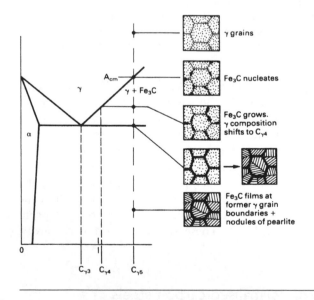

Figure 11.5   Microstructures during the slow cooling of a hypereutectoid steel. A_{cm} is the standard labelling for the temperature at which Fe_3C first appears. *Hyper*eutectoid means that the carbon content is *above* that of a eutectoid steel (in the sense that a hyperactive child has an above-normal activity!).

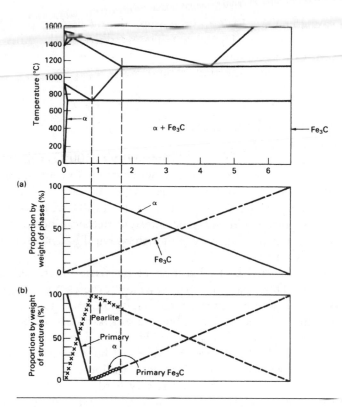

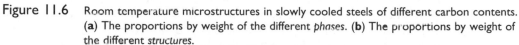

Figure 11.6   Room temperature microstructures in slowly cooled steels of different carbon contents. **(a)** The proportions by weight of the different *phases*. **(b)** The proportions by weight of the different *structures*.

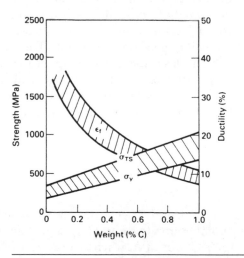

Figure 11.7   Mechanical properties of normalised carbon steels.

acts as a strengthening phase, and the proportion of Fe_3C in the steel is linear in carbon concentration (Fig. 11.6a). The ductility, on the other hand, falls rapidly as the carbon content goes up (Fig. 11.7) because the α-Fe_3C interfaces in pearlite are good at nucleating cracks.

Quenched and tempered carbon steels

We saw in Chapter 8 that, if we cool eutectoid γ to 500°C at about $200°C\,s^{-1}$, we will miss the nose of the C-curve. If we continue to cool below 280°C the unstable γ will begin to transform to martensite. At 220°C half the γ will have transformed to martensite. And at $-50°C$ the steel will have become completely martensitic. Hypoeutectoid and hypereutectoid steels can be quenched to give martensite in exactly the same way (although, as Fig. 11.8 shows, their C-curves are slightly different).

Figure 11.9 shows that the hardness of martensite increases rapidly with carbon content. This, again, is what we would expect. We saw in Chapter 8 that martensite is a supersaturated solid solution of C in Fe. Pure iron at room temperature would be b.c.c., but the supersaturated carbon distorts the lattice, making it tetragonal (Fig. 11.9). The distortion increases linearly with the amount of dissolved carbon (Fig. 11.9); and because the distortion is what gives martensite its hardness then this, too, must increase with carbon content.

Although 0.8% carbon martensite is very hard, it is also very *brittle*. You can quench a 3 mm rod of tool steel into cold water and then snap it like a carrot. But if you *temper* martensite (reheat it to 300–600°C) you can regain the lost toughness with only a moderate sacrifice in hardness. Tempering gives the carbon atoms enough thermal energy that they can diffuse out of supersaturated solution and react with iron to form small closely spaced precipitates of Fe_3C (Fig. 11.10). The lattice relaxes back to the undistorted b.c.c. structure of equilibrium α, and the ductility goes up as a result. The Fe_3C particles precipitation-harden the steel and keep the hardness up. If the steel is over-tempered, however, the Fe_3C particles *coarsen* (they get larger and further apart) and the hardness falls. Figure 11.11 shows the big improvements in yield and tensile strength that can be obtained by quenching and tempering steels in this way.

Cast irons

Alloys of iron containing more than 1.7 wt% carbon are called *cast irons*. Carbon lowers the melting point of iron (see Fig. 11.1): a medium-carbon steel must be heated to about 1500°C to melt it, whereas a 4% cast iron is molten at only 1160°C. This is why cast iron is called cast iron: it can be melted with primitive furnaces and can be cast into intricate shapes using very basic sand

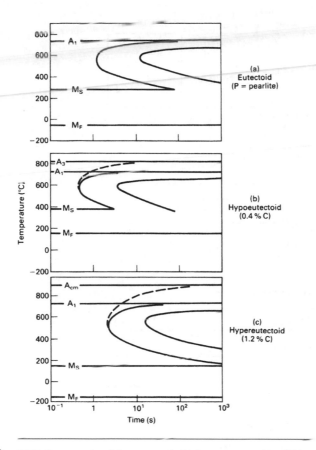

Figure 11.8   TTT diagrams for (**a**) eutectoid, (**b**) hypoeutectoid and (**c**) hypereutectoid steels. (**b**) and
(**c**) show (dashed lines) the C-curves for the formation of primary α and Fe_3C respectively.
Note that, as the carbon content increases, both M_S and M_F *decrease*.

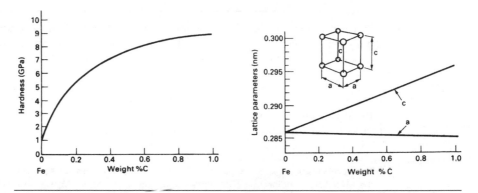

Figure 11.9   The hardness of martensite increases with carbon content because of the increasing
distortion of the lattice.

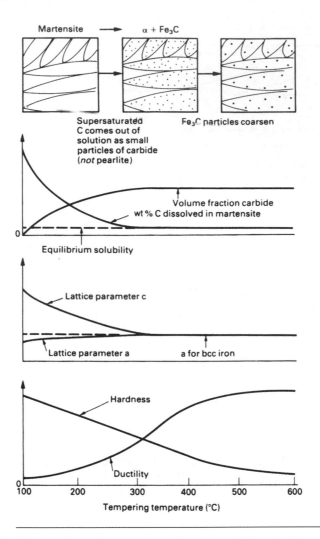

Figure 11.10   Changes during the tempering of martensite. There is a large driving force trying to make the martensite transform to the equilibrium phases of $\alpha + Fe_3C$. Increasing the temperature gives the atoms more thermal energy, allowing the transformation to take place.

casting technology. Cast iron castings have been made for hundreds of years.* The Victorians used cast iron for everything they could: bridges, architectural beams and columns, steam-engine cylinders, lathe beds, even garden furniture.

---

* The world's first iron bridge was put up in 1779 by the Quaker ironmaster Abraham Darby III. Spanning the River Severn in Shropshire the bridge is still there; the local village is now called Ironbridge. Another early ironmaster, the eccentric and ruthless "iron-mad" Wilkinson, lies buried in an iron coffin surmounted by an iron obelisk. He launched the world's first iron ship and invented the machine for boring the cylinders of James Watt's steam engines.

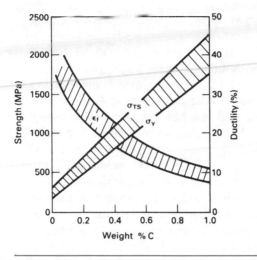

Figure 11.11   Mechanical properties of quenched-and-tempered steels. Compare with Fig. 11.7.

But most cast irons are brittle and should not be used where they are subjected to shock loading or high tensile stresses. When strong castings are needed, steel can be used instead. But it is only within the last 100 years that steel castings have come into use; and even now they are much more expensive than cast iron.

There are two basic types of cast iron: *white*, and *grey*. The phases in white iron are α and Fe_3C, and it is the large volume fraction of Fe_3C that makes the metal brittle. The name comes from the silvery appearance of the fracture surface, due to light being reflected from cleavage planes in the Fe_3C. In grey iron much of the carbon separates out as elemental carbon (graphite) rather than Fe_3C. Grey irons contain ≈ 2 wt% Si: this alters the thermodynamics of the system and makes iron–graphite more stable than iron–Fe_3C. If you cut a piece of grey iron with a hacksaw the graphite in the sawdust will turn your fingers black, and the cut surface will look dark as well, giving grey iron its name. It is the graphite that gives grey irons their excellent wear properties – in fact grey iron is the only metal which does not "scuff" or "pick up" when it runs on itself. The properties of grey iron depend strongly on the shape of the graphite phase. If it is in the form of large flakes, the toughness is low because the flakes are planes of weakness. If it is in the form of spheres (spheroidal-graphite, or "SG", iron) the toughness is high and the iron is surprisingly ductile. The graphite in grey iron is normally flaky, but SG irons can be produced if cerium or magnesium is added. Finally, some grey irons can be hardened by quenching and tempering in just the way that carbon steels can. The sliding surfaces of high-quality machine tools (lathes, milling machines, etc.) are usually hardened in this way, but in order to avoid distortion and cracking only the *surface* of the iron is heated to red heat (in a process called "induction hardening").

Some notes on the TTT diagram

The C-curves of TTT diagrams are determined by quenching a specimen to a given temperature, holding it there for a given time, and quenching to room temperature (Fig. 11.12). The specimen is then sectioned, polished and examined in the microscope. The percentage of Fe_3C present in the sectioned specimen allows one to find out how far the $\gamma \rightarrow \alpha + Fe_3C$ transformation has gone (Fig. 11.12). The complete set of C-curves can be built up by doing a large number of experiments at different temperatures and for different times. In order to get fast enough quenches, thin specimens are quenched into baths of molten salt kept at the various hold temperatures. A quicker alternative to quenching and sectioning is to follow the progress of the transformation with a high-resolution dilatometer: both α and Fe_3C are less dense than γ and the extent of the expansion observed after a given holding time tells us how far the transformation has gone.

When the steel transforms at a high temperature, with little undercooling, the pearlite in the steel is coarse – the plates in any nodule are relatively large and widely spaced. At slightly lower temperatures we get fine pearlite. Below the nose of the C-curve the transformation is too fast for the Fe_3C to grow in nice, tidy plates. It grows instead as isolated stringers to give a structure called "upper bainite" (Fig. 11.12). At still lower temperatures the Fe_3C grows as tiny rods and there is evidence that the α forms by a displacive transformation

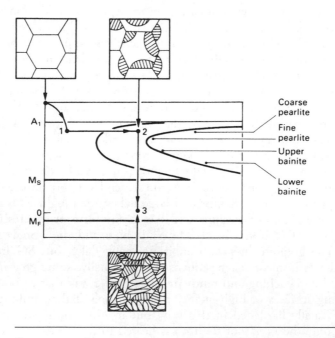

Figure 11.12   C-curves are determined using quench–hold–quench sequences.

("lower bainite"). The decreasing scale of the microstructure with increasing driving force (coarse pearlite → fine pearlite → upper bainite → lower bainite in Fig. 11.12) is an example of the general rule that, *the harder you drive a transformation, the finer the structure you get.*

Because C-curves are determined by quench–hold–quench sequences they can, strictly speaking, only be used to predict the microstructures that would be produced in a steel subjected to a quench–hold–quench heat treatment. But the curves do give a pretty good indication of the structures to expect in a steel that has been cooled *continuously.* For really accurate predictions, however, *continuous cooling diagrams* are available (see the literature of the major steel manufacturers).

The final note is that pearlite and bainite *only* form from undercooled γ. They *never* form from martensite. The TTT diagram *cannot* therefore be used to tell us anything about the rate of tempering in martensite.

Examples

11.1 The figure below shows the isothermal transformation diagram for a coarse-grained, plain-carbon steel of eutectoid composition. Samples of the steel are austenitised at 850°C and then subjected to the quenching treatments shown on the diagram. Describe the microstructure produced by each heat treatment.

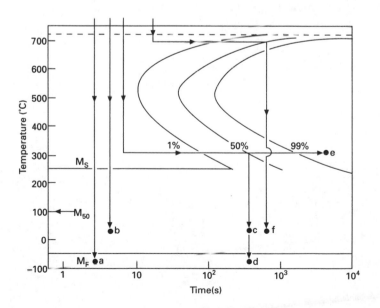

11.2   You have been given samples of the following materials:

(a) Pure iron.
(b) 0.3 wt% carbon steel.
(c) 0.8 wt% carbon steel.
(d) 1.2 wt% carbon steel.

Sketch the structures that you would expect to see if you looked at polished sections of the samples under a reflecting light microscope. Label the phases, and any other features of interest. You may assume that each specimen has been cooled moderately slowly from a temperature of 1100°C.

11.3   The densities of pure iron and iron carbide at room temperature are 7.87 and 8.15 $Mg\,m^{-3}$ respectively. Calculate the percentage by volume of a and Fe_3C in pearlite.

Answers

α, 88.9%; Fe_3C, 11.1%.

11.4   Samples of plain-carbon steel of eutectoid composition are to be heat-treated to produce the following metallurgical structures:

(a) pearlite;
(b) bainite;
(c) martensite;
(d) tempered martensite.

In each case, describe the structure (using simple sketches to illustrate your answer) and indicate a suitable heat treatment.

Chapter 12
Steels: II – alloy steels

Introduction

A small, but important, sector of the steel market is that of the alloy steels: the low-alloy steels, the high-alloy "stainless" steels and the tool steels. Alloying elements are added to steels with four main aims in mind:

(a) to improve the *hardenability* of the steel;
(b) to give *solution strengthening* and *precipitation hardening*;
(c) to give *corrosion resistance*;
(d) to *stabilise austenite*, giving a steel that is austenitic (f.c.c.) at room temperature.

Hardenability

We saw in the last chapter that carbon steels could be strengthened by *quenching and tempering*. To get the best properties we must quench the steel past the nose of the C-curve. The cooling rate that just misses the nose is called the *critical cooling rate* (CCR). If we cool at the critical rate, or faster, the steel will transform to 100% martensite.* The CCR for a plain carbon steel depends on two factors – carbon content and grain size. We have already seen (in Chapter 8) that adding carbon decreases the rate of the diffusive transformation by orders of magnitude: the CCR decreases from $\approx 10^5 {}^\circ C\,s^{-1}$ for pure iron to $\approx 200{}^\circ C\,s^{-1}$ for 0.8% carbon steel (see Fig. 12.1). We also saw in Chapter 8 that the rate of a diffusive transformation was proportional to the number of nuclei forming per m^3 per second. Since grain boundaries are favourite nucleation sites, a fine-grained steel should produce more nuclei than a coarse-grained one. The fine-grained steel will therefore transform more rapidly than the coarse-grained steel, and will have a higher CCR (Fig. 12.1).

Quenching and tempering is usually limited to steels containing more than about 0.1% carbon. Figure 12.1 shows that these must be cooled at rates ranging from 100 to 2000°C s^{-1} if 100% martensite is to be produced. There is no difficulty in transforming the *surface* of a component to martensite – we simply quench the red-hot steel into a bath of cold water or oil. But if the component is at all large, the surface layers will tend to insulate the

---

* Provided, of course, that we continue to cool the steel down to the martensite finish temperature.

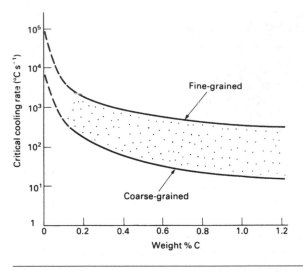

Figure 12.1   The effect of carbon content and grain size on the critical cooling rate.

bulk of the component from the quenching fluid. The bulk will cool more slowly than the CCR and will not harden properly. Worse, a rapid quench can create shrinkage stresses which are quite capable of cracking brittle, untempered martensite.

These problems are overcome by alloying. The entire TTT curve is shifted to the right by adding a small percentage of the right alloying element to the steel – usually molybdenum (Mo), manganese (Mn), chromium (Cr) or nickel (Ni) (Fig. 12.2). Numerous *low-alloy steels* have been developed with superior *hardenability* – the ability to form martensite in thick sections when

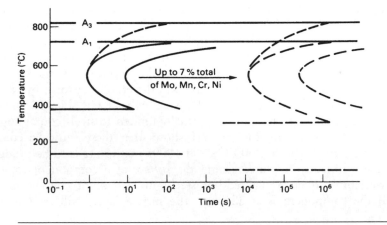

Figure 12.2   Alloying elements make steels more hardenable.

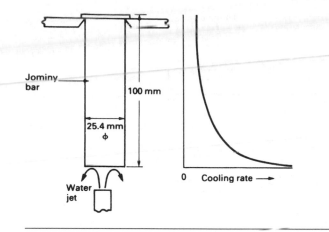

Figure 12.3   The Jominy end-quench test for hardenability.

quenched. This is one of the reasons for adding the 2–7% of alloying elements (together with 0.2–0.6% C) to steels used for things like crankshafts, high-tensile bolts, springs, connecting rods, and spanners. Alloys with lower alloy contents give martensite when quenched into oil (a moderately rapid quench); the more heavily alloyed give martensite even when cooled in air. Having formed martensite, the component is tempered to give the desired combination of strength and toughness.

Hardenability is so important that a simple test is essential to measure it. The Jominy end-quench test, though inelegant from a scientific standpoint, fills this need. A bar 100 mm long and 25.4 mm in diameter is heated and held in the austenite field. When all the alloying elements have gone into solution, a jet of water is sprayed onto one end of the bar (Fig. 12.3). The surface cools very rapidly, but sections of the bar behind the quenched surface cool progressively more slowly (Fig. 12.3). When the whole bar is cold, the hardness is measured along its length. A steel of high hardenability will show a uniform, high hardness along the whole length of the bar (Fig. 12.4). This is because the cooling rate, even at the far end of the bar, is greater than the CCR; and the whole bar transforms to martensite. A steel of medium hardenability gives quite different results (Fig. 12.5). The CCR is much higher, and is only exceeded in the first few centimetres of the bar. Once the cooling rate falls below the CCR the steel starts to transform to bainite rather than martensite, and the hardness drops off rapidly.

Solution hardening

The alloying elements in the *low-alloy steels* dissolve in the ferrite to form a substitutional solid solution. This solution strengthens the steel and gives useful additional strength. The *tool steels* contain large amounts of dissolved

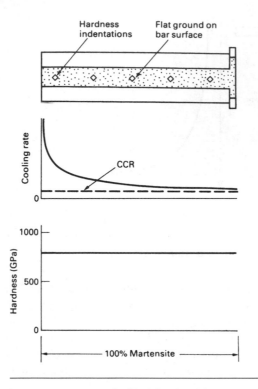

Figure 12.4   Jominy test on a steel of high hardenability.

tungsten (W) and cobalt (Co) as well, to give the maximum feasible solution strengthening. Because the alloying elements have large solubilities in both ferrite and austenite, no special heat treatments are needed to produce good levels of solution hardening. In addition, the solution-hardening component of the strength is not upset by overheating the steel. For this reason, low-alloy steels can be welded, and cutting tools can be run hot without affecting the solution-hardening contribution to their strength.

Precipitation hardening

The *tool steels* are an excellent example of how metals can be strengthened by precipitation hardening. Traditionally, cutting tools have been made from 1% carbon steel with about 0.3% of silicon (Si) and manganese (Mn). Used in the quenched and tempered state they are hard enough to cut mild steel and tough enough to stand up to the shocks of intermittent cutting. But they have one serious drawback. When cutting tools are in use they become hot: woodworking tools become warm to the touch, but metalworking tools can burn you. It is easy to "run the temper" of plain carbon metalworking tools, and the resulting drop in hardness will destroy the cutting edge. The problem can be overcome

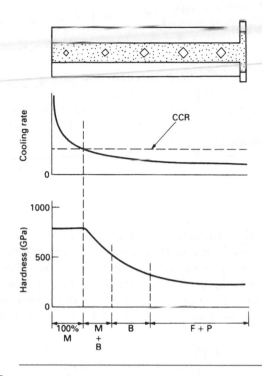

Figure 12.5   Jominy test on a steel of medium hardenability. M = martensite, B = bainite, F = primary ferrite, P = pearlite.

by using low cutting speeds and spraying the tool with cutting fluid. But this is an expensive solution – slow cutting speeds mean low production rates and expensive products. A better answer is to make the cutting tools out of *high-speed* steel. This is an alloy tool steel containing typically 1% C, 0.4% Si, 0.4% Mn, 4% Cr, 5% Mo, 6% W, 2% vanadium (V) and 5% Co. The steel is used in the quenched and tempered state (the Mo, Mn and Cr give good hardenability) and owes its strength to two main factors: the fine dispersion of Fe_3C that forms during tempering, and the solution hardening that the dissolved alloying elements give.

Interesting things happen when this high-speed steel is heated to 500–600°C. The Fe_3C precipitates dissolve and the carbon that they release combines with some of the dissolved Mo, W and V to give a fine dispersion of Mo_2C, W_2C and VC precipitates. This happens because Mo, W and V are strong *carbide formers*. If the steel is now cooled back down to room temperature, we will find that this *secondary hardening* has made it even stronger than it was in the quenched and tempered state. In other words, "running the temper" of a high-speed steel makes it harder, not softer; and tools made out of high-speed steel can be run at much higher cutting speeds (hence the name).

Corrosion resistance

Plain carbon steels rust in wet environments and oxidise if heated in air. But if chromium is added to steel, a hard, compact film of Cr_2O_3 will form on the surface and this will help to protect the underlying metal. The minimum amount of chromium needed to protect steel is about 13%, but up to 26% may be needed if the environment is particularly hostile. The iron–chromium system is the basis for a wide range of *stainless steels*.

Stainless steels

The simplest stainless alloy contains just iron and chromium (it is actually called stainless *iron*, because it contains virtually no carbon). Figure 12.6 shows the Fe–Cr phase diagram. The interesting thing about this diagram is that alloys containing $\geqslant$13% Cr have a b.c.c. structure all the way from 0 K to the melting point. They do not enter the f.c.c. phase field and cannot be quenched to form martensite. Stainless irons containing $\geqslant$13% Cr are therefore always *ferritic*.

Hardenable stainless steels usually contain up to 0.6% carbon. This is added in order to change the Fe–Cr phase diagram. As Fig. 12.7 shows, carbon expands the γ field so that an alloy of Fe–15% Cr, 0.6% C lies inside the γ field at 1000°C. This steel can be quenched to give martensite; and the martensite can be tempered to give a fine dispersion of alloy carbides.

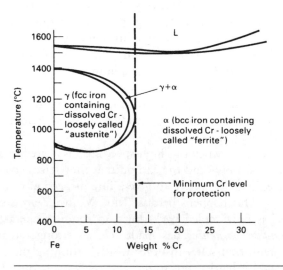

Figure 12.6   The Fe–Cr phase diagram.

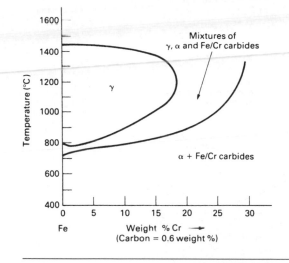

Figure 12.7   Simplified phase diagram for the Fe–Cr–0.6% C system.

These quenched and tempered stainless steels are ideal for things like non-rusting ball-bearings, surgical scalpels and kitchen knives.*

Many stainless steels, however, are *austenitic* (f.c.c.) at room temperature. The most common austenitic stainless, "18/8", has a composition Fe–0.1% C, 1% Mn, 18% Cr, 8% Ni. The chromium is added, as before, to give corrosion resistance. But *nickel* is added as well because it *stabilises austenite*. The Fe–Ni phase diagram (Fig. 12.8) shows why. Adding nickel lowers the temperature of the f.c.c.–b.c.c. transformation from 914°C for pure iron to 720°C for Fe–8% Ni. In addition, the Mn, Cr and Ni slow the diffusive f.c.c.–b.c.c. transformation down by orders of magnitude. 18/8 stainless steel can therefore be cooled in air from 800°C to room temperature without transforming to b.c.c. The austenite is, of course, unstable at room temperature. However, diffusion is far too slow for the metastable austenite to transform to ferrite by a diffusive mechanism. It is, of course, possible for the austenite to transform displacively to give martensite. But the large amounts of Cr and Ni lower the M_s temperature to ≈0°C. This means that we would have to cool the steel well below 0°C in order to lose much austenite.

Austenitic steels have a number of advantages over their ferritic cousins. They are tougher and more ductile. They can be formed more easily by stretching or deep drawing. Because diffusion is slower in f.c.c. iron than in b.c.c. iron, they have better creep properties. And they are non-magnetic, which makes them ideal for instruments like electron microscopes and mass spectrometers.

---

* Because both ferrite and martensite are magnetic, kitchen knives can be hung up on a strip magnet screwed to the kitchen wall.

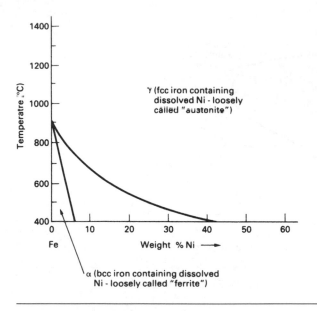

Figure 12.8   The Fe–Ni phase diagram.

But one drawback is that austenitic steels work harden very rapidly, which makes them rather difficult to machine.

Examples

12.1   Explain the following.

(a) The critical cooling rate (CCR) is approximately $700°C\,s^{-1}$ for a fine-grained 0.6% carbon steel, but is only around $30°C\,s^{-1}$ for a coarse-grained 0.6% carbon steel.

(b) A stainless steel containing 18% Cr has a bcc structure at room temperature, whereas a stainless steel containing 18% Cr plus 8% Ni has an fcc structure at room temperature.

(c) High-speed steel cutting tools retain their hardness to well above the temperature at which the initial martensitic structure has become over-tempered.

12.2   A steel shaft 40 mm in diameter is to be hardened by austenitising followed by quenching into cold oil. The centre of the bar must be 100% martensite. The following table gives the cooling rate at the centre of an oil quenched bar as a function of bar diameter.

Bar diameter (mm)	Cooling rate (°C s^{-1})
500	0.17
100	2.5
20	50
5	667

It is proposed to make the shaft from a NiCrMo low-alloy steel. The critical cooling rates of NiCrMo steels are given quite well by the empirical equation

$$\log_{10}(\text{CCR in °C s}^{-1}) = 4.3 - 3.27 \text{ C} - \frac{(\text{Mn} + \text{Cr} + \text{Mo} + \text{Ni})}{1.6}$$

where the symbol given for each element denotes its weight percentage. Which of the following steels would be suitable for this application?

Steel	Weight percentages				
	C	Mn	Cr	Mo	Ni
A	0.30	0.80	0.50	0.20	0.55
B	0.40	0.60	1.20	0.30	1.50
C	0.36	0.70	1.50	0.25	1.50
D	0.40	0.60	1.20	0.15	1.50
E	0.41	0.85	0.50	0.25	0.55
F	0.40	0.65	0.75	0.25	0.85
G	0.40	0.60	0.65	0.55	2.55

[Hint: there is a log–log relationship between bar diameter and cooling rate.]

Answer

Steels B, C, D, G.

Chapter 13
Case studies in steels

Metallurgical detective work after a boiler explosion

The first case study shows how a knowledge of steel microstructures can help us trace the chain of events that led to a damaging engineering failure.

The failure took place in a large water-tube boiler used for generating steam in a chemical plant. The layout of the boiler is shown in Fig. 13.1. At the bottom of the boiler is a cylindrical pressure vessel – the mud drum – which contains water and sediments. At the top of the boiler is the steam drum, which contains water and steam. The two drums are connected by 200 tubes through which the water circulates. The tubes are heated from the outside by the flue gases from a coal-fired furnace. The water in the "hot" tubes moves upwards from the mud drum to the steam drum, and the water in the "cool" tubes moves downwards from the steam drum to the mud drum. A convection circuit is therefore set up where water circulates around the boiler and picks up heat in the process. The water tubes are 10 m long, have an outside diameter of 100 mm and are 5 mm thick in the wall. They are made from a steel of composition Fe–0.18% C, 0.45% Mn, 0.20% Si. The boiler operates with a working pressure of 50 bar and a water temperature of 264°C.

In the incident some of the "hot" tubes became overheated, and started to bulge. Eventually one of the tubes burst open and the contents of the boiler were discharged into the environment. No one was injured in the explosion, but it took several months to repair the boiler and the cost was heavy. In order to prevent another accident, a materials specialist was called in to examine the failed tube and comment on the reasons for the failure.

Figure 13.2 shows a schematic diagram of the burst tube. The first operation was to cut out a 20 mm length of the tube through the centre of the failure. One of the cut surfaces of the specimen was then ground flat and tested for hardness. Figure 13.3 shows the data that were obtained. The hardness of most of the section was about 2.2 GPa, but at the edges of the rupture the hardness went up to 4 GPa. This indicates (see Fig. 13.3) that the structure at the rupture edge is mainly martensite. However, away from the rupture, the structure is largely bainite. Hardness tests done on a spare boiler tube gave only 1.5 GPa, showing that the failed tube would have had a ferrite + pearlite microstructure to begin with.

In order to produce martensite and bainite the tube must have been overheated to at least the A_3 temperature of 870°C (Fig. 13.4). When the rupture occurred the rapid outrush of boiler water and steam cooled the steel rapidly

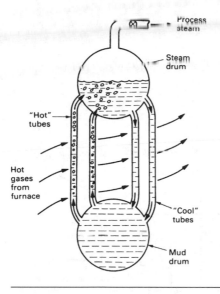

Figure 13.1   Schematic of water-tube boiler.

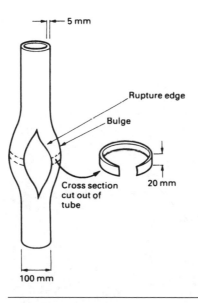

Figure 13.2   Schematic of burst tube.

down to 264°C. The cooling rate was greatest at the rupture edge, where the section was thinnest: high enough to quench the steel to martensite. In the main bulk of the tube the cooling rate was less, which is why bainite formed instead.

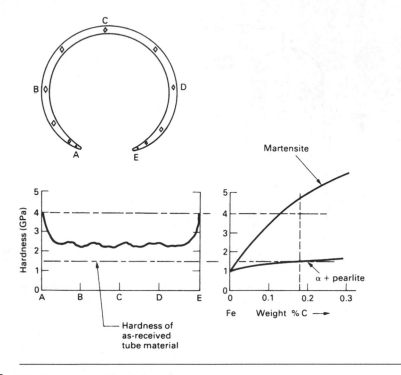

Figure 13.3   The hardness profile of the tube.

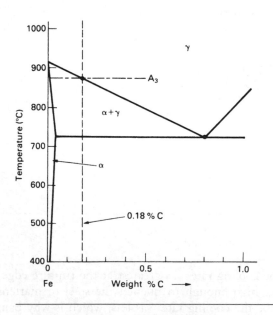

Figure 13.4   Part of the iron–carbon phase diagram.

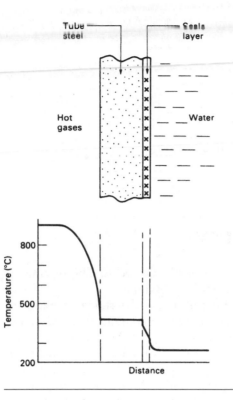

Figure 13.5   Temperature distribution across the water-tube wall.

The hoop stress in the tube under the working pressure of 50 bar (5 MPa) is 5 MPa × 50 mm/5 mm = 50 MPa. Creep data indicate that, at 900°C and 50 MPa, the steel should fail after only 15 minutes or so. In all probability, then, the failure occurred by creep rupture during a short temperature excursion to at least 870°C.

How was it that water tubes reached such high temperatures? We can give two probable reasons. The first is that "hard" feed water will – unless properly treated – deposit scale inside the tubes (Fig. 13.5). This scale will help to insulate the metal from the boiler water and the tube will tend to overheat. Secondly, water circulation in a natural convection boiler can be rather hit-and-miss; and the flow in some tubes can be very slow. Under these conditions a stable layer of dry steam can form next to the inner wall of the tube and this, too, can be a very effective insulator.

Welding steels together safely

Many steel structures – like bridges, storage tanks, and ships – are held together by welds. And when incidents arise from fast fractures or fatigue failures they

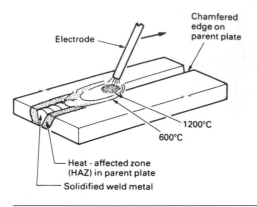

Figure 13.6   Schematic of a typical welding operation.

can often be traced to weaknesses in the welds. The sinking of the Alexander Keilland oil platform in 1980 is an example.

Figure 13.6 is a schematic of a typical welding operation. An electric arc is struck between the electrode (which contains filler metal and flux) and the parent plates. The heat from the arc melts the filler metal which runs into the gap to form a molten pool. Although the pool loses heat to the surrounding cold metal this is replaced by energy from the arc. In fact some of the parent material is melted back as well. But as the arc moves on, the molten steel that it leaves behind solidifies rapidly, fusing the two plates together. Figure 13.6 shows how the temperature in the metal varies with distance from the weld pool. Because the electrode moves along the weld, the isotherms bunch up in front of the pool like the bow wave of a ship; behind the pool they are spaced more widely. The passage of this thermal "wave" along the weld leads to very rapid *heating* of the metal in front of the weld pool and slightly less rapid *cooling* of the metal behind (Fig. 13.7). The section of the plate that is heated above $\approx 650°C$ has its mechanical properties changed as a result. For this reason it is called the *heat-affected zone* (HAZ).

The most critical changes occur in the part of the HAZ that has been heated above the A_3 temperature (Fig. 13.7). As the arc moves on, this part of the HAZ can cool as quickly as $100°C \, s^{-1}$. With a fine-grained carbon steel this should not be a problem: the CCR is more than $400°C \, s^{-1}$ (Fig. 12.1) and little, if any, martensite should form. But some of the steel in the HAZ goes up to temperatures as high as $1400°C$. At such temperatures diffusion is extremely rapid, and in only a few seconds significant *grain growth* will take place (see Chapter 5). The CCR for this grain-growth zone will be reduced accordingly (Fig. 12.1). In practice, it is found that martensite starts to appear in the HAZ when the carbon content is more than 0.5 to 0.6%.

The last thing that is wanted in a weld is a layer of hard, brittle martensite. It can obviously make the weld as a whole more brittle. And it can encourage *hydrogen cracking*. All welds contain quantities of atomic hydrogen (the molten

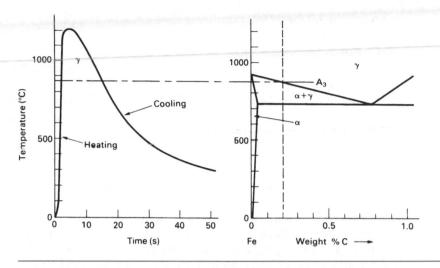

Figure 13.7 The left-hand graph shows how the temperature at one point in the parent plate changes as the welding arc passes by. The point chosen here is quite close to the edge of the plate, which is why it reaches a high peak temperature. A point further away from the edge would not reach such a high peak temperature.

steel in the weld pool rapidly reduces atmospheric moisture to give iron oxide and hydrogen). Because hydrogen is such a small atom it can diffuse rapidly through steel (even at room temperature) and coalesce to give voids which, in a brittle material like martensite, will grow into cracks. These cracks can then extend (e.g. by fatigue) until they are long enough to cause fast fracture.

Example 1: Weldable structural steel to BS 4360 grade 43A

BS 4360 is the structural steel workhorse. Grade 43A has the following specifications: C $\leqslant$ 0.25%, Mn $\leqslant$ 1.60%, Si $\leqslant$ 0.50%; σ_{TS} 430 to 510 MPa; $\sigma_y \geqslant$ 240 MPa.

The maximum carbon content of 0.25% is well below the 0.5 to 0.6% that may give HAZ problems. But, in common with all "real" carbon steels, 4360 contains manganese. This is added to react with harmful impurities like sulphur. Any unreacted manganese dissolves in ferrite where it contributes solid-solution strengthening and gives increased hardenability. But how do we know whether our 1.6% of manganese is likely to give HAZ problems? Most welding codes assess the effect of alloying elements from the empirical formula

$$CE = C + \frac{Mn}{6} + \frac{Cr + Mo + V}{5} + \frac{Ni + Cu}{15} \tag{13.1}$$

where CE is the *carbon equivalent* of the steel. The 1.6% Mn in our steel would thus be equivalent to 1.6%/6 = 0.27% carbon in its contribution to martensite formation in the HAZ. The total carbon equivalent is $0.25 + 0.27 = 0.52$. The cooling rate in most welding operations is unlikely to be high enough to quench this steel to martensite. But poorly designed welds – like a small weld bead laid on a massive plate – can cool very quickly. For this reason failures often initiate not in the main welds of a structure but in small welds used to attach ancillary equipment like ladders and gangways. (It was a small weld of this sort which started the crack which led to the Alexander Keilland failure.)

Example 2: Pressure vessel steel to A 387 grade 22 class 2

A 387–22(2) is a creep-resistant steel which can be used at about 450°C. It is a standard material for pipes and pressure vessels in chemical plants and oil refineries. The specification is: C $\leqslant$ 0.15%, Mn 0.25 to 0.66%, Si $\leqslant$ 0.50%, Cr 1.88 to 2.62%, Mo 0.85 to 1.15%; σ_{TS} 515 to 690 MPa; $\sigma_y \geqslant$ 310 MPa.

The carbon equivalent for the maximum composition figures given is

$$CE = 0.15 + \frac{0.66}{6} + \frac{2.62 + 1.15}{5} = 1.01. \tag{13.2}$$

The Mn, Cr and Mo in the steel have increased the hardenability considerably, and martensite is likely to form in the HAZ unless special precautions are taken. But the cooling rate can be decreased greatly if the parent plate is pre-heated to $\approx$350°C before welding starts, and this is specified in the relevant welding code. Imagine yourself, though, as a welder working inside a large pre-heated pressure vessel: clad in an insulating suit and fed with cool air from outside you can only stand 10 minutes in the heat before making way for somebody else!

The case of the broken hammer

The father of one of the authors was breaking up some concrete slabs with an engineers' hammer when a large fragment of metal broke away and narrowly missed his eye. A $1\frac{1}{2}$ lb hammer is not heavy enough for breaking up slabs, and he should have been wearing eye protection, but that is another matter. We examined the hammer to see whether the fracture might have been caused by faulty heat treatment.

Figure 13.8(a) shows the general shape of the hammer head, and marks the origin of the offending fragment. Figure 13.8(b) is a close-up of the crater that the fragment left behind when it broke away.

The first thing that we did was to test the hardness of the hammer. The results are shown in Fig. 13.9. The steel is very hard near the striking face

(a)   (b)

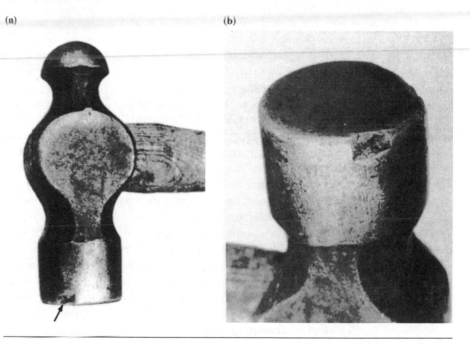

Figure 13.8   (a) General view of hammer head. The origin of the fragment is marked with an arrow.
(b) Close-up of crater left by fragment.

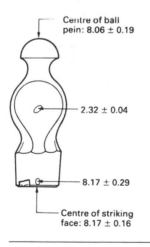

Centre of ball
pein: 8.06 ± 0.19

2.32 ± 0.04

8.17 ± 0.29

Centre of striking
face: 8.17 ± 0.16

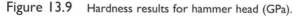

Figure 13.9   Hardness results for hammer head (GPa).

and the ball pein, but the main body of the head is much softer. A typical standard for hammers (BS 876, 1995) states that hammers shall be made from a medium-carbon forging steel with the following limits on composition: 0.5 to 0.6% C, 0.5 to 0.9% Mn and 0.1 to 0.4% Si. We would expect this steel to have a hardness of between 1.8 and 2.2 GPa in the as-received state. This is very close to the figure of 2.32 GPa that we found for the bulk of the head, but much less than the hardness of face and pein. The bulk of the hammer was thus in the as-forged state, but both face and pein must have been hardened by quenching.

We can see from Fig. 11.9 that untempered 0.55% carbon martensite should have a hardness of 8.0 GPa. This is essentially identical to the hardnesses that we found on the striking face and the ball pein, and suggests strongly that the face had not been tempered at all. In fact, the Standard says that faces and peins must be tempered to bring the hardness down into the range 5.1 to 6.6 GPa. Presumably experience has shown that this degree of tempering makes the steel tough enough to stand up to normal wear and tear. Untempered martensite is far too brittle for hammer heads.

Having solved the immediate problem to our satisfaction we still wondered how the manufacturer had managed to harden the striking faces without affecting the bulk of the hammer head. We contacted a reputable maker of hammers who described the sequence of operations that is used. The heads are first shaped by hot forging and then allowed to cool slowly to room temperature. The striking face and ball pein are finished by grinding. The striking faces are austenitised as shown in Fig. 13.10 and are then quenched into cold water. The only parts of the hammers to go above 723°C are the ends: so only these parts can be hardened by the quench. The quenched heads are then dried and the ends are tempered by completely immersing the heads in a bath of molten salt at 450°C. Finally, the heads are removed from the tempering bath and washed in cold water. The rather complicated austenitising treatment is needed because the Standard insists that the hardened zone must not extend into the

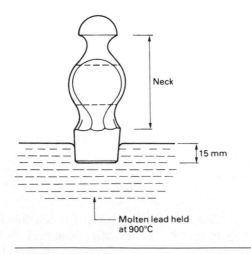

Figure 13.10   Austenitising a striking face.

neck of the hammer (Fig. 13.10). You can imagine how dangerous it would be if a hammer broke clean in two. The neck does not need to be hard, but it *does* need to be tough. Although there are standards for hammers, there is often no legislation which compels retailers to supply only standard hammers. It is, in fact, quite difficult to get standard hammers "over the counter". But reputable makers spot check their hammers, because they will not knowingly sell improperly heat-treated hammers.

Examples

13.1 The heat exchanger in a reformer plant consisted of a bank of tubes made from a low-alloy ferritic steel containing 0.2 weight% carbon. The tubes contained hydrocarbon gas at high pressure and were heated from the outside by furnace gases. The tubes had an internal diameter of 128 mm and a wall thickness of 7 mm. Owing to a temperature overshoot, one of the tubes fractured and the resulting gas leak set the plant on fire.

When the heat exchanger was stripped down it was found that the tube wall had bulged over a distance of about 600 mm. In the most expanded region of the bulge, the tube had split longitudinally over a distance of about 300 mm. At the edge of the fracture the wall had thinned down to about 3 mm. Metallurgical sections were cut from the tube at two positions: (i) immediately next to the fracture surface half-way along the length of the split, (ii) 100 mm away from the end of the split in the part of the tube which, although slightly expanded, was otherwise intact.

The microstructure at position (ii) consisted of grains of ferrite and colonies of pearlite. It was noticed that the pearlite had started to "spheroidise" (see Example 5.2). The microstructure at position (i) consisted of grains of ferrite and grains of lower bainite in roughly equal proportions. Estimate the temperatures to which the tube been heated at positions (i) and (ii). Explain the reasoning behind your answers.

Answers

(i) 830°C; (ii) 700°C.

13.2 In 1962 a span of Kings Bridge (Melbourne, Australia) collapsed by brittle fracture. The fracture started from a crack in the heat-affected zone (HAZ) of a transverse fillet weld, which had been used to attach a reinforcing plate to the underside of a main structural I-beam (see the diagram on the next page). The concentrations of the alloying elements in the steel (in weight%) were: C, 0.26; Mn, 1.80; Cr, 0.25. The welding was done by hand, without any special precautions. The welding electrodes had become damp before use.

Account for the HAZ cracking. After the collapse, the other transverse welds in the bridge were milled-out and rewelded. What procedures would you specify to avoid a repeat of the HAZ cracking?

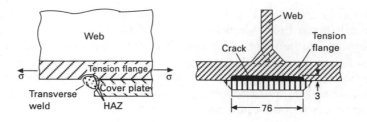

13.3 Steels for railroad rails typically contain 0.80 weight% carbon, 0.3 weight% silicon and 1.0 weight% manganese. The steel is processed to give a fine-grained pearlitic structure with a hardness of approximately 2.8 GPa. However, after a period in service, it is commonly found that a thin, hard layer (the "white layer") forms in patches on the top (running) surface of the rail. The microhardness of this white layer is typically around 8 GPa. Given that frictional heating between the wheels of rail vehicles and the running surface of the rail can raise the temperature at the interface to 800°C, explain why the white layer forms and account for its high hardness.

13.4 A rotating steel shaft from a high-speed textile machine was repaired with a circumferential surface weld. The shaft failed shortly after being put back into service by fatigue, which initiated in HAZ cracks. The composition of the steel was (wt%): C, 0.40; Mn, 0.72; Cr, 0.96; Mo, 0.22; Ni, 0.23. Explain why the HAZ cracked.

Chapter 14
Production, forming and joining of metals

Introduction

Figure 14.1 shows the main routes that are used for processing raw metals into finished articles. Conventional forming methods start by *melting* the basic metal and then *casting* the liquid into a mould. The casting may be a large prism-shaped ingot, or a continuously cast "strand", in which case it is *worked* to standard sections (e.g. sheet, tube) or *forged* to shaped components. Shaped components are also made from standard sections by *machining* or *sheet metal-working*. Components are then assembled into finished articles by *joining* operations (e.g. welding) which are usually carried out in conjunction with *finishing* operations (e.g. grinding or painting). Alternatively, the casting can be made to the final shape of the component, although some light machining will usually have to be done on it.

Increasing use is now being made of alternative processing routes. In *powder metallurgy* the liquid metal is atomised into small droplets which solidify to a fine powder. The powder is then *hot pressed* to shape (as we shall see in Chapter 19, hot-pressing is the method used for shaping high-technology ceramics). *Melt spinning* (Chapter 9) gives high cooling rates and is used to make amorphous alloys. Finally, there are a number of specialised processes in which components are formed directly from metallic compounds (e.g. *electro forming* or *chemical vapour deposition*).

It is not our intention here to give a comprehensive survey of the forming processes listed in Fig. 14.1. This would itself take up a whole book, and details can be found in the many books on production technology. Instead, we look at the underlying principles, and relate them to the characteristics of the materials that we are dealing with.

Casting

We have already looked at casting structures in Chapter 9. Ingots tend to have the structure shown in Fig. 14.2. When the molten metal is poured into the mould, *chill* crystals nucleate on the cold walls of the mould and grow inwards. But the chill crystals are soon overtaken by the much larger columnar grains. Finally, nuclei are swept into the remaining liquid and these grow to produce *equiaxed* grains at the centre of the ingot. As the crystals grow they reject dissolved impurities into the remaining liquid, causing *segregation*. This can

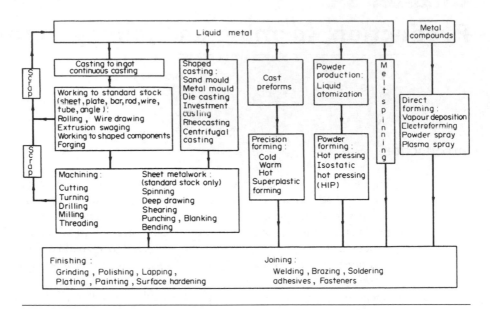

Figure 14.1   Processing routes for metals.

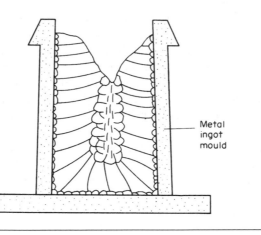

Figure 14.2   Typical ingot structure.

lead to bands of solid impurities (e.g. iron sulphide in steel) or to gas bubbles (e.g. from dissolved nitrogen). And because most metals contract when they solidify, there will be a substantial *contraction cavity* at the top of the ingot as well (Fig. 14.2).

These *casting defects* are not disastrous in an ingot. The top, containing the cavity, can be cut off. And the gas pores will be squashed flat and welded solid when the white-hot ingot is sent through the rolling mill. But there are still a

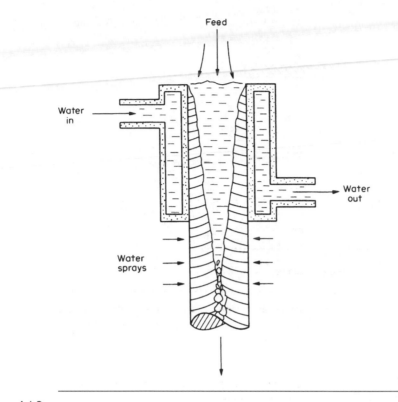

Feed

Water
in

Water
out

Water
sprays

Figure 14.3   Continuous casting.

number of disadvantages in starting with ingots. Heavy segregation may persist through the rolling operations and can weaken the final product.* And a great deal of work is required to roll the ingot down to the required section.

Many of these problems can be solved by using *continuous casting* (Fig. 14.3). Contraction cavities do not form because the mould is continuously topped up with liquid metal. Segregation is reduced because the columnar grains grow over smaller distances. And, because the product has a small cross-section, little work is needed to roll it to a finished section.

Shaped castings must be poured with much more care than ingots. Whereas the structure of an ingot will be greatly altered by subsequent working operations, the structure of a shaped casting will directly determine the strength of the finished article. Gas pores should be avoided, so the liquid metal must be *degassed* to remove dissolved gases (either by adding reactive chemicals or – for high-technology applications – casting in a vacuum). *Feeders* must

* Welded joints are usually in a state of high *residual stress*, and this can tear a steel plate apart if it happens to contain layers of segregated impurity.

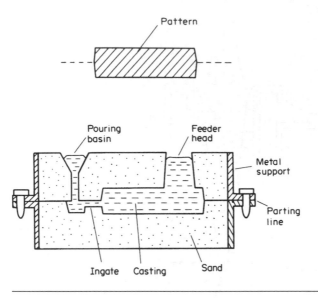

Figure 14.4   Sand casting. When the casting has solidified it is removed by destroying the sand mould. The casting is then "fettled" by cutting off the ingate and the feeder head.

be added (Fig. 14.4) to make up the contraction. And inoculants should be added to *refine* the grain size (Chapter 9). This is where powder metallurgy is useful. When atomised droplets solidify, contraction is immaterial. Segregation is limited to the size of the powder particles (2 to 150 μm); and the small powder size will give a small grain size in the hot-pressed product.

Shaped castings are usually poured into moulds of sand or metal (Fig. 14.4). The first operation in sand casting is to make a *pattern* (from wood, metal or plastic) shaped like the required article. Sand is rammed around the pattern and the mould is then split to remove the pattern. Passages are cut through the sand for ingates and risers. The mould is then re-assembled and poured. When the casting has gone solid it is removed by destroying the mould. Metal moulds are machined from the solid. They must come apart in enough places to allow the casting to be removed. They are costly, but can be used repeatedly; and they are ideal for pressure die casting (Fig. 14.5), which gives high production rates and improved accuracy. Especially intricate castings cannot be made by these methods: it is impossible to remove a complex pattern from a sand mould, and impossible to remove a complex casting from a metal one! This difficulty can be overcome by using *investment casting* (Fig. 14.6). A wax pattern is coated with a ceramic slurry. The slurry is dried to give it strength, and is then fired (as Chapter 19 explains, this is just how we make ceramic cups and plates). During firing the wax burns out of the ceramic mould to leave a perfectly shaped mould cavity.

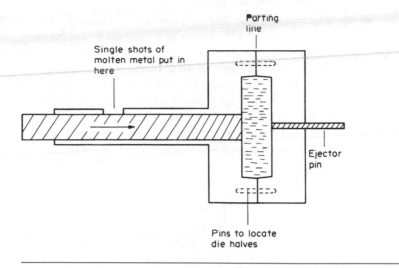

Figure 14.5   Pressure die casting.

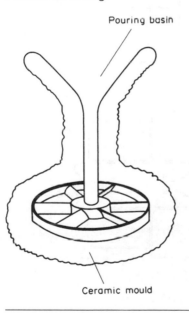

Figure 14.6   Investment casting.

Working processes

The working of metals and alloys to shape relies on their great *plasticity*: they can be deformed by large percentages, especially in compression, without breaking. But the *forming pressures* needed to do this can be large – as high as $3\sigma_y$ or even more, depending on the geometry of the process.

We can see where these large pressures come from by modelling a typical forging operation (Fig. 14.7). In order to calculate the forming pressure at a given position x we apply a force f to a movable section of the forging die. If we break the forging up into four separate pieces we can arrange for it to deform when the movable die sections are pushed in. The sliding of one piece over another requires a shear stress k (the shear yield stress). Now the work needed to push the die sections in must equal the work needed to shear the pieces of the forging over one another. The work done on each die section is $f \times u$, giving a total work input of $2fu$. Each sliding interface has area $\sqrt{2}(d/2)L$.

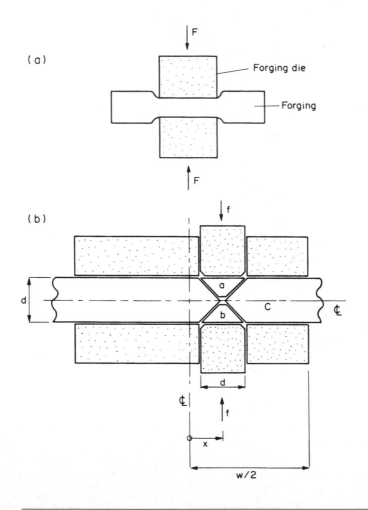

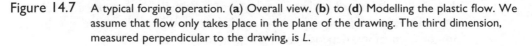

Figure 14.7   A typical forging operation. (a) Overall view. (b) to (d) Modelling the plastic flow. We assume that flow only takes place in the plane of the drawing. The third dimension, measured perpendicular to the drawing, is L.

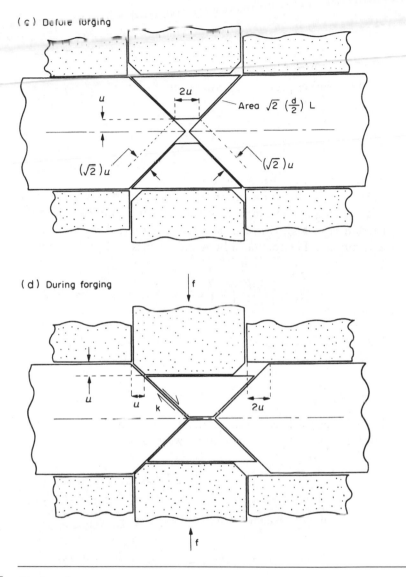

(c) Before forging

2u

u

Area $\sqrt{2}\left(\frac{d}{2}\right)$ L

$(\sqrt{2})u$

$(\sqrt{2})u$

(d) During forging

f

u

u

k

2u

f

Figure 14.7 (Continued)

The sliding force at each interface is thus $\sqrt{2}(d/2)L \times k$. Each piece slides a distance $(\sqrt{2})u$ relative to its neighbour. The work absorbed at each interface is thus $\sqrt{2}(d/2)Lk(\sqrt{2})u$; and there are four interfaces. The work balance thus gives

$$2fu = 4\sqrt{2}(d/2)Lk(\sqrt{2})u = 4dLku, \tag{14.1}$$

or

$$f = 2dLk. \tag{14.2}$$

The forming pressure, p_f, is then given by

$$p_f = \frac{f}{dL} = 2k = \sigma_y \tag{14.3}$$

which is just what we would expect.

We get a quite different answer if we include the friction between the die and the forging. The extreme case is one of sticking friction: the coefficient of friction is so high that a shear stress k is needed to cause sliding between die and forging. The total area between the dies and piece c is given by

$$2\left\{\left(\frac{W}{2}\right) - \left(x + \frac{d}{2}\right)\right\}L = (w - 2x - d)L. \tag{14.4}$$

Piece c slides a distance $2u$ relative to the die surfaces, absorbing work of amount

$$(w - 2x - d)Lk2u. \tag{14.5}$$

Pieces a and b have a total contact area with the dies of $2dL$. They slide a distance u over the dies, absorbing work of amount

$$2dLku. \tag{14.6}$$

The overall work balance is now

$$2fu = 4dLku + 2(w - 2x - d)Lku + 2dLku \tag{14.7}$$

or

$$f = 2Lk\left(d + \frac{w}{2} - x\right). \tag{14.8}$$

The forming pressure is then

$$p_f = \frac{f}{dL} = \sigma_y\left\{1 + \frac{(w/2) - x}{d}\right\}. \tag{14.9}$$

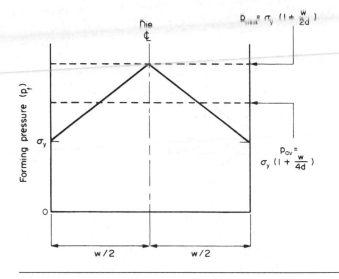

$$p_{\text{max}} = \sigma_y \left(1 + \frac{w}{2d}\right)$$

$$p_{\text{av}} = \sigma_y \left(1 + \frac{w}{4d}\right)$$

Figure 14.8   How the forming pressure varies with position in the forging.

This equation is plotted in Fig. 14.8: p_f increases linearly from a value of σ_y at the edge of the die to a maximum of

$$p_{\text{max}} = \sigma_y \left(1 + \frac{w}{2d}\right) \tag{14.10}$$

at the centre.

It is a salutary exercise to put some numbers into eqn. (14.10): if $w/d = 10$, then $p_{\text{max}} = 6\sigma_y$. Pressures of this magnitude are likely to deform the metal-forming tools themselves – clearly an undesirable state of affairs. The problem can usually be solved by heating the workpiece to $\approx 0.7\ T_m$ before forming, which greatly lowers σ_y. Or it may be possible to change the geometry of the process to reduce w/d. *Rolling* is a good example of this. From Fig. 14.9 we can write

$$(r - b)^2 + w^2 = r^2. \tag{14.11}$$

Provided b $\ll 2r$ this can be expanded to give

$$w = \sqrt{2rb}. \tag{14.12}$$

Thus

$$\frac{w}{d} = \frac{\sqrt{2rb}}{d} = \left(\frac{2r}{d}\right)^{1/2} \left(\frac{b}{d}\right)^{1/2}. \tag{14.13}$$

Well-designed rolling mills therefore have rolls of small diameter. However, as Fig. 14.9 shows, these may need to be supported by additional secondary rolls which do not touch the workpiece. In fact, aluminium cooking foil is rolled by primary rolls the diameter of a pencil, backed up by a total of 18 secondary rolls.

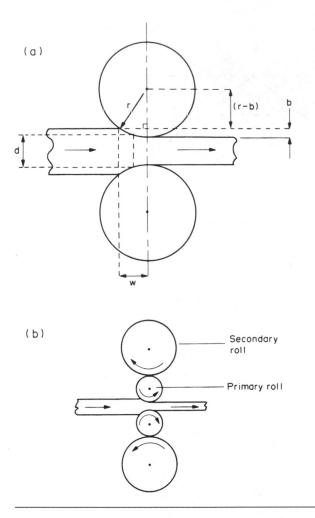

Figure 14.9   **(a)** In order to minimise the effects of friction, rolling operations should be carried out with minimum values of w/d. **(b)** Small rolls give small w/d values, but they may need to be supported by additional secondary rolls.

Recovery and recrystallisation

When metals are forged, or rolled, or drawn to wire, they *work-harden*. After a deformation of perhaps 80% a limit is reached, beyond which the metal cracks or fractures. Further rolling or drawing is possible if the metal is *annealed* (heated to about $0.6\ T_m$). During annealing, old, deformed grains are replaced by new, undeformed grains, and the working can be continued for a further 80% or so.

Figure 14.10 shows how the microstructure of a metal changes during plastic working and annealing. If the metal has been annealed to begin with

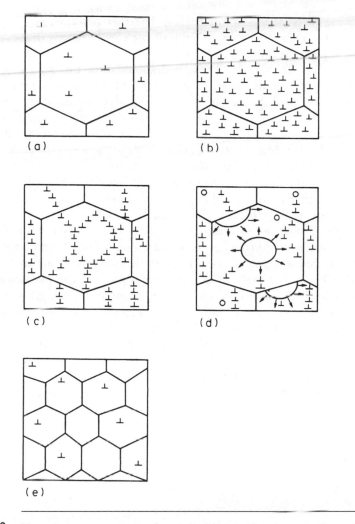

Figure 14.10   How the microstructure of a metal is changed by plastic working and annealing. (a) If the starting metal has already been annealed it will have a comparatively low dislocation density. (b) Plastic working greatly increases the dislocation density. (c) Annealing leads initially to recovery – dislocations move to low-energy positions. (d) During further annealing new grains nucleate and grow. (e) The fully recrystallised metal consists entirely of new undeformed grains.

(Fig. 14.10a) it will have a comparatively low dislocation density (about 10^{12} m^{-2}) and will be relatively soft and ductile. Plastic working (Fig. 14.10b) will greatly increase the dislocation density (to about 10^{15} m^{-2}). The metal will work-harden and will lose ductility. Because each dislocation strains the lattice the deformed metal will have a large strain energy (about 2 MJ m^{-3}). Annealing gives the atoms enough thermal energy that they can move under the driving force of this strain energy. The first process to occur is *recovery* (Fig. 14.10c). Because the strain fields of the closely spaced dislocations interact, the total strain energy can be reduced by rearranging the dislocations into low-angle grain boundaries. These boundaries form the surfaces of irregular *cells* – small volumes which are relatively free of dislocations. During recovery the dislocation density goes down only slightly: the hardness and ductility are almost unchanged. The major changes come from *recrystallisation*. New grains nucleate and grow (Fig. 14.10d) until the whole of the metal consists of undeformed grains (Fig. 14.10e). The dislocation density returns to its original value, as do the values of the hardness and ductility.

Recrystallisation is not limited just to getting rid of work-hardening. It is also a powerful way of controlling the grain size of worked metals. Although single crystals are desirable for a few specialised applications (see Chapter 9) the metallurgist almost always seeks a fine grain size. To begin with, fine-grained metals are stronger and tougher than coarse-grained ones. And large grains can be undesirable for other reasons. For example, if the grain size of a metal sheet is comparable to the sheet thickness, the surface will rumple when the sheet is pressed to shape; and this makes it almost impossible to get a good surface finish on articles such as car-body panels or spun aluminium saucepans.

The ability to control grain size by recrystallisation is due to the general rule (e.g. Chapter 11) that the harder you drive a transformation, the finer the structure you get. In the case of recrystallisation this means that the greater the prior plastic deformation (and hence the stored strain energy) the finer the recrystallised grain size (Fig. 14.11). To produce a fine-grained sheet, for

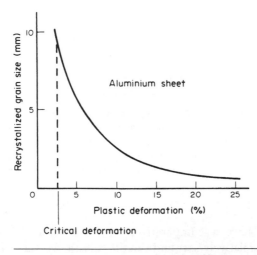

Figure 14.11   Typical data for recrystallised grain size as a function of prior plastic deformation. Note that, below a critical deformation, there is not enough strain energy to nucleate the new strain-free grains. This is just like the critical undercooling needed to nucleate a solid from its liquid (see Fig. 7.4).

example, we simply reduce the thickness by about 50% in a cold rolling operation (to give the large stored strain energy) and then anneal the sheet in a furnace (to give the fine recrystallised structure).

Machining

Most engineering components require at least some machining: turning, drilling, milling, shaping, or grinding. The cutting tool (or the abrasive particles of the grinding wheel) parts the chip from the workpiece by a process of plastic shear (Fig. 14.12). Thermodynamically, all that is required is the energy of the two new surfaces created when the chip peels off the surface; in reality, the work done in the plastic shear (a strain of order 1) greatly exceeds this minimum necessary energy. In addition, the friction is very high ($\mu \approx 0.5$) because the chip surface which bears against the tool is freshly formed, and free from adsorbed films which could reduce adhesion. This friction can be reduced by generous lubrication with water-soluble *cutting fluids*, which also cool the tool. *Free cutting* alloys have a built-in lubricant which smears across the tool face as the chip forms: lead in brass, manganese sulphide in steel.

Machining is expensive – in energy, wasted material and time. Forming routes which minimise or avoid machining result in considerable economies.

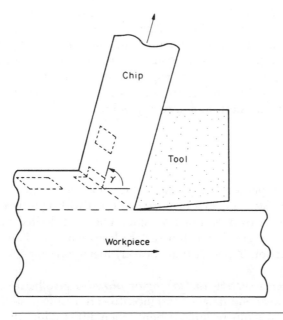

Figure 14.12   Machining.

Joining

Many of the processes used to join one metal to another are based on casting. We have already looked at *fusion welding* (Fig. 13.6). The most widely used welding process is arc welding: an electric arc is struck between an electrode of filler metal and the workpieces, providing the heat needed to melt the filler and fuse it to the parent plates. The electrode is coated with a *flux* which melts and forms a protective cover on the molten metal. In *submerged arc* welding, used for welding thick sections automatically, the arc is formed beneath a pool of molten flux. In *gas welding* the heat source is an oxyacetylene flame. In *spot welding* the metal sheets to be joined are pressed together between thick copper electrodes and fused together locally by a heavy current. Small, precise welds can be made using either an electron beam or a laser beam as the heat source.

Brazing and soldering are also fine-scale casting processes. But they use filler metals which melt more easily than the parent metal (see Table 4.1). The filler does not join to the parent metal by fusion (melting together). Instead, the filler spreads over, or wets, the solid parent metal and, when it solidifies, becomes firmly stuck to it. True metal-to-metal contact is essential for good wetting. Before brazing, the parent surfaces are either mechanically abraded or acid pickled to remove as much of the surface oxide film as possible. Then a flux is added which chemically reduces any oxide that forms during the heating cycle. Specialised brazing operations are done in a vacuum furnace which virtually eliminates oxide formation.

Adhesives, increasingly used in engineering applications, do not necessarily require the application of heat. A thin film of epoxy, or other polymer, is spread on the surfaces to be joined, which are then brought together under pressure for long enough for the adhesive to polymerise or set. Special methods are required with adhesives, but they offer great potential for design.

Metal parts are also joined by a range of *fasteners*: rivets, bolts, or tabs. In using them, the stress concentration at the fastener or its hole must be allowed for: fracture frequently starts at a fastening point.

Surface engineering

Often it is the properties of a surface which are critical in an engineering application. Examples are components which must withstand wear; or exhibit low friction; or resist oxidation or corrosion. Then the desired properties can often be achieved by creating a thin surface layer with good (but expensive) properties on a section of poorer (but cheaper) metal, offering great economies of production.

Surface treatments such as *carburising* or *nitriding* give hard surface layers, which give good wear and fatigue resistance. In carburising, a steel component is heated into the austenite region. Carbon is then diffused into the surface until its concentration rises to 0.8% or more. Finally the component is quenched into oil, transforming the surface into hard martensite. Steels for nitriding contain aluminium: when nitrogen is diffused into the surface it reacts to form aluminium nitride, which hardens the surface by precipitation hardening. More recently *ion implantation* has been used: foreign ions are accelerated in a strong electric field and are implanted into the surface. Finally, laser heat treatment has been developed as a powerful method for producing hard surfaces. Here the surface of the steel is scanned with a laser beam. As the beam passes over a region of the surface it heats it into the austenite region. When the beam passes on, the surface it leaves behind is rapidly quenched by the cold metal beneath to produce martensite.

Energy-efficient forming

Many of the processes used for working metals are energy-intensive. Large amounts of energy are needed to melt metals, to roll them to sections, to machine them or to weld them together. Broadly speaking, the more steps there are between raw metal and finished article (see Fig. 14.1) then the greater is the cost of production. There is thus a big incentive to minimise the number of processing stages and to maximise the efficiency of the remaining operations. This is not new. For centuries, lead sheet for organ pipes has been made in a single-stage casting operation. The Victorians were the pioneers of pouring intricate iron castings which needed the minimum of machining. Modern processes which are achieving substantial energy savings include the single-stage casting of thin wires or ribbons (melt spinning, see Chapter 9) or the spray deposition of "atomised" liquid metal to give semi-finished seamless tubes. But modifications of conventional processes can give useful economies too. In examining a production line it is always worth questioning whether a change in processing method could be introduced with economic benefits.

Examples

14.1   Estimate the percentage volume contraction due to solidification in pure copper. Use the following data: $T_m = 1083°C$; density of solid copper at $20°C = 8.96\,\mathrm{Mg\,m^{-3}}$; average coefficient of thermal expansion in the range 20 to $1083°C = 20.6\,\mathrm{M\,K^{-1}}$; density of liquid copper at $T_m = 8.00\,\mathrm{Mg\,m^{-3}}$.

Answer

5%.

14.2   A silver replica of a holly leaf is to be made by investment casting. (A natural leaf is coated with ceramic slurry which is then dried and fired. During firing the leaf burns away, leaving a mould cavity.) The thickness of the leaf is

0.4 mm. Calculate the liquid head needed to force the molten silver into the mould cavity. It can be assumed that molten silver does not wet the mould walls.

[Hint: the pressure needed to force a non-wetting liquid into a parallel-sided cavity of thickness t is given by

$$p = \frac{T}{(t/2)}$$

where T is the surface tension of the liquid.] The density and surface tension of molten silver are $9.4 \, Mg \, m^{-3}$ and $0.90 \, Nm^{-1}$.

Answer

49 mm.

14.3 Aluminium sheet is to be rolled according to the following parameters: starting thickness 1 mm, reduced thickness 0.8 mm, yield strength 100 MPa. What roll radius should be chosen to keep the forming pressure below 200 MPa?

Answer

16.2 mm, or less.

14.4 Aluminium sheet is to be rolled according to the following parameters: sheet width 300 mm, starting thickness 1 mm, reduced thickness 0.8 mm, yield strength 100 MPa, maximum forming pressure 200 MPa, roll radius 16.2 mm, roll length 300 mm. Calculate the force F that the rolling pressure will exert on each roll.

[Hint: use the average forming pressure, p_{av}, shown in Fig. 14.8.]

The design states that the roll must not deflect by more than 0.01 mm at its centre. To achieve this bending stiffness, each roll is to be backed up by one secondary roll as shown in Fig. 14.9(b). Calculate the secondary roll radius needed to meet the specification. The central deflection of the secondary roll is given by

$$\delta = \frac{5FL^3}{384 \, EI}$$

where L is the roll length and E is the Young's modulus of the roll material. I, the second moment of area of the roll section, is given by

$$I = \pi r_s^4 / 4$$

where r_s is the secondary roll radius. The secondary roll is made from steel, with $E = 210 \, GPa$. You may neglect the bending stiffness of the primary roll.

Answers

$F = 81 \, \text{kN}; \, r_s = 64.5 \, \text{mm}.$

14.5 Copper capillary fittings are to be used to solder copper water pipes together as shown below:

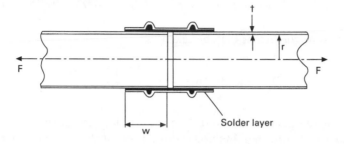

The joint is designed so that the solder layer will yield in shear at the same axial load F that causes the main tube to fail by tensile yield. Estimate the required value of W, given the following data: $t = 1 \, \text{mm}$; σ_y (copper) $= 120 \, \text{MPa}$; σ_y (solder) $= 10 \, \text{MPa}$.

Answer

24 mm.

14.6 A piece of plain carbon steel containing 0.2 wt% carbon was case-carburised to give a case depth of 0.3 mm. The carburising was done at a temperature of 1000°C. The Fe–C phase diagram shows that, at this temperature, the iron can dissolve carbon to a maximum concentration of 1.4 wt%. Diffusion of carbon into the steel will almost immediately raise the level of carbon in the steel to a constant value of 1.4 wt% just beneath the surface of the steel. However, the concentration of carbon well below the surface will increase more slowly toward the maximum value of 1.4 wt% because of the time needed for the carbon to diffuse into the interior of the steel.

The diffusion of carbon into the steel is described by the time-dependent diffusion equation.

$$C(x, t) = (C_s - C_0) \left\{ 1 - \text{erf}\left(\frac{x}{2\sqrt{Dt}}\right) \right\} + C_0.$$

The symbols have the meanings: C, concentration of carbon at a distance x below the surface after time t; C_s, 1.4 wt% C; C_0, 0.2 wt% C; D, diffusion coefficient for carbon in steel. The "error function", erf(y), is given by

$$\text{erf}(y) = \frac{2}{\sqrt{\pi}} \int_0^y e^{-Z^2} \, dZ.$$

The following table gives values for this integral.

y	0	0.1	0.2	0.3	0.4	0.5	0.6	0.7	
erf(y)	0	0.11	0.22	0.33	0.43	0.52	0.60	0.68	

y	0.8	0.9	1.0	1.1	1.2	1.3	1.4	1.5	∞
erf(y)	0.74	0.80	0.84	0.88	0.91	0.93	0.95	0.97	1.00

The diffusion coefficient may be taken as

$$D = 9 \times 10^{-6}\,\text{m}^2\,\text{s}^{-1}\,\exp\left\{\frac{-125\,\text{kJ}\,\text{mol}^{-1}}{RT}\right\}$$

where R is the gas constant and T is the absolute temperature.
    Calculate the time required for carburisation, if the depth of the case is taken be the value of x for which $C = 0.5$ wt% carbon.

Answer

8.8 minutes.

14.7 Using the equations and tabulated error function data from Example 14.6, show that the expression $x = \sqrt{Dt}$ gives the distance over which the concentration of the diffusion profile halves.

14.8 Describe the processes of recovery and recrystallisation that occur during the high-temperature annealing of a work-hardened metal. How does the grain size of the fully recrystallised metal depend on the initial amount of work hardening? Mention some practical situations in which recrystallisation is important.

14.9 A bar of cold-drawn copper had a yield strength of 250 MPa. The bar was later annealed at 600°C for 5 minutes. The yield strength after annealing was 50 MPa. Explain this change. [Both yield strengths were measured at 20°C.]

B. Ceramics and glasses

Chapter 15
Ceramics and glasses

Introduction

If you have ever dropped a plate on the kitchen floor and seen it disintegrate, you might question whether ceramics have a role as load-bearing materials in engineering. But any friend with a historical perspective will enlighten you. Ceramic structures are larger and have survived longer than any other works. The great pyramid of Giza is solid ceramic (nearly 1,000,000 tonnes of it); so is the Parthenon, the Forum, the Great Wall of China. The first cutting tools and weapons were made of flint – a glass; and pottery from 5000 BC survives to the present day. Ceramics may not be as tough as metals, but for resistance to corrosion, wear, decay and corruption, they are unsurpassed.

Today, cement and concrete replace stone in most large structures. But cement, too, is a ceramic: a complicated but fascinating one. The understanding of its structure, and how it forms, is better now than it used to be, and has led to the development of special high-strength cement pastes which can compete with polymers and metals in certain applications.

But the most exciting of all is the development, in the past 20 years, of a range of high-performance engineering ceramics. They can replace, and greatly improve on, metals in many very demanding applications. Cutting tools made of sialons or of dense alumina can cut faster and last longer than the best metal tools. Engineering ceramics are highly wear-resistant: they are used to clad the leading edges of agricultural machinery like harrows, increasing the life by 10 times. They are inert and biocompatible, so they are good for making artificial joints (where wear is a big problem) and other implants. And, because they have high melting points, they can stand much higher temperatures than metals can: vast development programs in Japan, the US and Europe aim to put increasing quantities of ceramics into reciprocating engines, turbines and turbochargers. In the next decade the potential market is estimated at $1 billion per year. Even the toughness of ceramics has been improved: modern body-armour is made of plates of boron carbide or of alumina, sewn into a fabric vest.

The next six chapters of this book focus on ceramics and glasses: non-metallic, inorganic solids. Five classes of materials are of interest to us here:

(a) *Glasses*, all of them based on silica (SiO_2), with additions to reduce the melting point, or give other special properties.
(b) The traditional *vitreous ceramics*, or clay products, used in vast quantities for plates and cups, sanitary ware, tiles, bricks, and so forth.
(c) The new *high-performance ceramics*, now finding application for cutting tools, dies, engine parts and wear-resistant parts.
(d) *Cement and concrete*: a complex ceramic with many phases, and one of three essential bulk materials of civil engineering.
(e) *Rocks and minerals*, including ice.

As with metals, the number of different ceramics is vast. But there is no need to remember them all: the generic ceramics listed below (and which you *should* remember) embody the important features; others can be understood in terms of these. Although their properties differ widely, they all have one feature in common: they are intrinsically brittle, and it is this that dictates the way in which they can be used.

They are, potentially or actually, cheap. Most ceramics are compounds of oxygen, carbon or nitrogen with metals like aluminium or silicon; all five are among the most plentiful and widespread elements in the Earth's crust. The processing costs may be high, but the ingredients are almost as cheap as dirt: dirt, after all, is a ceramic.

The generic ceramics and glasses

Glasses

Glasses are used in enormous quantities: the annual tonnage is not far below that of aluminium. As much as 80% of the surface area of a modern office block can be glass; and glass is used in a load-bearing capacity in car windows, containers, diving bells and vacuum equipment. All important glasses are based on silica (SiO_2). Two are of primary interest: common window glass, and the temperature-resisting borosilicate glasses. Table 15.1 gives details.

Vitreous ceramics

Potters have been respected members of society since ancient times. Their products have survived the ravages of time better than any other; the pottery of an era or civilisation often gives the clearest picture of its state of development and its customs. Modern pottery, porcelain, tiles, and structural and refractory bricks are made by processes which, though automated, differ very

Table 15.1   Generic glasses

Glass	Typical composition (wt%)	Typical uses
Soda-lime glass	70 SiO_2, 10 CaO, 15 Na_2O	Windows, bottles, etc.; easily formed and shaped.
Borosilicate glass	80 SiO_2, 15 B_2O_3, 5 Na_2O	Pyrex; cooking and chemical glassware; high-temperature strength, low coefficient of expansion, good thermal shock resistance.

little from those of 2000 years ago. All are made from clays, which are formed in the wet, plastic state and then dried and fired. After firing, they consist of crystalline phases (mostly silicates) held together by a glassy phase based, as always, on silica (SiO_2). The glassy phase forms and melts when the clay is fired, and spreads around the surface of the inert, but strong, crystalline phases, bonding them together. The important information is summarised in Table 15.2.

High-performance engineering ceramics

Diamond, of course, is the ultimate engineering ceramic; it has for many years been used for cutting tools, dies, rock drills, and as an abrasive. But it is expensive. The strength of a ceramic is largely determined by two characteristics: its *toughness* (K_c), and the size distribution of *microcracks* it contains. A new class of fully dense, high-strength ceramics is now emerging which combine a higher K_c with a much narrower distribution of smaller microcracks, giving properties which make them competitive with metals, cermets, even with diamond, for cutting tools, dies, implants and engine parts. And (at least potentially) they are cheap. The most important are listed in Table 15.3.

Table 15.2   Generic vitreous ceramics

Ceramic	Typical composition	Typical uses
Porcelain China Pottery Brick	Made from clays: hydrous alumino-silicate such as $Al_2(Si_2O_5)(OH)_4$ mixed with other inert minerals.	Electrical insulators. Artware and tableware tiles. Construction; refractory uses.

Table 15.3   Generic high-performance ceramics

Ceramic	Typical composition	Typical uses
Dense alumina	Al_2O_3	Cutting tools, dies; wear-resistant
Silicon carbide, nitride	SiC, Si_3N_4	surfaces, bearings; medical implants;
Sialons	e.g. Si_2AlON_3	engine and turbine parts; armour.
Cubic zirconia	$ZrO_2 + 5wt\%\ MgO$	

Cement and concrete

Cement and concrete are used in construction on an enormous scale, equalled only by structural steel, brick and wood. *Cement* is a mixture of a combination of lime (CaO), silica (SiO_2) and alumina (Al_2O_3), which sets when mixed with water. Concrete is sand and stones (aggregate) held together by a cement. Table 15.4 summarises the most important facts.

Natural ceramics

Stone is the oldest of all construction materials and the most durable. The pyramids are 5000 years old; the Parthenon 2200. Stone used in a load-bearing capacity behaves like any other ceramic; and the criteria used in design with stone are the same. One natural ceramic, however, is unique. Ice forms on the Earth's surface in enormous volumes: the Antarctic ice cap, for instance, is up to 3 km thick and almost 3000 km across; something like 10^{13} m^3 of pure ceramic. The mechanical properties are of primary importance in some major engineering problems, notably ice breaking, and the construction of offshore oil rigs in the Arctic. Table 15.5 lists the important natural ceramics.

Ceramic composites

The great stiffness and hardness of ceramics can sometimes be combined with the toughness of polymers or metals by making composites. Glass- and carbon-

Table 15.4   Generic cements and concretes

Cement	Typical composition	Uses
Portland cement	$CaO + SiO_2 + Al_2O_3$	Cast facings, walkways, etc. and as component of concrete. General construction.

Table 15.5   Generic natural ceramics

Ceramic	Composition	Typical uses
Limestone (marble)	Largely CaCO$_3$	
Sandstone	Largely SiO$_2$	Building foundations, construction.
Granite	Aluminium silicates	
Ice	H$_2$O	Arctic engineering.

Table 15.6   Ceramics composites

Ceramic composite	Components	Typical uses
Fibre glass	Glass–polymer	
CFRP	Carbon–polymer	High-performance structures.
Cermet	Tungsten carbide–cobalt	Cutting tools, dies.
Bone	Hydroxyapatite–collagen	Main structural material of animals.
New ceramic composites	Alumina–silicon carbide	High temperature and high toughness applications.

fibre reinforced plastics are examples: the glass or carbon fibres stiffen the rather floppy polymer; but if a fibre fails, the crack runs out of the fibre and blunts in the ductile polymer without propagating across the whole section. Cermets are another example: particles of hard tungsten carbide bonded by metallic cobalt, much as gravel is bonded with tar to give a hard-wearing road surface (another ceramic-composite). Bone is a natural ceramic-composite: particles of hydroxyapatite (the ceramic) bonded together by collagen (a polymer). Synthetic ceramic–ceramic composites (like glass fibres in cement, or silicon carbide fibres in silicon carbide) are now under development and may have important high-temperature application in the next decade. The examples are summarised in Table 15.6.

Data for ceramics

Ceramics, without exception, are hard, brittle solids. When designing with metals, failure by plastic collapse and by fatigue are the primary considerations. For ceramics, plastic collapse and fatigue are seldom problems; it is brittle failure, caused by direct loading or by thermal stresses, that is the overriding consideration.

Because of this, the data listed in Table 15.7 for ceramic materials differ in emphasis from those listed for metals. In particular, the Table shows the *modulus of rupture* (the maximum surface stress when a beam breaks in bending)

Table 15.7 Properties of ceramics

Ceramic	Cost (UK£ (US$) tonne^{-1})	Density (Mg m^{-3})	Young's modulus (GPa)	Compressive strength (MPa)	Modulus of rupture (MPa)	Weibull exponent m	Time exponent n	Fracture toughness (MPa m$^{1/2}$)	Melting (softening) temperature (K)	Specific heat (J kg^{-1} K^{-1})	Thermal conductivity (W m^{-1} K^{-1})	Thermal expansion coefficient (MK^{-1})	Thermal shock resistance (K)
Glasses													
Soda glass	700 (1000)	2.48	74	1000	50	Assume 10 in design	10	0.7	(1000)	990	1	8.5	84
Borosilicate glass	1000 (1400)	2.23	65	1200	55		10	0.8	(1100)	800	1	4.0	280
Pottery, etc.													
Porcelain	260–1000 (360–1400)	2.3–2.5	70	350	45	–	–	1.0	(1400)	800	1	3	220
High-performance engineering ceramics													
Diamond	4 × 10⁸ (6 × 10⁸)	3.52	1050	5000	–	–	–	–	–	510	70	1.2	1000
Dense alumina	Expensive at present.	3.9	380	3000	300–400	10	10	3–5	2323 (1470)	795	25.6	8.5	150
Silicon carbide	Potentially	3.2	410	2000	200–500	10	40	–	3110	1422	84	4.3	300
Silicon nitride	350–1000	3.2	310	1200	300–850	–	40	4	2173	627	17	3.2	500
Zirconia	(490–1400)	5.6	200	2000	200–500	10–21	10	4–12	2843	670	1.5	8	500
Sialons		3.2	300	2000	500–830	15	10	5	–	710	20–25	3.2	510
Cement, etc.													
Cement	52 (73)	2.4–2.5	20–30	50	7	12	40	0.2	–	–	1.8	10–14	<50
Concrete	26 (36)	2.4	30–50	50	7	12	40	0.2	–	–	2	10–14	
Rocks and ice													
Limestone	Cost of mining and transport	2.7	63	30–80	20	–	–	0.9	–	–	–	8	≈100
Granite		2.6	60–80	65–150	23	–	–	–	–	–	–	8	
Ice		0.92	9.1	6	1.7	–	–	0.12	273 (250)	–	–	–	

and the *thermal shock resistance* (the ability of the solid to withstand sudden changes in temperature). These, rather than the yield strength, tend to be the critical properties in any design exercise.

As before, the data presented here are approximate, intended for the first phase of design. When the choice has narrowed sufficiently, it is important to consult more exhaustive data compilations (see Appendix 3); and then to obtain detailed specifications from the supplier of the material you intend to use. Finally, if the component is a critical one, you should conduct your own tests. The properties of ceramics are more variable than those of metals: the same material, from two different suppliers, could differ in toughness and strength by a factor of two.

There are, of course, many more ceramics available than those listed here: alumina is available in many densities, silicon carbide in many qualities. As before, the structure-insensitive properties (density, modulus and melting point) depend little on quality – they do not vary by more than 10%. But the structure-sensitive properties (fracture toughness, modulus of rupture and some thermal properties including expansion) are much more variable. For these, it is essential to consult manufacturers' data sheets or conduct your own tests.

Examples

15.1 What are the five main generic classes of ceramics and glasses? For each generic class:

(a) give one example of a specific component made from that class;
(b) indicate why that class was selected for the component.

15.2 How do the unique characteristics of ceramics and glasses influence the way in which these materials are used?

15.3 The glass walls of the Sydney Opera House are constructed from glass panels butted together with a 10 mm expansion gap between adjacent panels. The gap is injected with a flexible polymer which is cured in-situ and also sticks to the edge of the glass (see photographs). Taking a notional panel width of 3 m, and a maximum temperature variation of 40°C, calculate the minimum strain capacity required from the polymer. See Table 15.7 for the thermal expansion coefficient of soda glass. Apart from tensile fracture of the polymer, what additional mechanical failure mechanism is possible in the jointing system?

Answer

10%

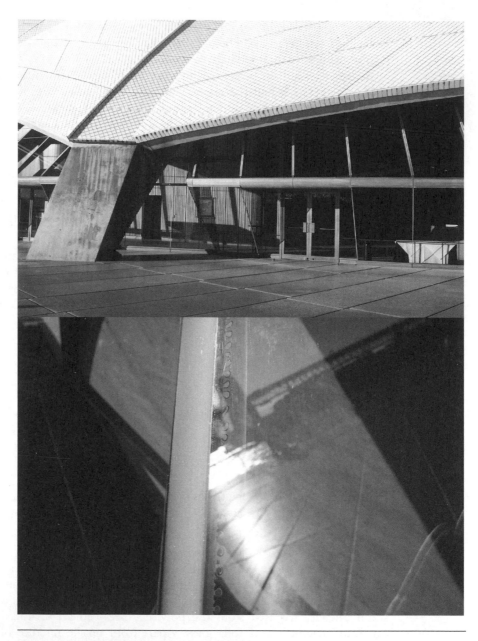

Glass walls of Sydney Opera House.

Chapter 16
Structure of ceramics

Introduction

A ceramic, like a metal, has structure at the atomic scale: its crystal structure (if it is crystalline), or its amorphous structure (if it is glassy). It has structure at a larger scale too: the shape and arrangement of its grains or phases; the size and volume fraction of pores it contains (most ceramics are porous). Ceramic structures differ from those of metals. Their intricate details (and they can be pretty intricate) are the province of the professional ceramist; but you need a working knowledge of their basic features to understand the processing and the engineering uses of ceramics and to appreciate the recently developed high-strength ceramics.

Ionic and covalent ceramics

One critical distinction should be drawn right at the beginning. It is that between predominantly *ionic* ceramics and those which are predominantly *covalent* in their bonding.

Ionic ceramics are, typically, compounds of a metal with a non-metal; sodium chloride, NaCl; magnesium oxide, MgO; alumina Al_2O_3; zirconia ZrO_2. The metal and non-metal have unlike electric charges: in sodium chloride, for instance, the sodium atoms have one positive charge and the chlorine atoms have one negative charge each. The electrostatic attraction between the unlike charges gives most of the bonding. So the ions pack *densely* (to get as many plus and minus charges close to each other as possible), but with the *constraint* that ions of the same type (and so with the same charge) must not touch. This leads to certain basic ceramic structures, typified by rock salt, NaCl, or by alumina Al_2O_3, which we will describe later.

Covalent ceramics are different. They are compounds of two non-metals (like silica SiO_2), or, occasionally, are just pure elements (like diamond, C, or silicon, Si). An atom in this class of ceramic bonds by sharing electrons with its neighbours to give a fixed number of directional bonds. Covalent atoms are a bit like the units of a child's construction kit which snap together: the position and number of neighbours are rigidly fixed by the number and position of the connectors on each block. The resulting structures are quite different from those given by ionic bonding; and as we will see later, the mechanical properties are different too. The energy is minimised, not by dense packing, but by forming *chains, sheets,* or *three-dimensional networks.* Often these are non-crystalline; all commercial glasses, for instance, are three-dimensional amorphous networks based on silica, SiO_2.

We will first examine the simple structures given by ionic and covalent bonding, and then return to describe the microstructures of ceramics.

Simple ionic ceramics

The archetype of the ionic ceramic is sodium chloride ("rocksalt"), NaCl, shown in Fig. 16.1(a). Each sodium atom loses an electron to a chlorine atom; it is the electrostatic attraction between the Na^+ ions and the Cl^- ions that holds the crystal together. To achieve the maximum electrostatic interaction, each Na^+ has 6 Cl^- neighbours and no Na^+ neighbours (and vice versa); there is no way of arranging single-charged ions that does better than this. So most of the simple ionic ceramics with the formula AB have the rocksalt structure.

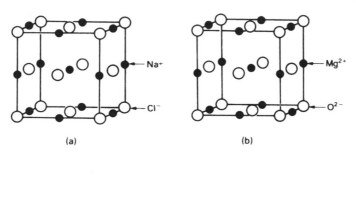

(a)   (b)

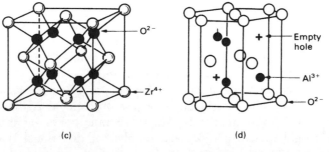

(c)   (d)

Figure 16.1   Ionic ceramics. **(a)** The rocksalt, or NaCl, structure. **(b)** Magnesia, MgO, has the rocksalt structure. It can be thought of as an f.c.c. packing with Mg ions in the octahedral holes. **(c)** Cubic zirconia ZrO_2: an f.c.c. packing of Zr with O in the tetrahedral holes. **(d)** Alumina, Al_2O_3: a c.p.h. packing of oxygen with Al in two-thirds of the octahedral holes.

Magnesia, MgO, is an example (Fig. 16.1b). It is an engineering ceramic, used as a refractory in furnaces, and its structure is exactly the same as that of rocksalt: the atoms pack to maximise the density, with the constraint that like ions are not nearest neighbours.

But there is another way of looking at the structure of MgO, and one which greatly simplifies the understanding of many of the more complex ceramic structures. Look at Fig. 16.1(b) again: the oxygen ions (open circles) form an f.c.c. packing. Figure 16.2 shows that the f.c.c. structure contains two sorts of interstitial holes: the larger octahedral holes, of which there is one for each oxygen atom; and the smaller tetrahedral holes, of which there are two for each oxygen atom. Then the structure of MgO can be described as a *face-centred cubic packing of oxygen with an Mg ion squeezed into each octahedral hole*. Each Mg^{2+} has six O^{2-} as immediate neighbours, and vice versa: the coordination number is 6. The Mg ions wedge the oxygens apart, so that they do not touch. Then the attraction between the Mg^{2+} and the O^{2-} ions greatly outweighs the repulsion between the O^{2-} ions, and the solid is very strong and stable (its melting point is over 2000°C).

This "packing" argument may seem an unnecessary complication. But its advantage comes now. Consider cubic zirconia, ZrO_2, an engineering ceramic of growing importance. The structure (Fig. 16.1c) looks hard to describe, but it isn't. It is simply an *f.c.c. packing of zirconium with the O^{2-} ions in the tetrahedral holes*. Since there are two tetrahedral holes for each atom of the f.c.c. structure, the formula works out at ZrO_2.

Alumina, Al_2O_3 (Fig. 16.1d), is a structural ceramic used for cutting tools and grinding wheels, and a component in brick and pottery. It has a structure which can be understood in a similar way. The oxygen ions are close-packed, but this time in the c.p.h. arrangement, like zinc or titanium. The hexagonal structure (like the f.c.c. one) has one octahedral hole and two tetrahedral holes per atom. In Al_2O_3 the Al^{3+} ions are put into the octahedral interstices, so that each is surrounded by six O^{2-} ions. But if the charges are to balance (as they must) there are only enough Al ions to fill two-thirds of the sites. So one-third of the sites, in an ordered pattern, remain empty. This introduces

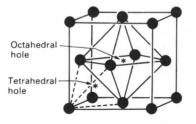

Octahedral hole

Tetrahedral hole

Figure 16.2   Both the f.c.c. and the c.p.h. structures are close-packed. Both contain one octahedral hole per atom, and two tetrahedral holes per atom. The holes in the f.c.c. structures are shown here.

a small distortion of the original hexagon, but from our point of view this is unimportant.

There are many other ionic oxides with structures which are more complicated than these. We will not go into them here. But it is worth knowing that most can be thought of as a dense (f.c.c. or c.p.h.) packing of oxygen, with various metal ions arranged, in an orderly fashion, in the octahedral or the tetrahedral holes.

Simple covalent ceramics

The ultimate covalent ceramic is diamond, widely used where wear resistance or very great strength are needed: the diamond stylus of a pick-up, or the diamond anvils of an ultra-high pressure press. Its structure, shown in Fig. 16.3(a), shows the 4 coordinated arrangement of the atoms within the cubic unit cell: each atom is at the centre of a tetrahedron with its four bonds directed to the four corners of the tetrahedron. It is not a close-packed structure (atoms in close-packed structures have 12, not four, neighbours) so its density is low.

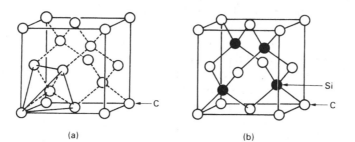

(a)   (b)

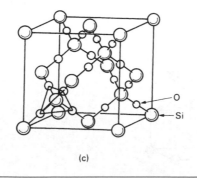

(c)

Figure 16.3   Covalent ceramics. (a) The diamond-cubic structure; each atom bonds to four neighbours. (b) Silicon carbide: the diamond cubic structure with half the atoms replaced by silicon. (c) Cubic silica: the diamond cubic structure with an SiO_4 tetrahedron on each atom site.

The very hard structural ceramics silicon carbide, SiC, and silicon nitride, Si_3N_4 (used for load-bearing components such as high-temperature bearings and engine parts) have a structure closely related to that of diamond. If, in the diamond cubic structure, every second atom is replaced by silicon, we get the *sphalerite* structure of SiC, shown in Fig. 16.3(b). Next to diamond, this is one of the hardest of known substances, as the structural resemblance would suggest.

Silica and silicates

The earth's crust is largely made of silicates. Of all the raw materials used by man, silica and its compounds are the most widespread, plentiful and cheap.

Silicon atoms bond strongly with four oxygen atoms to give a tetrahedral unit (Fig. 16.4a). This stable *tetrahedron* is the basic unit in all silicates, including that of pure silica (Fig. 16.3c); note that it is just the diamond cubic structure with every C atom replaced by an SiO_4 unit. But there are a number of other, quite different, ways in which the tetrahedra can be linked together.

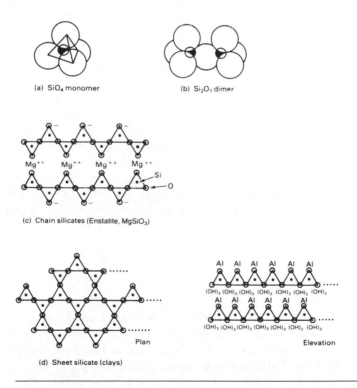

(a) SiO_4 monomer

(b) Si_2O_7 dimer

(c) Chain silicates (Enstatite, $MgSiO_3$)

(d) Sheet silicate (clays)

Plan

Elevation

Figure 16.4   Silicate structures. (a) The SiO_4 monomer. (b) The Si_2O_7 dimer with a bridging oxygen. (c) A chain silicate. (d) A sheet silicate. Each triangle is the projection of an SiO_4 monomer.

The way to think of them all is as SiO_4 tetrahedra (or, in polymer terms, *monomers*) linked to each other either directly or via a metal ion (M) link. When silica is combined with metal oxides like MgO, CaO or Al_2O_3 such that the ratio MO/SiO_2 is 2/1 or greater, then the resulting silicate is made up of separated SiO_4 monomers (Fig. 16.4a) linked by the MO molecules. (Olivene, the dominant material in the Earth's upper mantle, is a silicate of this type.)

When the ratio MO/SiO_2 is a little less than 2/1, silica *dimers* form (Fig. 16.4b). One oxygen is shared between two tetrahedra; it is called a *bridging* oxygen. This is the first step in the polymerisation of the monomer to give chains, sheets and networks.

With decreasing amounts of metal oxide, the degree of polymerisation increases. *Chains* of linked tetrahedra form, like the long chain polymers with a —C—C—back-bone, except that here the backbone is an —Si—O—Si—O—Si— chain (Fig. 16.4c). Two oxygens of each tetrahedron are shared (there are two bridging oxygens). The others form ionic bonds between chains, joined by the MO. These are weaker than the —Si—O—Si— bonds which form the backbone, so these silicates are fibrous; asbestos, for instance, has this structure.

If three oxygens of each tetrahedron are shared, *sheet structures* form (Fig. 16.4d). This is the basis of clays and micas. The additional M attaches itself preferentially to one side of the sheet – the side with the spare oxygens on it. Then the sheet is polarised: it has a net positive charge on one surface and a negative charge on the other. This interacts strongly with water, attracting a layer of water between the sheets. This is what makes clays plastic: the sheets of silicate slide over each other readily, lubricated by the water layer. As you might expect, sheet silicates are very strong in the plane of the sheet, but cleave or split easily between the sheets: think of mica and talc.

Pure silica contains no metal ions and every oxygen becomes a bridge between two silicon atoms giving a *three-dimensional network*. The high-temperature form, shown in Fig. 16.3(c), is cubic; the tetrahedra are stacked in the same way as the carbon atoms in the diamond-cubic structure. At room temperature the stable crystalline form of silica is more complicated but, as before, it is a three-dimensional network in which all the oxygens bridge silicons.

Silicate glasses

Commercial glasses are based on silica. They are made of the same SiO_4 tetrahedra on which the crystalline silicates are based, but they are arranged in a non-crystalline, or *amorphous*, way. The difference is shown schematically in Fig. 16.5. In the glass, the tetrahedra link at the corners to give a random (rather than a periodic) network. Pure silica forms a glass with a high softening temperature (about 1200°C). Its great strength and stability, and its low thermal expansion, suit it for certain special applications, but it is hard to work with because its viscosity is high.

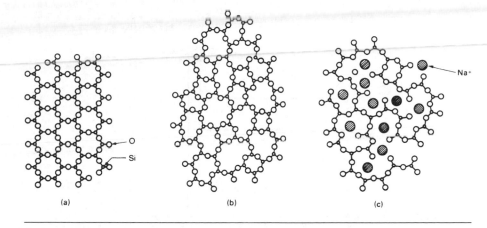

Figure 16.5   Glass formation. A 3-co-ordinated crystalline network is shown at (a). But the bonding requirements are still satisfied if a random (or glassy) network forms, as shown at (b). The network is broken up by adding network modifiers, like Na_2O, which interrupt the network as shown at (c).

This problem is overcome in commercial glasses by introducing *network modifiers* to reduce the viscosity. They are metal oxides, usually Na_2O and CaO, which add positive ions to the structure, and break up the network (Fig. 16.5c). Adding one molecule of Na_2O, for instance, introduces two Na^+ ions, each of which attaches to an oxygen of a tetrahedron, making it non-bridging. This reduction in cross-linking softens the glass, reducing its *glass temperature* T_G (the temperature at which the viscosity reaches such a high value that the glass is a solid). Glance back at the table in Chapter 15 for generic glasses; common window glass is only 70% SiO_2: it is heavily modified, and easily worked at 700°C. Pyrex is 80% SiO_2; it contains less modifier, has a much better thermal shock resistance (because its thermal expansion is lower), but is harder to work, requiring temperatures above 800°C.

Ceramic alloys

Ceramics form alloys with each other, just as metals do. But the reasons for alloying are quite different: in metals it is usually to increase the yield strength, fatigue strength or corrosion resistance; in ceramics it is generally to allow sintering to full density, or to improve the fracture toughness. But for the moment this is irrelevant; the point here is that one deals with ceramic alloys just as one did with metallic alloys. Molten oxides, for the most part, have large solubilities for other oxides (that is why they make good *fluxes*, dissolving undesirable impurities into a harmless slag). On cooling, they solidify as one or more phases: solid solutions or new compounds. Just as for metals, the *constitution of a ceramic alloy* is described by the appropriate phase diagram.

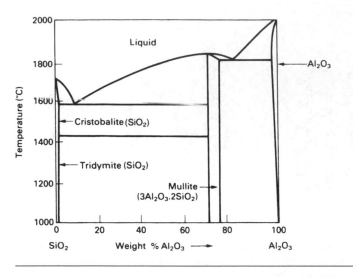

Figure 16.6   A typical ceramic phase diagram: that for alloys of SiO_2 with Al_2O_3. The intermediate compound $3Al_2O_3 SiO_2$ is called mullite.

Take the silica-alumina system as an example. It is convenient to treat the components as the two pure oxides SiO_2 and Al_2O_3 (instead of the three elements Si, Al and O). Then the phase diagram is particularly simple, as shown in Fig. 16.6. There is a compound, *mullite*, with the composition $(SiO_2)_2$ $(Al_2O_3)_3$, which is slightly more stable than the simple solid solution, so the alloys break up into mixtures of mullite and alumina, or mullite and silica. The phase diagram has two eutectics, but is otherwise straightforward.

The phase diagram for MgO and Al_2O_3 is similar, with a central compound, *spinel*, with the composition $MgOAl_2O_3$. That for MgO and SiO_2, too, is simple with a compound, *forsterite*, having the composition $(MgO)_2 SiO_2$. Given the composition, the equilibrium constitution of the alloy is read off the diagram in exactly the way described in Chapter 3.

The microstructure of ceramics

Crystalline ceramics form polycrystalline microstructures, very like those of metals (Fig. 16.7). Each grain is a more or less perfect crystal, meeting its neighbours at grain boundaries. The structure of ceramic grain boundaries is obviously more complicated than those in metals: ions with the same sign of charge must still avoid each other and, as far as possible, valency requirements must be met in the boundary, just as they are within the grains. But none of this is visible at the microstructural level, which for a pure, dense ceramic, looks just like that of a metal.

Many ceramics are not fully dense. Porosities as high as 20% are a common feature of the microstructure (Fig. 16.7). The pores weaken the material, though

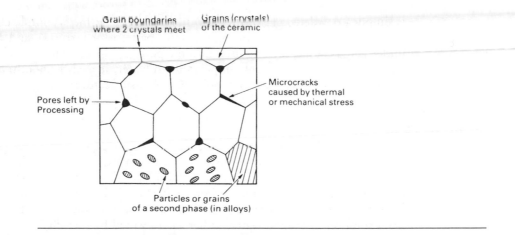

Grain boundaries
where 2 crystals meet

Grains (crystals)
of the ceramic

Microcracks
caused by thermal
or mechanical stress

Pores left by
Processing

Particles or grains
of a second phase (in alloys)

Figure 16.7   Microstructural features of a crystalline ceramic: grains, grain boundaries, pores, microcracks and second phases.

if they are well rounded, the stress concentration they induce is small. More damaging are cracks; they are much harder to see, but they are nonetheless present in most ceramics, left by processing, or nucleated by differences in thermal expansion or modulus between grains or phases. These, as we shall see in the next chapter, ultimately determine the strength of the material. Recent developments in ceramic processing aim to reduce the size and number of these cracks and pores, giving ceramic bodies with tensile strengths as high as those of high-strength steel (more about that in Chapter 18).

Vitreous ceramics

Pottery and tiles survive from 5000 BC, evidence of their extraordinary corrosion resistance and durability. Vitreous ceramics are today the basis of an enormous industry, turning out bricks, tiles and white-ware. All are made from clays: sheet silicates such as the hydrated alumino-silicate *kaolin*, $Al_2(Si_2O_5)(OH)_4$. When wet, the clay draws water between the silicate sheets (because of its polar layers), making it plastic and easily worked. It is then dried to the green state, losing its plasticity and acquiring enough strength to be handled for firing. The firing – at a temperature between 800 and 1200°C – drives off the remaining water, and causes the silica to combine with impurities like CaO to form a liquid glass which wets the remaining solids. On cooling, the glass solidifies (but is still a glass), giving strength to the final composite of crystalline silicates bonded by vitreous bonds. The amount of glass which forms during firing has to be carefully controlled: too little, and the bonding is poor; too much, and the product slumps, or melts completely.

As fired, vitreous ceramics are usually porous. To seal the surface, a glaze is applied, and the product refired at a lower temperature than before. The glaze is simply a powdered glass with a low melting point. It melts completely, flows over the surface (often producing attractive patterns or textures), and wets the underlying ceramic, sucking itself into the pores by surface tension. When cold again, the surface is not only impervious to water, it is also smooth, and free of the holes and cracks which would lead to easy fracture.

Stone or rock

Sedimentary rocks (like sandstone) have a microstructure rather like that of a vitreous ceramic. Sandstone is made of particles of silica, bonded together either by more silica or by calcium carbonate ($CaCO_3$). Like pottery, it is porous. The difference lies in the way the bonding phase formed: it is precipitated from solution in ground water, rather than formed by melting.

Igneous rocks (like granite) are much more like the SiO_2–Al_2O_3 alloys described in the phase diagram of Fig. 16.6. These rocks have, at some point in their history, been hot enough to have melted. Their structure can be read from the appropriate phase diagram: they generally contain several phases and, since they have melted, they are fully dense (though they still contain cracks nucleated during cooling).

Ceramic composites

Most successful composites combine the stiffness and hardness of a ceramic (like glass, carbon, or tungsten carbide) with the ductility and toughness of a polymer (like epoxy) or a metal (like cobalt). You will find all you need to know about them in Chapter 25.

Examples

16.1 Describe, in a few words, with an example or sketch as appropriate, what is meant by each of the following:

(a) an ionic ceramic;
(b) a covalent ceramic;
(c) a chain silicate;
(d) a sheet silicate;
(e) a glass;
(f) a network modifier;
(g) the glass temperature;
(h) a vitreous ceramic;
(i) a glaze;
(j) a sedimentary rock;
(k) an igneous rock;

Chapter 17
The mechanical properties of ceramics

Introduction

A Ming vase could, one would hope, perform its primary function – that of pleasing the eye – without being subjected to much stress. Much glassware, vitreous ceramic and porcelain fills its role without carrying significant direct load, though it must withstand thermal shock (if suddenly heated or cooled), and the wear and tear of normal handling. But others, such as brick, refractories and structural cement, are deliberately used in a load-bearing capacity; their strength has a major influence on the design in which they are incorporated. And others still – notably the high-performance engineering ceramics and abrasives – are used under the most demanding conditions of stress and temperature.

In this chapter we examine the mechanical properties of ceramics and, particularly, what is meant by their "strength".

The elastic moduli

Ceramics, like metals (but unlike polymers) have a well-defined Young's modulus: the value does not depend significantly on loading time (or, if the loading is cyclic, on frequency). Ceramic moduli are generally larger than those of metals, reflecting the greater stiffness of the ionic bond in simple oxides, and of the covalent bond in silicates. And since ceramics are largely composed of light atoms (oxygen, carbon, silicon, aluminium) and their structures are often not close-packed, their densities are low. Because of this their specific moduli (E/ρ) are attractively high. Table 17.1 shows that alumina, for instance, has a specific modulus of 100 (compared to 27 for steel). This is one reason ceramic or glass fibres are used in composites: their presence raises the specific stiffness of the composite enormously. Even cement has a reasonable specific stiffness – high enough to make boats out of it.

Strength, hardness and the lattice resistance

Ceramics are the hardest of solids. Corundum (Al_2O_3), silicon carbide (SiC) and, of course, diamond (C) are used as abrasives: they will cut, or grind, or polish almost anything – even glass, and glass is itself a very hard solid. Table 17.2 gives some feel for this: it lists the hardness H, normalised by the

Table 17.1   Specific moduli: ceramics compared to metals

Material	Modulus E (GPa)	Density ρ (Mg m^{-3})	Specific modulus E/ρ (GPa/Mg m^{-3})
Steels	210	7.8	27
Al alloys	70	2.7	26
Alumina, Al$_2$O$_3$	390	3.9	100
Silica, SiO$_2$	69	2.6	27
Cement	45	2.4	19

Table 17.2   Normalised hardness of pure metals, alloys and ceramics

Pure metal	H/E	Metal alloy	H/E	Ceramic	H/E
Copper	1.2×10^{-3}	Brass	9×10^{-3}	Diamond	1.5×10^{-1}
Aluminium	1.5×10^{-3}	Dural (Al 4% Cu)	1.5×10^{-2}	Alumina	4×10^{-2}
Nickel	9×10^{-4}	Stainless steel	6×10^{-3}	Zirconia	6×10^{-2}
Iron	9×10^{-4}	Low alloy steel	1.5×10^{-2}	Silicon carbide	6×10^{-2}
Mean, metals	1×10^{-3}	Mean, alloys	1×10^{-2}	Mean, ceramics	8×10^{-2}

Young's modulus E, for a number of pure metals and alloys, and for four pure ceramics. Pure metals (first column of Table 17.2) have a very low hardness and yield strength (remember $H \approx 3\sigma_y$). The main purpose of alloying is to raise it. The second column shows that this technique is very successful: the hardness has been increased from around $10^{-3}E$ to about $10^{-2}E$. But now look at the third column: even pure, unalloyed ceramics have hardnesses which far exceed even the best metallic alloys. Why is this?

When a material yields in a tensile test, or when a hardness indenter is pressed into it, dislocations move through its structure. Each test, in its own way, measures the difficulty of moving dislocations in the material. Metals are *intrinsically soft*. When atoms are brought together to form a metal, each loses one (or more) electrons to the gas of free electrons which moves freely around the ion cores (Fig. 17.1a). The binding energy comes from the general electrostatic interaction between the positive ions and the negative electron gas, and the bonds are not localised. If a dislocation passes through the structure, it displaces the atoms above its slip plane over those which lie below, but this has only a small effect on the electron–ion bonding. Because of this, there is a slight *drag* on the moving dislocation; one might liken it to wading through tall grass.

Most ceramics are *intrinsically hard*; ionic or covalent bonds present an enormous *lattice resistance* to the motion of a dislocation. Take the covalent bond first. The covalent bond is localised; the electrons which form the bond are concentrated in the region between the bonded atoms; they behave like little elastic struts joining the atoms (Fig. 17.1b). When a dislocation moves through the structure it must break and reform these bonds as it moves: it is like traversing a forest by uprooting and then replanting every tree in your path.

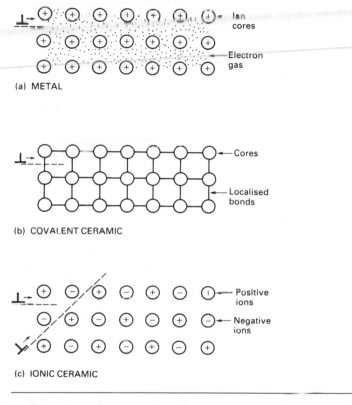

(a) METAL

(b) COVALENT CERAMIC

(c) IONIC CERAMIC

Figure 17.1   (a) Dislocation motion is intrinsically easy in pure metals – though alloying to give solid solutions or precipitates can make it more difficult. (b) Dislocation motion in covalent solids is intrinsically difficult because the interatomic bonds must be broken and reformed. (c) Dislocation motion in ionic crystals is easy on some planes, but hard on others. The hard systems usually dominate.

Most ionic ceramics are hard, though for a slightly different reason. The ionic bond, like the metallic one, is electrostatic: the attractive force between a sodium ion (Na^+) and a chlorine ion (Cl^-) is simply proportional to q^2/r where q is the charge on an electron and r the separation of the ions. If the crystal is sheared on the 45° plane shown in Fig. 17.1(c) then like ions remain separated: Na^+ ions do not ride over Na^+ ions, for instance. This sort of shear is relatively easy – the lattice resistance opposing it is small. But look at the other shear – the horizontal one. This *does* carry Na^+ ions over Na^+ ions and the electrostatic repulsion between like ions opposes this strongly. The lattice resistance is high. In a polycrystal, you will remember, many slip systems are necessary, and some of them are the hard ones. So the hardness of a polycrystalline ionic ceramic is usually high (though not as high as a covalent ceramic), even though a single crystal of the same material might – if loaded in the right way – have a low yield strength.

So ceramics, at room temperature, generally have a very large lattice resistance. The stress required to make dislocations move is a large fraction of Young's modulus: typically, around $E/30$, compared with $E/10^3$ or less for the soft metals like copper or lead. This gives to ceramics yield strengths which are of order 5 GPa – so high that the only way to measure them is to indent the ceramic with a diamond and measure the hardness.

This enormous hardness is exploited in grinding wheels which are made from small particles of a high-performance engineering ceramic (Table 15.3) bonded with an adhesive or a cement. In design with ceramics it is never necessary to consider plastic collapse of the component: fracture always intervenes first. The reasons for this are as follows.

Fracture strength of ceramics

The penalty that must be paid for choosing a material with a large lattice resistance is brittleness: the fracture toughness is low. Even at the tip of a crack, where the stress is intensified, the lattice resistance makes slip very difficult. It is the crack-tip plasticity which gives metals their high toughness: energy is absorbed in the plastic zone, making the propagation of the crack much more difficult. Although some plasticity can occur at the tip of a crack in a ceramic too, it is very limited; the energy absorbed is small and the fracture toughness is low.

The result is that ceramics have values of K_c which are roughly one-fiftieth of those of good, ductile metals. In addition, they almost always contain cracks and flaws (see Fig. 16.7). The cracks originate in several ways. Most commonly the production method (see Chapter 19) leaves small holes: sintered products, for instance, generally contain angular pores on the scale of the powder (or grain) size. Thermal stresses caused by cooling or thermal cycling can generate small cracks. Even if there are no processing or thermal cracks, corrosion (often by water) or abrasion (by dust) is sufficient to create cracks in the surface of any ceramic. And if they do not form any other way, cracks appear during the loading of a brittle solid, nucleated by the elastic anisotropy of the grains, or by easy slip on a single slip system.

The design strength of a ceramic, then, is determined by its low fracture toughness and by the lengths of the microcracks it contains. If the longest microcrack in a given sample has length $2a_m$ then the tensile strength is simply

$$\sigma_{TS} = \frac{K_C}{\sqrt{\pi a_m}}. \tag{17.1}$$

Some engineering ceramics have tensile strengths about half that of steel – around 200 MPa. Taking a typical toughness of 2 MPa m$^{1/2}$, the largest microcrack has a size of 60 μm, which is of the same order as the original particle size. (For reasons given earlier, particle-size cracks commonly pre-exist in dense ceramics.) Pottery, brick and stone generally have tensile strengths which are much lower than this – around 20 MPa. These materials are full of cracks and voids left by the manufacturing process (their porosity is, typically, 5–20%). Again, it is the size of the longest crack – of the order of 2 mm for a typical toughness of 1 MPa m$^{1/2}$ – which determines the strength. The tensile strength of cement and concrete is even lower – 2 MPa in large sections – implying the presence of at least one crack 6 mm or more in length for a toughness of 0.2 MPa m$^{1/2}$.

As we shall see, there are two ways of improving the strength of ceramics: decreasing a_m by careful quality control, and increasing K_c by alloying, or by making the ceramic into a composite. But first, we must examine how strength is measured.

The common tests are shown in Fig. 17.2. The obvious one is the simple tensile test (Fig. 17.2a). It measures the stress required to make the longest crack in the sample propagate unstably in the way shown in Fig. 17.3(a). But it is hard to do tensile tests on ceramics – they tend to break in the grips. It is much easier to measure the force required to break a beam in bending (Fig. 17.2b). The maximum tensile stress in the surface of the beam when it breaks is called the modulus of rupture, σ_r; for an elastic beam it is related to the maximum moment in the beam, M_r, by

$$\sigma_r = \frac{6M_r}{bd^2} \qquad (17.2)$$

where d is the depth and b the width of the beam. You might think that σ_r (which is listed in Table 15.7) should be equal to the tensile strength σ_{TS}. But it is actually a little larger (typically 1.7 times larger), for reasons which we will get to when we discuss the statistics of strength in the next chapter.

The third test shown in Fig. 17.2 is the compression test. For metals (or any plastic solid) the strength measured in compression is the same as that measured

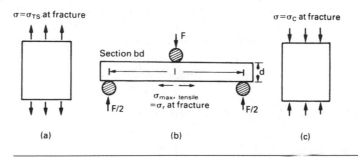

σ = σ$_{TS}$ at fracture   σ = σ$_C$ at fracture

Section bd

F

σ$_{max,\ tensile}$ = σ$_r$ at fracture

F/2   F/2

(a)   (b)   (c)

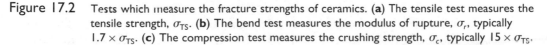

Figure 17.2   Tests which measure the fracture strengths of ceramics. (a) The tensile test measures the tensile strength, σ_{TS}. (b) The bend test measures the modulus of rupture, σ_r, typically 1.7 × σ_{TS}. (c) The compression test measures the crushing strength, σ_c, typically 15 × σ_{TS}.

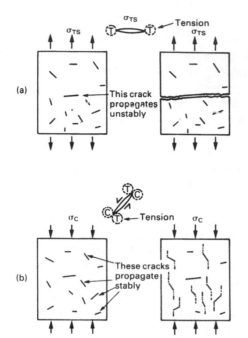

Figure 17.3   **(a)** In tension the largest flaw propagates unstably. **(b)** In compression, many flaws propagate stably to give general crushing.

in tension. But for brittle solids this is not so; for these, the compressive strength is roughly 15 times larger, with

$$\sigma_C \approx 15\sigma_{TS}. \tag{17.3}$$

The reason for this is explained by Fig. 17.3(b). Cracks in compression propagate *stably*, and twist out of their original orientation to propagate *parallel to the compression axis*. Fracture is not caused by the rapid unstable propagation of one crack, but the slow extension of many cracks to form a crushed zone. It is not the size of the largest crack (a_m) that counts, but that of the average $\bar{a}$. The compressive strength is still given by a formula like eqn. (17.1), with

$$\sigma_C = C \frac{K_c}{\sqrt{\pi\bar{a}}} \tag{17.4}$$

but the constant C is about 15, instead of 1.

Thermal shock resistance

When you pour boiling water into a cold bottle and discover that the bottom drops out with a smart pop, you have re-invented the standard test for thermal shock resistance. Fracture caused by sudden changes in temperature is a problem with ceramics. But while some (like ordinary glass) will only take a temperature "shock" of 80°C before they break, others (like silicon nitride) will stand a sudden change of 500°C, and this is enough to fit them for use in environments as violent as an internal combustion engine.

One way of measuring *thermal shock resistance* is to drop a piece of the ceramic, heated to progressively higher temperatures, into cold water. The maximum temperature drop ΔT (in K) which it can survive is a measure of its thermal shock resistance. If its coefficient of expansion is α then the quenched surface layer suffers a shrinkage strain of $\alpha \Delta T$. But it is part of a much larger body which is still hot, and this constrains it to its original dimensions: it then carries an elastic tensile stress $E\alpha\Delta T$. If this tensile stress exceeds that for tensile fracture, σ_{TS}, the surface of the component will crack and ultimately spall off. So the maximum temperature drop ΔT is given by

$$E\alpha\Delta T = \sigma_{TS}. \qquad (17.5)$$

Values of ΔT are given in Table 15.7. For ordinary glass, α is large and ΔT is small – about 80°C, as we have said. But for most of the high-performance engineering ceramics, α is small and σ_{TS} is large, so they can be quenched suddenly through several hundred degrees without fracturing.

Creep of ceramics

Like metals, ceramics creep when they are hot. The creep curve (Fig. 17.4) is just like that for a metal (see Book 1, Chapter 20). During primary creep, the strain-rate decreases with time, tending towards the steady state creep rate

$$\dot{\varepsilon}_{ss} = A\sigma^n \exp\left(-Q/RT\right). \qquad (17.6)$$

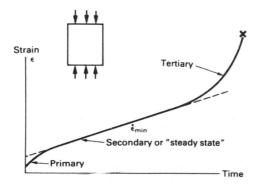

Figure 17.4   A creep curve for a ceramic.

Here σ is the stress, A and n are creep constants and Q is the activation energy for creep. Most engineering design against creep is based on this equation. Finally, the creep rate accelerates again into *tertiary creep* and *fracture*.

But what is "hot"? Creep becomes a problem when the temperature is greater than about $\frac{1}{3}T_m$. The melting point T_m of engineering ceramics is high – over $2000°C$ – so creep is design limiting only in very high-temperature applications (refractories, for instance). There is, however, one important ceramic – ice – which has a low melting point and creeps extensively, following eqn. (17.6). The sliding of glaciers, and even the spreading of the Antarctic ice-cap, are controlled by the creep of the ice; geophysicists who model the behaviour of glaciers use eqn. (17.6) to do so.

Examples

17.1 Explain why the yield strengths of ceramics can approach the ideal strength $\tilde{\sigma}$, whereas the yield strengths of metals are usually much less than $\tilde{\sigma}$. How would you attempt to measure the yield strength of a ceramic, given that the fracture strengths of ceramics in tension are usually much less than the yield strengths?

17.2 Why are ceramics usually much stronger in compression than in tension?

Al_2O_3 has a fracture toughness K_c of about $3\,MPa\,m^{1/2}$. A batch of Al_2O_3 samples is found to contain surface flaws about $30\,\mu m$ deep. Estimate (a) the tensile strength and (b) the compressive strength of the samples.

Answers

(a) 309 MPa, (b) 4635 MPa.

17.3 Modulus-of-rupture tests are carried out using the arrangement shown in Fig. 17.2. The specimens break at a load F of about $330\,N$. Find the modulus of rupture, given that $l = 50\,mm$, and that $b = d = 5\,mm$.

Answer

198 MPa.

17.4 Estimate the thermal shock resistance ΔT for the ceramics listed in Table 15.7. Use the data for Young's modulus E, modulus of rupture σ_r and thermal expansion coefficient α given in Table 15.7. How well do your calculated estimates of ΔT agree with the values given for ΔT in Table 15.7?

[Hints: (a) assume that $\sigma_{TS} \approx \sigma_r$ for the purposes of your estimates; (b) where there is a spread of values for E, σ_r or α, use the average values for your calculation.]

17.5 The photograph shows part of the cloisters of a monastery near the city of Huelva, Andalucia, Spain. Explain in a qualitative way why none of the elements in the structure experience tensile loading. Why is this important in masonry structures? [This place is of great historical importance, because it was from here in August 1492 that Christopher Columbus set sail for his epic voyage of discovery to North America.]

Monastery cloisters.

Chapter 18
The statistics of brittle fracture and case study

Introduction

The chalk with which I write on the blackboard when I teach is a brittle solid. Some sticks of chalk are weaker than others. On average, I find (to my slight irritation), that about 3 out of 10 sticks break as soon as I start to write with them; the other 7 survive. The failure probability, P_f, for this chalk, loaded in bending under my (standard) writing load is 3/10, that is

$$P_f = 0.3. \tag{18.1}$$

When you write on a blackboard with chalk, you are not unduly inconvenienced if 3 pieces in 10 break while you are using it; but if 1 in 2 broke, you might seek an alternative supplier. So the failure probability, P_f, of 0.3 is acceptable (just barely). If the component were a ceramic cutting tool, a failure probability of 1 in 100 ($P_f = 10^{-2}$) might be acceptable, because a tool is easily replaced. But if it were the window of a vacuum system, the failure of which can cause injury, one might aim for a P_f of 10^{-6}; and for a ceramic protective tile on the re-entry vehicle of a space shuttle, when one failure in any one of 10,000 tiles could be fatal, you might calculate that a P_f of 10^{-8} was needed.

When using a brittle solid under load, it is not possible to be certain that a component will not fail. But if an acceptable risk (the failure probability) can be assigned to the function filled by the component, then it *is* possible to design so that this acceptable risk is met. This chapter explains why ceramics have this dispersion of strength; and shows how to design components so they have a given probability of survival. The method is an interesting one, with application beyond ceramics to the malfunctioning of any complex system in which the breakdown of one component will cause the entire system to fail.

The statistics of strength and the Weibull distribution

Chalk is a porous ceramic. It has a fracture toughness of $0.9\,\mathrm{MPa\,m^{1/2}}$ and, being poorly consolidated, is full of cracks and angular holes. The average tensile strength of a piece of chalk is 15 MPa, implying an average length for the longest crack of about 1 mm (calculated from eqn. 17.1). But the chalk

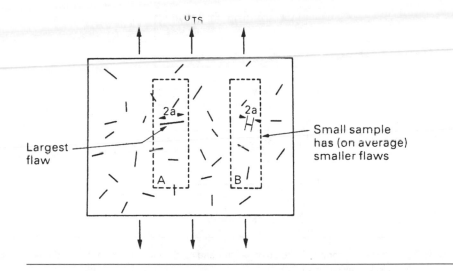

Figure 18.1 If small samples are cut from a large block of a brittle ceramic, they will show a dispersion of strengths because of the dispersion of flaw sizes. The average strength of the small samples is greater than that of the large sample.

itself contains a distribution of crack lengths. Two nominally identical pieces of chalk can have tensile strengths that differ greatly – by a factor of 3 or more. This is because one was cut so that, by chance, all the cracks in it are small, whereas the other was cut so that it includes one of the longer flaws of the distribution. Figure 18.1 illustrates this: if the block of chalk is cut into pieces, piece A will be weaker than piece B because it contains a larger flaw. It is inherent in the strength of ceramics that there will be a statistical variation in strength. There is no single "tensile strength"; but there is a certain, definable, *probability* that a given sample will have a given strength.

The distribution of crack lengths has other consequences. A large sample will fail at a lower stress than a small one, on average, because it is more likely that it will contain one of the larger flaws (Fig. 18.1). So there is a *volume dependence* of the strength. For the same reason, a ceramic rod is stronger in bending than in simple tension: in tension the entire sample carries the tensile stress, while in bending only a thin layer close to one surface (and thus a relatively smaller volume) carries the peak tensile stress (Fig. 18.2). That is why the modulus of rupture (Chapter 17, eqn. 17.2) is larger than the tensile strength.

The Swedish engineer, Weibull, invented the following way of handling the statistics of strength. He defined the *survival probability* $P_s(V_0)$ as the fraction of identical samples, each of volume V_0, which survive loading to a tensile stress σ. He then proposed that

$$P_s(V_0) = \exp\left\{-\left(\frac{\sigma}{\sigma_0}\right)^m\right\} \qquad (18.2)$$

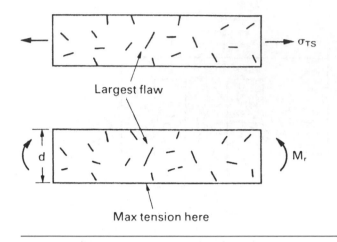

Max tension here

Figure 18.2   Ceramics appear to be stronger in bending than in tension because the largest flaw may not be near the surface.

where σ_0 and m are constants. This equation is plotted in Fig. 18.3(a). When $\sigma = 0$ all the samples survive, of course, and $P_s(V_0) = 1$. As σ increases, more and more samples fail, and $P_s(V_0)$ decreases. Large stresses cause virtually all the samples to break, so $P_s(V_0) \rightarrow 0$ and $\sigma \rightarrow \infty$.

If we set $\sigma = \sigma_0$ in eqn. (18.2) we find that $P_s(V_0) = 1/e \, (=0.37)$. So σ_0 is simply the tensile stress that allows 37% of the samples to survive. The constant m tells us how rapidly the strength falls as we approach σ_0 (see Fig. 18.3b). It is called the *Weibull modulus*. The lower m, the greater the *variability* of strength. For ordinary chalk, m is about 5, and the variability is great. Brick, pottery and cement are like this too. The engineering ceramics (e.g. SiC, Al_2O_3 and Si_3N_4) have values of m of about 10; for these, the strength varies rather less. Even steel shows some variation in strength, but it is small: it can be described by a Weibull modulus of about 100. Figure 18.3(b) shows that, for $m \approx 100$, a material can be treated as having a single, well-defined failure stress.

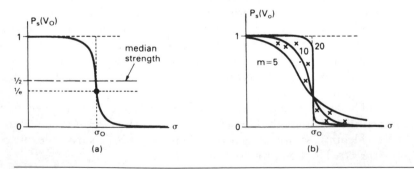

(a)   (b)

Figure 18.3   **(a)** The Weibull distribution function. **(b)** When the modulus, m, changes, the survival probability changes as shown.

σ_0 and m can be found from experiment. A batch of samples, each of volume V_0, is tested at a stress σ_1 and the fraction $P_{s1}(V_0)$ that survive is determined. Another batch is tested at σ_2 and so on. The points are then plotted on Fig. 18.3(b). It is easy to determine σ_0 from the graph, but m has to be found by curve-fitting. There is a better way of plotting the data which allows m to be determined more easily. Taking natural logs in eqn. (18.2) gives

$$\ln \left\{ \frac{1}{P_s(V_0)} \right\} = \left(\frac{\sigma}{\sigma_0} \right)^m. \tag{18.3}$$

And taking logs again gives

$$\ln \left\{ \ln \left(\frac{1}{P_s(V_0)} \right) \right\} = m \ln \left(\frac{\sigma}{\sigma_0} \right). \tag{18.4}$$

Thus a plot of $\ln\{\ln(1/P_s(V_0))\}$ against $\ln(\sigma/\sigma_0)$ is a straight line of slope m. Weibull-probability graph paper does the conversion for you (see Fig. 18.4).

So much for the stress dependence of P_s. But what of its volume dependence? We have already seen that the probability of one sample surviving a stress σ is $P_s(V_0)$. The probability that a batch of n such samples all survive the stress is just $\{P_s(V_0)\}^n$. If these n samples were stuck together to give a single sample of volume $V = nV_0$ then its survival probability would still be $\{P_s(V_0)\}^n$. So

$$P_s(V) = \{P_s(V_0)\}^n = \{P_s(V_0)\}^{V/V_0}. \tag{18.5}$$

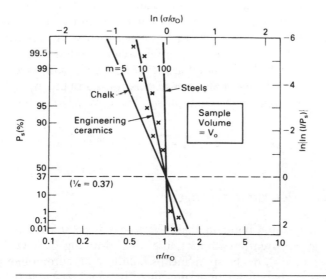

Figure 18.4   Survival probability plotted on "Weibull probability" axes for samples of volume V_0. This is just Fig. 18.3(b) plotted with axes that straighten out the lines of constant m.

This is equivalent to

$$\ln P_s(V) = \frac{V}{V_0} \ln P_s(V_0) \tag{18.6}$$

or

$$P_s(V) = \exp\left\{ \frac{V}{V_0} \ln P_s(V_0) \right\}. \tag{18.7}$$

The Weibull distribution (eqn. 18.2) can be rewritten as

$$\ln P_s(V_0) = -\left(\frac{\sigma}{\sigma_0} \right)^m. \tag{18.8}$$

If we insert this result into eqn. (18.7) we get

$$P_s(V) = \exp\left\{ -\frac{V}{V_0} \left(\frac{\sigma}{\sigma_0} \right)^m \right\}, \tag{18.9}$$

or

$$\ln P_s(V) = -\frac{V}{V_0} \left(\frac{\sigma}{\sigma_0} \right)^m.$$

This, then, is our final design equation. It shows how the survival probability depends on both the stress σ and the volume V of the component. In using it, the first step is to fix on an acceptable failure probability, P_f: 0.3 for chalk, 10^{-2} for the cutting tool, 10^{-6} for the vacuum-chamber window. The survival probability is then given by $P_s = 1 - P_f$. (It is useful to remember that, for small P_f, $\ln P_s = \ln(1 - P_f) \approx -P_f$.) We can then substitute suitable values of σ_0, m and V/V_0 into the equation to calculate the design stress.

Note that eqn. (18.9) assumes that the component is subjected to a *uniform* tensile stress σ. In many applications, σ is not constant, but instead varies with position throughout the component. Then, we rewrite eqn. (18.9) as

$$P_s(V) = \exp\left\{ -\frac{1}{\sigma_0^m V_0} \int_V \sigma^m dV \right\}. \tag{18.10}$$

The time-dependence of ceramic strength

Most people, at some point in their lives, have been startled by the sudden disintegration, apparently without cause, of a drinking glass (a "toughened" glass, almost always), or the spontaneous failure of an automobile windshield. These poltergeist-like happenings are caused by *slow crack growth*.

To be more specific, if a glass rod at room temperature breaks under a stress σ in a short time t, then an identical rod stressed at 0.75σ will break in a time of order $10t$. Most oxides behave like this, which is something which must be taken into account in engineering design. Carbides and nitrides (e.g. SiC or Si_3N_4) do not suffer from this *time-dependent failure* at room temperature, although at high temperatures they may do so. Its origin is the slow growth of surface microcracks caused by a chemical interaction between the ceramic and the water in its environment. Water or water vapour reaching the crack tip reacts chemically with molecules there to form a hydroxide, breaking the Si—O—Si or M—O—M bonds (Fig. 18.5). When the crack has grown to the critical length for failure at that stress level (eqn. 17.1) the part fails suddenly, often after a long period. Because it resembles fatigue failure, but under static load, it is sometimes called "static fatigue". Toughened glass is particularly prone to this sort of failure because it contains internal stresses which can drive the slow crack growth, and which drive spontaneous fast fracture when the crack grows long enough.

Fracture mechanics can be applied to this problem, much as it is to fatigue. We use only the final result, as follows. If the standard test which was used to measure σ_{TS} takes a time $t(\text{test})$, then the stress which the sample will support safely for a time t is

$$\left(\frac{\sigma}{\sigma_{TS}}\right)^n = \frac{t(\text{test})}{t} \tag{18.11}$$

where n is the slow crack-growth exponent. Its value for oxides is between 10 and 20 at room temperature. When $n = 10$, a factor of 10 in time reduces the strength by 20%. For carbides and nitrides, n can be as large as 100; then a factor of 10 in time reduces the strength by only 2%. (Data for n are included

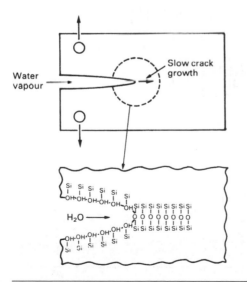

Figure 18.5   Slow crack growth caused by surface hydration of oxide ceramics.

in Table 15.7.) An example of the use of this equation is given in the following case study.

CASE STUDY: THE DESIGN OF PRESSURE WINDOWS

Glass can support large static loads for long times. Aircraft windows support a pressure difference of up to 1 atmosphere. Windows of tall buildings support wind loads; diving bells have windows which support large water pressures; glass vacuum equipment carries stress due to the pressure differences at which it operates. In Cambridge (UK) there is a cake shop with glass shelves, simply supported at both ends, which on weekdays are so loaded with cakes that the centre deflects by some centimetres. The owners say that they have loaded them like this, without mishap, for decades. But what about cake-induced slow crack growth? In this case study, we analyse safe design with glass under load.

Consider the design of a glass window for a vacuum chamber (Fig. 18.6). It is a circular glass disc of radius R and thickness t, freely supported in a rubber seal around its periphery and subjected to a uniform pressure difference $\Delta p = 0.1\,\text{MPa}$ (1 atmosphere). The pressure bends the disc. We shall simply quote the result of the stress analysis of such a disc: it is that the peak tensile stress is on the low-pressure face of the window and has magnitude

$$\sigma_{max} = \frac{3(3+v)}{8}\,\Delta p\,\frac{R^2}{t^2}. \tag{18.12}$$

Poisson's ratio v for ceramics is close to 0.3 so that

$$\sigma_{max} \approx \Delta p\,\frac{R^2}{t^2}. \tag{18.13}$$

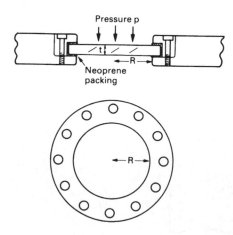

Figure 18.6   A flat-faced pressure window. The pressure difference generates tensile stresses in the low-pressure face.

Table 18.1   Properties of soda glass

Modulus E (GPa)	74
Compressive strength σ_c (MPa)	1000
Modulus of rupture σ_r (MPa)	50
Weibull modulus m	10
Time exponent n	10
Fracture toughness K_c (MPa m$^{1/2}$)	0.7
Thermal shock resistance ΔT (K)	84

The material properties of window glass are summarised in Table 18.1. To use these data to calculate a safe design load, we must assign an acceptable failure probability to the window, and decide on its design life. Failure could cause injury, so the window is a critical component: we choose a failure probability of 10^{-6}. The vacuum system is designed for intermittent use and is seldom under vacuum for more than 1 hour, so the design life under load is 1000 hours.

The modulus of rupture ($\sigma_r = 50\,\text{MPa}$) measures the mean strength of the glass in a short-time bending test. We shall assume that the test sample used to measure σ_r had dimensions similar to that of the window (otherwise a correction for volume is necessary) and that the test time was 10 minutes. Then the Weibull equation (eqn. 18.9) for a failure probability of 10^{-6} requires a strength-reduction factor of 0.25. And the static fatigue equation (eqn. 18.11) for a design life of 1000 hours [$t/t(\text{test}) \approx 10^4$] requires a reduction factor of 0.4. For this critical component, a design stress $\sigma = 50\,\text{MPa} \times 0.25 \times 0.4 = 5.0\,\text{MPa}$ meets the requirements. We apply a further safety factor of $S = 1.5$ to allow for uncertainties in loading, unforeseen variability and so on.

We may now specify the dimensions of the window. Inverting eqn. (18.13) gives

$$\frac{t}{R} = \left(\frac{S\Delta p}{\sigma} \right)^{1/2} = 0.17. \tag{18.14}$$

A window designed to these specifications should withstand a pressure difference of 1 atmosphere for 1000 hours with a failure probability of better than 10^{-6} – provided, of course, that it is not subject to thermal stresses, impact loads, stress concentrations or contact stresses. The commonest mistake is to overtighten the clamps holding the window in place, generating contact stresses: added to the pressure loading, they can lead to failure. The design shown in Fig. 18.6 has a neoprene gasket to distribute the clamping load, and a large number of clamping screws to give an even clamping pressure.

If, for reasons of weight, a thinner window is required, two options are open to the designer. The first is to select a different material. Thermally toughened glass (quenched in such a way as to give compressive surface stress) has a modulus of rupture which is 3 times greater than that of ordinary glass,

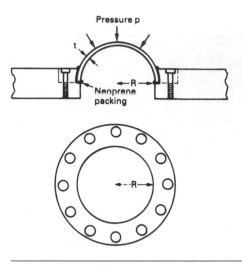

Figure 18.7   A hemispherical pressure window. The shape means that the glass is everywhere in compression.

allowing a window $\sqrt{3}$ times thinner than before. The second is to *redesign* the window itself. If it is made in the shape of a hemisphere (Fig. 18.7) the loading in the glass caused by a pressure difference is purely compressive ($\sigma_{max} = [\Delta p R/2t]$). Then we can utilise the enormous compressive strength of glass (1000 MPa) to design a window for which t/R is 7×10^{-5} with the same failure probability and life.

There is, of course, a way of cheating the statistics. If a batch of components has a distribution of strengths, it is possible to weed out the weak ones by loading them all up to a *proof stress* (say σ_0); then all those with big flaws will fail, leaving the fraction which were stronger than σ_0. Statistically speaking, proof testing selects and rejects the low-strength tail of the distribution. The method is widely used to reduce the probability of failure of critical components, but its effectiveness is undermined by slow crack growth which lets a small, harmless, crack grow with time into a large, dangerous one. The only way out is to proof test regularly throughout the life of the structure – an inconvenient, often impractical procedure. Then design for long-term safety is essential.

Examples

18.1   In order to test the strength of a ceramic, cylindrical specimens of length 25 mm and diameter 5 mm are put into axial tension. The tensile stress σ which causes 50% of the specimens to break is 120 MPa. Cylindrical ceramic components of length 50 mm and diameter 11 mm are required to withstand an axial tensile stress σ^1 with a survival probability of 99%. Given that $m = 5$, use eqn. (18.9) to determine σ^1.

Answer

32.6 MPa.

18.2 Modulus-of-rupture tests were carried out on samples of silicon carbide using the three-point bend test geometry shown in Fig. 17.2. The samples were 100 mm long and had a 10 mm by 10 mm square cross section. The median value of the modulus of rupture was 400 MPa. Tensile tests were also carried out using samples of identical material and dimensions, but loaded in tension along their lengths. The median value of the tensile strength was only 230 MPa. Account in a qualitative way for the difference between the two measures of strength.

Answer

In the tensile test, the whole volume of the sample is subjected to a tensile stress of 230 MPa. In the bend test, only the lower half of the sample is subjected to a tensile stress. Furthermore, the average value of this tensile stress is considerably less than the peak value of 400 MPa (which is only reached at the underside of the sample beneath the central loading point). The probability of finding a fracture-initiating defect in the small volume subjected to the highest stresses is small.

18.3 Modulus-of-rupture tests were done on samples of ceramic with dimensions $l = 100$ mm, $b = d = 10$ mm. The median value of σ_r (i.e. σ_r for $P_s = 0.5$) was 300 MPa. The ceramic is to be used for components with dimensions $l = 50$ mm, $b = d = 5$ mm loaded in simple tension along their length. Calculate the tensile stress σ that will give a probability of failure, P_f, of 10^{-6}. Assume that $m = 10$. Note that, for $m = 10$, $\sigma_{TS} = \sigma_r/1.73$.

Answer

55.7 MPa.

18.4 The diagram is a schematic of a stalactite, a cone-shaped mineral deposit hanging downwards from the roof of a cave. Its failure due to self weight loading is to be modelled using Weibull statistics. The geometry of the stalactite is idealised as a cone of length L and semiangle α. The cone angle is assumed small so that the base radius equals αL. The stalactite density is ρ.

(a) Show that the variation of tensile stress σ with height x is given by $\sigma = \frac{1}{3}\rho g x$. You may assume that the volume of a cone is given by $(\pi/3) \times$ (base radius)$^2 \times$ height.

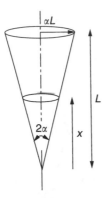

(b) Use equation (18.10) for Weibull statistics with a varying stress to show that the probability of survival $P_s(L)$ for a stalactite of length L is given by

$$P_s(L) = \exp\left\{-\left(\frac{\rho g}{3\sigma_0}\right)^m \frac{\pi\alpha^2 L^{m+3}}{(m+3)V_0}\right\}$$

Explain why there is a dependency on cone angle α, even though the stress variation up the stalactite is independent of α.

18.5 (a) Cylinderical samples of radius r and length l taken from stalactites are tested in uniform tension. If the stress level which gives a 50 per cent failure rate for specimens is equal to σ_t, derive an expression for the length of stalactite which can be expected to have a 50 per cent survival rate.

(b) Comment on possible practical difficulties with the sample tests.

Answer

$$L = \left\{\frac{lr^2(m+3)}{\alpha^2}\left(\frac{3\sigma_t}{\rho g}\right)^m\right\}^{1/(m+3)}$$

Chapter 19
Production, forming and joining of ceramics

Introduction

When you squeeze snow to make a snowball, you are hot-pressing a ceramic. Hot-pressing of powders is one of several standard *sintering methods* used to form ceramics which require methods appropriate to their special properties.

Glass, it is true, becomes liquid at a modest temperature (1000°C) and can be cast like a metal. At a lower temperature (around 700°C) it is very viscous, and can again be formed by the methods used for metals: rolling, pressing and forging. But the engineering ceramics have high melting points – typically 2000°C – precluding the possibility of melting and casting. And they lack the plasticity which allows the wide range of secondary forming processes used for metals: forging, rolling, machining and so forth. So most ceramics are made from *powders* which are pressed and fired, in various ways, to give the final product shape.

Vitreous ceramics are different. Clay, when wet, is *hydroplastic*: the water is drawn between the clay particles, lubricating their sliding, and allowing the clay to be formed by hand or with simple machinery. When the shaped clay is dried and fired, one component in it melts and spreads round the other components, bonding them together.

Low-grade ceramics – stone, and certain refractories – are simply mined and shaped. We are concerned here not with these, but with the production and shaping of high-performance engineering ceramics, clay products and glasses. Cement and concrete are discussed separately in Chapter 20. We start with engineering ceramics.

The production of engineering ceramics

Alumina powder is made from bauxite, a hydrated aluminium oxide with the formula $Al(OH)_3$, of which there are large deposits in Australia, the Caribbean and Africa. After crushing and purification, the bauxite is heated at 1150°C to decompose it to alumina, which is then milled and sieved

$$2Al(OH)_3 = Al_2O_3 + 3H_2O. \qquad (19.1)$$

Zirconia, ZrO_2, is made from the natural hydrated mineral, or from zircon, a silicate. Silicon carbide and silicon nitride are made by reacting silicon with carbon or nitrogen. Although the basic chemistry is very simple, the processes are complicated by the need for careful quality control, and the goal of producing fine (<1 μm) powders which, almost always, lead to a better final product.

These powders are then consolidated by one of a number of methods.

Forming of engineering ceramics

The surface area of fine powders is enormous. A cupful of alumina powder with a particle size of 1 μm has a surface area of about $10^3 \, m^2$. If the surface energy of alumina is $1 \, J \, m^{-2}$, the surface energy of the cupful of powder is 1 kJ.

This energy drives *sintering* (Fig. 19.1). When the powder is packed together and heated to a temperature at which diffusion becomes very rapid (generally, to around $\frac{2}{3} T_m$), the particles *sinter*, that is, they bond together to form small necks which then grow, reducing the surface area, and causing the powder to densify. Full density is not reached by this sort of sintering, but the residual porosity is in the form of small, rounded holes which have only a small effect on mechanical strength.

Figure 19.2 shows, at a microscopic level, what is going on. Atoms *diffuse* from the grain boundary which must form at each neck (since the particles which meet there have different orientations), and deposit in the pore, tending to fill it up. The atoms move by *grain boundary diffusion* (helped a little by lattice diffusion, which tends to be slower). The reduction in surface area drives the process, and the rate of diffusion controls its rate. This immediately tells us the two most important things we need to know about solid state sintering:

(a) Fine particles sinter much faster than coarse ones because the surface area (and thus the driving force) is higher, and because the diffusion distances are smaller.

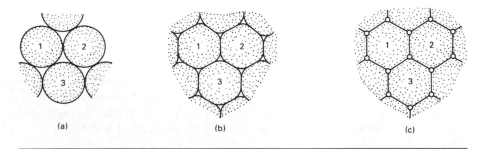

(a)   (b)   (c)

Figure 19.1   Powder particles pressed together at (a) sinter, as shown at (b), reducing the surface area (and thus energy) of the pores; the final structure usually contains small, nearly spherical pores (c).

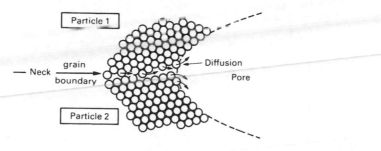

Figure 19.2 The microscopic mechanism of sintering. Atoms leave the grain boundary in the neck between two particles and diffuse into the pore, filling it up.

(b) The rate of sintering varies with temperature in exactly the same way as the diffusion coefficient. Thus the rate of densification is given by

$$\frac{\mathrm{d}\rho}{\mathrm{d}t} = \frac{C}{a^n} \exp\left(-Q/RT\right). \tag{19.2}$$

Here ρ is the density, a is the particle size, C and n are constants, Q is the activation energy for sintering, R is the gas constant and T is the absolute temperature. n is typically about 3, and Q is usually equal to the activation energy for grain boundary diffusion.

The sintering of powder is a production method used not only for ceramics but for metals and polymers too (see Chapter 14). In practice, the powder is first pressed to an initial shape in a die, mixing it with a binder, or relying on a little plasticity, to give a "green compact" with just enough strength to be moved into a sintering furnace. Considerable shrinkage occurs, of course, when the compact is fired. But by mixing powders of different sizes to get a high density to start with, and by allowing for the shrinkage in designing the die, a product can be produced which requires the minimum amount of finishing by machining or grinding. The final microstructure shows grains with a distribution of small, nearly spherical pores at the edges of the grains (see Fig. 16.7). The pore size and spacing are directly proportional to the original particle size, so the finer the particles, the smaller are these defects, and the better the mechanical strength (see Chapter 17). During sintering the grains in the ceramic grow, so the final grain size is often much larger than the original particle size (see Chapter 5).

Higher densities and smaller grains are obtained by *hot-pressing*: the simultaneous application of pressure and temperature to a powder. The powder is squeezed in a die (die pressing, Fig. 19.3), or in a pressure vessel which is pumped up to a high gas pressure (hot-isostatic pressing, or "HIPing", Fig. 19.4). At the same time the powder is heated to the sintering temperature. The pressure adds to the surface energy to drive sintering more quickly than before. The rate is still controlled by diffusion, and so it still varies with temperature according to eqn. (19.2). But the larger driving force shortens the

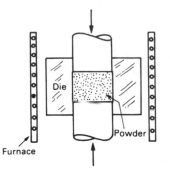

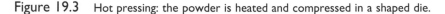

Figure 19.3 Hot pressing: the powder is heated and compressed in a shaped die.

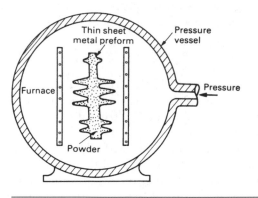

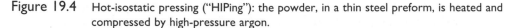

Figure 19.4 Hot-isostatic pressing ("HIPing"): the powder, in a thin steel preform, is heated and compressed by high-pressure argon.

sintering time from hours to minutes, and increases the final density. Indeed, full density can only be reached by pressure sintering, and the short time gives no opportunity for grain growth, so the mechanical properties of the product are good. Die pressing allows such precision that no subsequent finishing processes are necessary; but the dies, and thus the process, are expensive.

Full density can be reached by another route – though with some loss of mechanical strength. Small amounts of additive, such as MgO in the sintering of Al_2O_3 or Si_3N_4, greatly increase the rate of sintering. The additive reacts with the powder (and any impurities it may contain) to form a low-melting point glass which flows between the powder particles at the sintering temperature. Diffusional transport through the melt is high – it is like squeezing wet sugar – and the rate of sintering of the solid is increased. As little as 1% of glass is all that is needed, but it remains at the boundaries of the grains in the final product, and (because it melts again) drastically reduces their high-temperature strength. This process of *liquid phase sintering* (Fig. 19.5) is widely used to

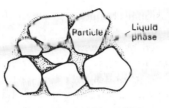

Figure 19.5   Liquid phase sintering: a small amount of additive forms a liquid which accelerates sintering and gives fully dense products but with some loss of high-temperature strength.

produce dense ceramics. It can be applied to metals too. The unhappy reader with bad teeth will know this only too well: it is the way dental amalgam works (silver, sintered at 36.9°C in the presence of a liquid phase, mercury).

There are two further processes. Silicon-based ceramics can be fabricated by sintering or by hot-pressing. But a new route, *reaction bonding* (Fig. 19.6), is cheaper and gives good precision. If pure silicon powder is heated in nitrogen gas, or a mixture of silicon and carbon powders is sintered together, then the reactions

$$3Si + 2N_2 = Si_3N_4 \qquad (19.3)$$

and

$$Si + C = SiC \qquad (19.4)$$

occur during the sintering, and bonding occurs simultaneously. In practice silicon, or the silicon–carbon mixture, is mixed with a polymer to make it plastic, and then formed by rolling, extrusion or pressing, using the methods which are normally used for polymer forming (Chapter 24): thin shells and complicated shapes can be made in this way. The polymer additive is then burnt out and the temperature raised so that the silicon and carbon react. The final porosity is high (because nitrogen or carbon must be able to penetrate

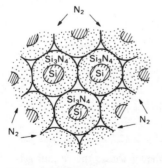

Figure 19.6   Silicon ceramics (SiC, Si$_3$N$_4$) can be shaped by reaction bonding.

through the section), but the dimensional change is so small (0.1%) that no further finishing operations need be necessary.

Finally, some ceramics can be formed by *chemical–vapour deposition* (CVD) processes. Silicon nitride is an example: Si_3N_4 can be formed by reacting suitable gases in such a way that it deposits on (or in) a former to give a shell or a solid body. When solids grow from the vapour they usually have a structure like that of a casting: columnar grains grow from the original surface, and may extend right through the section. For this reason, CVD products often have poor mechanical properties.

Production and forming of glass

Commercial glasses are based on silica, SiO_2, with additives: 30% of additives in a soda glass, about 20% in high-temperature glass like Pyrex. The additives, as you will remember from Chapter 15, lower the viscosity by breaking up the network. Raw glasses are produced, like metals, by melting the components together and then casting them.

Glasses, like metals, are formed by deformation. Liquid metals have a low viscosity (about the same as that of water), and transform discontinuously to a solid when they are cast and cooled. The viscosity of glasses falls slowly and continuously as they are heated. Viscosity is defined in the way shown in Fig. 19.7. If a shear stress σ_s is applied to the hot glass, it shears at a shear strain rate $\dot{\gamma}$. Then the viscosity, η, is defined by

$$\eta = \frac{\sigma_s}{10\dot{\gamma}}. \tag{19.5}$$

It has units of poise (P) or 10^{-1} Pa s. Glasses are worked in the temperature range in which their viscosity is between 10^4 and 10^7 poise (Fig. 19.8).

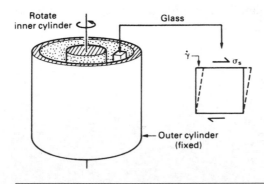

Figure 19.7   A rotation viscometer. Rotating the inner cylinder shears the viscous glass. The torque (and thus the shear stress σ_s) is measured for a given rotation rate (and thus shear strain rate $\dot{\gamma}$).

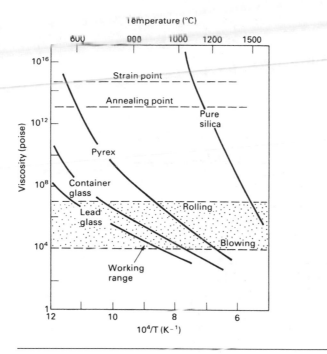

Figure 19.8   The variation of glass viscosity with temperature. It follows an Arrhenius law
($\eta \propto \exp(Q/RT)$) at high temperature.

Viscous flow is a thermally activated process. For flow to take place, the network must break and reform locally. Below the *glass temperature*, T_G there is insufficient thermal energy to allow this breaking and reforming to occur, and the glass ceases to flow; it is convenient to define this as the temperature at which the viscosity reaches 10^{17} P. (At T_G, it would take a large window 10,000 years to deform perceptibly under its own weight. The story that old church windows do so at room temperature is a myth.) Above T_G, the thermal energy of the molecules is sufficient to break and remake bonds at a rate which is fast enough to permit flow. As with all simple thermally activated processes, the rate of flow is given by

$$\text{rate of flow} \propto \exp\left(-Q/RT\right) \qquad (19.6)$$

where Q (this time) is the activation energy for viscous flow. The viscosity, η, is proportional to (rate of flow)$^{-1}$, so

$$\eta \propto \exp\left(Q/RT\right). \qquad (19.7)$$

The figure shows how the viscosity of three sorts of glass, and of silica itself, vary with temperature. On it, log η is plotted against $1/T$, to give lines with slope Q/R. The viscosity corresponding to the glass temperature is at the top

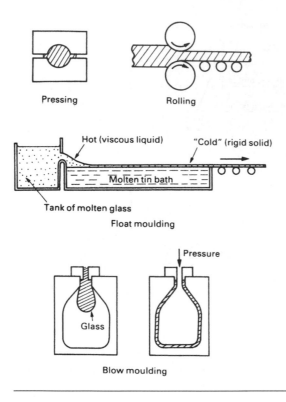

Pressing   Rolling

Hot (viscous liquid)   "Cold" (rigid solid)

Molten tin bath

Tank of molten glass

Float moulding

Pressure

Glass

Blow moulding

Figure 19.9   Forming methods for glass: pressing, rolling, float-moulding and blow-moulding.

of the figure. The working range is shown as a shaded band: it is wide because working procedures vary. Typical processes, shown in Fig. 19.9, are:

(a) Hot-pressing, in which a slug of hot glass is pressed between dies (like the forging of metals); it is used to make heavy glass insulators, and requires a high viscosity.
(b) Rolling, to produce a glass sheet; again, requires a high viscosity.
(c) Float moulding, to produce optically smooth window glass; needs a *low* viscosity.
(d) Blow moulding, to produce bottles or the thin envelopes for light bulbs, at rates of several thousand per hour; requires a low viscosity.

Two other temperatures are important in the working of glass. At the *annealing point* ($\eta = 10^{13}$ poise) there is still enough fluidity to relax internal stresses in about 15 minutes. Most glass products are held briefly at this temperature to remove tensile stresses that might otherwise induce fracture. At the *strain point* ($\eta = 10^{14}$ poise) atom motion in the glass is so sluggish that rapid cooling from this temperature does not introduce new stresses. So, in processing, the product is cooled slowly from the annealing point to the strain point and faster from there to room temperature.

Residual tensile stresses, as we have seen, are a problem. But compressive residual stresses, in the right place, can be used to advantage. *Toughened glass* is made by heating the product above its annealing point, and then cooling rapidly. The surface contracts and hardens while the interior is still hot and more fluid; it deforms, allowing the tensile stress in the surface to relax. Then the interior cools and contracts. But the surface is below its strain point; it cannot flow, so it is put into compression by the contracting interior. With the surface in compression, the glass is stronger, because the microcracks which initiate failure in a glass are always in the surface (caused by abrasion or corrosion). The interior, of course, is in tension; and if a crack should penetrate through the protective compressive layer it is immediately unstable and the toughened glass shatters spontaneously.

The production and forming of pottery, porcelain and brick

Pottery is one of the oldest materials. Clay artefacts as old as the pyramids (5000 BC) are sophisticated in their manufacture and glazing; and shards of pottery of much earlier date are known. Then, as now, the clay was *mined* from sites where weathering had deposited them, *hydroplastically formed*, *fired* and then *glazed*.

Clays have plate-like molecules with charges on their surfaces (Chapter 16). The charges draw water into the clay as a thin lubricating layer between the plates. With the right moisture content, clays are *plastic*: they can be moulded, extruded, turned or carved. But when they are dried, they have sufficient strength to be handled and stacked in kilns for firing.

In *slip casting* a thin slurry, or suspension, of clay in water is poured into a porous mould. Water is absorbed into the mould wall, causing a layer of clay to form and adhere to it. The excess slurry is tipped out of the mould and the slip-cast shell, now dry enough to have strength, is taken out and fired. The process allows intricate shapes (like plates, cups, vases) to be made quickly and accurately.

When a clay is *fired*, the water it contains is driven off and a silicate glass forms by reaction between the components of the clay. The glass melts and is drawn by surface tension into the interstices between the particles of clay, like water into a sponge. Clays for brick and pottery are usually a blend of three constituents which occur together naturally: pure clay, such as the $Al_2O_3 2SiO_2 2H_2O$ (kaolinite) described in Chapter 16; a flux (such as feldspar) which contains the Na or K used to make the glass; and a filler such as quartz sand, which reduces shrinkage but otherwise plays no role in the firing. Low-fire clays contain much flux and can be fired at 1000°C. High-fire clays have less, and require temperatures near 1200°C. The final microstructure shows particles of filler surrounded by particles of mullite (the reaction product of SiO_2 and Al_2O_3 in the clay) all bonded together by the glass.

Vitreous ceramics are made waterproof and strengthened by *glazing*. A slurry of powdered glass is applied to the surface by spraying or dipping, and the part is refired at a lower temperature (typically 800°C). The glass melts, flows over the surface, and is drawn by capillary action into pores and microcracks, sealing them.

Improving the performance of ceramics

When we speak of the "strength" of a metal, we mean its yield strength or tensile strength; to strengthen metals, they are alloyed in such a way as to obstruct dislocation motion, and thus raise the yield strength. By contrast, the "strength" of a ceramic is its fracture strength; to strengthen ceramics, we must seek ways of making fracture more difficult.

There are two, and they are complementary. The tensile fracture strength (Chapter 17) is roughly

$$\sigma_{TS} = \frac{K_c}{\sqrt{\pi a}} \tag{19.8}$$

and the compressive strength is about 15 times this value. First, we can seek to reduce the inherent flaw size, a; and second (though this is more difficult) we can seek to increase the fracture toughness, K_c.

Most ceramics (as we have seen) contain flaws: holes and cracks left by processing, cracks caused by thermal stress, corrosion or abrasion. Even if there are no cracks to start with, differences in elastic moduli between phases will nucleate cracks on loading. And most of these flaws have a size which is roughly that of the powder particles from which the ceramic was made. If the flaw size can be reduced, or if samples containing abnormally large flaws can be detected and rejected, the mean strength of the ceramic component is increased.

This is largely a problem of *quality control*. It means producing powders of a controlled, small size; pressing and sintering them under tightly controlled conditions to avoid defects caused by poor compaction, or by grain growth; and careful monitoring of the product to detect any drop in standard. By these methods, the modulus of rupture for dense Al_2O_3 and silicon carbide can be raised to 1000 MPa, making them as strong in tension as a high-strength steel; in compression they are 15 times stronger again.

The other alternative is to attempt to increase K_c. Pure ceramics have a fracture toughness between 0.2 and 2 MPa m$^{1/2}$. A dispersion of *particles of a second phase* can increase this a little: the advancing crack is pinned by the particles and bows between them, much as a dislocation is pinned by strong second phase particles (Chapter 10).

A more complicated, and more effective, mechanism operates in *partially stabilised zirconia* (PSZ), which has general application to other ceramics. Consider the analogy of a chocolate bar. Chocolate is a brittle solid and because of this it is notch-sensitive: notches are moulded into chocolate to help you break it in a fair, controlled way. Some chocolate bars have raisins and nuts in them, and they are less brittle: a crack, when it runs into a raisin, is arrested; and more energy is needed to break the bar in half. PSZ works in rather the same way. When ZrO_2 is alloyed with MgO, a structure can be created which has small particles of tetragonal zirconia (the raisins). When a crack approaches a particle, the particle transforms by a displacive transformation to a new (monoclinic) crystal structure, and this process absorbs energy. The details are complicated, but the result is simple: the toughness is increased from 2 to $8\,MPa\,m^{1/2}$. This may not seem much compared with $100\,MPa\,m^{1/2}$ for a tough steel, but it is big for a ceramic, dramatically increasing its strength and resistance to thermal shock, and opening up new applications for it.

Ceramics can be *fibre-strengthened* to improve their toughness. The plaster in old houses contains horse hair; and from the earliest times straw has been put into mud brick, in both cases to increase the toughness. In Arctic regions, ice is used for aircraft runways; the problem is that heavy aircraft knock large chips out of the brittle surface. One solution is to spread sawdust or straw onto the surface, flood it with water, and refreeze it; the fibres toughen the ice and reduce cracking. More recently, methods have been developed to toughen cement with glass fibres to produce high-strength panels and pipes. The details of the toughening mechanisms are the same as those for fibre-reinforced polymers, which we will discuss in Chapter 25. The effect can be spectacular: toughnesses of over $10\,MPa\,m^{1/2}$ are possible.

An older and successful way of overcoming the brittleness of ceramics is to make a sort of composite called a *cermet*. The best example is the cemented carbide used for cutting tools. Brittle particles of tungsten carbide (WC) are bonded together with a film of cobalt (Co) by sintering the mixed powders. If a crack starts in a WC particle, it immediately runs into the ductile cobalt film, which deforms plastically and absorbs energy (Fig. 19.10). The composite has a fracture toughness of around $15\,MPa\,m^{1/2}$, even though that of the WC is only $1\,MPa\,m^{1/2}$.

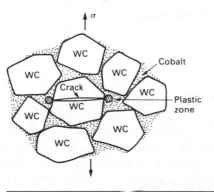

Figure 19.10   A cermet is a particulate composite of a ceramic (WC) in a metal (Co). A crack in the ceramic is arrested by plasticity in the cobalt.

Table 19.1   Applications of high-performance ceramics

Application	Property	Material
Cutting tools	Hardness, toughness	Alumina, sialons
Bearings, liners, seals	Wear resistance	Alumina, zirconia
Agricultural machinery	Wear resistance	Alumina, zirconia
Engine and turbine parts, burner nozzles	Heat and wear resistance	SiC, Si_3N_4, alumina, sialons, ceramic–ceramic composites
Shielding, armour	Hardness, toughness	Alumina, boron carbide
High-performance windows	Translucence and strength	Alumina, magnesia
Artificial bone, teeth, joints	Wear resistance, strength	Zirconia, alumina
Integrated circuit substrates	Insulation, heat resistance	Alumina, magnesia

The combination of better processing to give smaller flaws with alloying to improve toughness is a major advance in ceramic technology. The potential, not yet fully realised, appears to be enormous. Table 19.1 lists some of the areas in which ceramics have, or may soon replace other materials.

Joining of ceramics

Ceramics cannot be bolted or riveted: the contact stresses would cause brittle failure. Instead, ceramic components are bonded to other ceramic or metal parts by techniques which avoid or minimise stress concentrations.

Two such techniques are *diffusion bonding* and *glaze bonding* (Fig. 19.11). In diffusion bonding, the parts are heated while being pressed together; then, by processes like those which give sintering, the parts bond together. Even dissimilar materials can be bonded in this way. In glaze bonding the parts are coated with a low-melting (600°C) glass; the parts are placed in contact and heated above the melting point of the glass.

Ceramics are joined to metals by *metal coating and brazing*, and by the use of *adhesives*. In metal coating, the mating face of the ceramic part is coated in a thin film of a refractory metal such as molybdenum (usually applied as a powder and then heated). The metal film is then electroplated with copper, and the metal part brazed to the copper plating. Adhesives, usually epoxy resins, are used to join parts at low temperatures. Finally, ceramic parts can be clamped together, provided the clamps avoid stress concentrations, and are provided with soft (e.g. rubber) packing to avoid contact stresses.

The forming and joining of ceramics is summarised in the flowchart of Table 19.2.

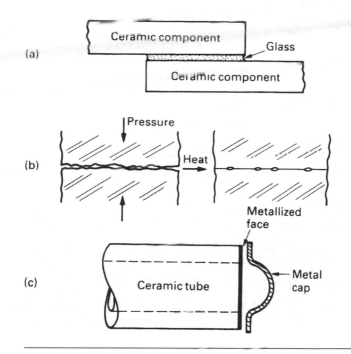

Figure 19.11   Joining methods for ceramics: (a) glaze bonding, (b) diffusion bonding, (c) metallisation plus brazing. In addition, ceramics can be clamped, and can be joined with adhesives.

Table 19.2   Forming and joining of ceramics

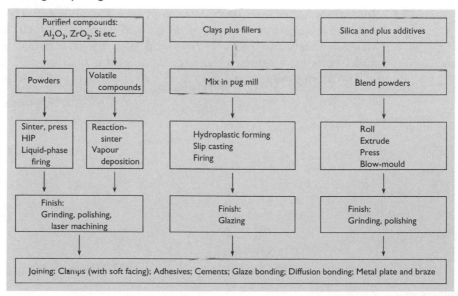

Examples

19.1   You have been given samples of the following ceramics.

(a) A hot-pressed thermocouple sheath of pure alumina.
(h) A piece of window glass.
(c) An unglazed fired clay pot
(d) A tungsten-carbide/cobalt cutting tool.

Sketch the structures that you would expect to see if you looked at polished sections of the samples under a reflecting light microscope. Label the phases and any other features of interest.

19.2   Describe briefly how the tensile strength of ceramic materials is determined by their microstructures. How may the tensile strength of ceramics be improved?

19.3   Describe the stages which might typically be followed in producing a small steel gear wheel by powder processing. Discuss the relative advantages and disadvantages of producing the gear wheel by powder processing or machining.

19.4   Why are special precautions necessary when joining ceramic components to metal components? What methods are available for the satisfactory joining of ceramics to metals?

19.5   The window glass in old buildings often has an uneven surface, with features which look like flow marks. The common explanation is that the glass has suffered creep deformation over the years under its own weight. Explain why this scenario is complete rubbish (why do you think the glass does appear to have "flow marks"?).

19.6   In addition to its dominant use for windows, glass is increasingly being used in innovative architectural applications such as bathroom panels, balcony walls, staircase treads and roofs. How would you design a bolted joint type for attaching a glass panel to the steel framing of a balcony wall? Indicate how and why your joint type differs from that used to attach a metal panel to the framing.

Chapter 20
Special topic: cements and concretes

Introduction

Concrete is a *particulate composite* of stone and sand, held together by an adhesive. The adhesive is usually a *cement paste* (used also as an adhesive to join bricks or stones), but asphalt or even polymers can be used to give special concretes. In this chapter we examine three cement pastes: the primitive pozzolana; the widespread Portland cement; and the newer, and somewhat discredited, high-alumina cement. And we consider the properties of the principal cement-based composite, concrete. The chemistry will be unfamiliar, but it is not difficult. The properties are exactly those expected of a ceramic containing a high density of flaws.

Chemistry of cements

Cement, of a sort, was known to the ancient Egyptians and Greeks. Their lime-cement was mixed with volcanic ash by the Romans to give a lime mortar; its success can be judged by the number of Roman buildings still standing 2000 years later. In countries which lack a sophisticated manufacturing and distribution system, these *pozzolana cements* are widely used (they are named after Pozzuoli, near Naples, where the ash came from, and which is still subject to alarming volcanic activity). To make them, chalk is heated at a relatively low temperature in simple wood-fired kilns to give lime

$$\text{Chalk } (CaCO_3) \xrightarrow[600°C]{\text{Heat}} \text{Lime } (CaO). \qquad (20.1)$$

The lime is mixed with water and volcanic ash and used to bond stone, brick, or even wood. The water reacts with lime, turning it into $Ca(OH)_2$; but in doing so, a surface reaction occurs with the ash (which contains SiO_2) probably giving a small amount of $(CaO)_3(SiO_2)_2(H_2O)_3$ and forming a strong bond. Only certain volcanic ashes have an active surface which will bond in this way; but they are widespread enough to be readily accessible.

The chemistry, obviously, is one of the curses of the study of cement. It is greatly simplified by the use of a *reduced nomenclature*. The four ingredients that matter in any cement are, in this nomenclature

$$\begin{array}{lll} \text{Lime} & CaO & = C \\ \text{Alumina} & Al_2O_3 & = A \end{array}$$

$$\text{Silica}\quad SiO_2 = S$$
$$\text{Water}\quad H_2O = H.$$

The key product, which bonds everything together, is

$$\text{Tobomorite gel } (CaO)_3(SiO_2)_2(H_2O)_3 = C_3S_2H_3.$$

In this terminology, pozzolana cement is C mixed with a volcanic ash which has active S on its surface. The reactions which occur when it sets (Fig. 20.1) are

$$C + H \rightarrow CH \text{ (in the bulk)} \tag{20.2}$$

and

$$3C + 2S + 3H \rightarrow C_3S_2H_3 \text{ (on the pozzolana surface).} \tag{20.3}$$

The tobomorite gel bonds the hydrated lime (CH) to the pozzolana particles. These two equations are all you need to know about the chemistry of pozzolana cement. Those for other cements are only slightly more complicated.

The world's construction industry thrived on lime cements until 1824, when a Leeds entrepreneur, Jo Aspdin, took out a patent for "a cement of superior quality, resembling Portland stone" (a white limestone from the island of Portland). This *Portland cement* is prepared by firing a controlled mixture of chalk ($CaCO_3$) and clay (which is just S_2AH_2) in a kiln at 1500°C (a high temperature, requiring special kiln materials and fuels, so it is a technology adapted to a developed country). Firing gives three products

$$\text{Chalk} + \text{Clay} \xrightarrow[1500°C]{\text{Heat}} C_3A + C_2S + C_3S. \tag{20.4}$$

When Portland cement is mixed with water, it hydrates, forming *hardened cement paste* ("h.c.p."). All cements harden by reaction, not by drying; indeed, it is important to keep them wet until full hardness is reached. Simplified a bit, two groups of reactions take place during the hydration of Portland cement.

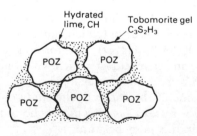

Figure 20.1   A pozzolana cement. The lime (C) reacts with silica (S) in the ash to give a bonding layer of tobomorite gel $C_3S_2H_3$.

The first is fast, occurring in the first 4 hours, and causing the cement to *set*. It is the hydration of the C_3A

$$C_3A + 6H \rightarrow C_3AH_6 + \text{heat}. \tag{20.5}$$

The second is slower, and causes the cement to *harden*. It starts after a delay of 10 hours or so, and takes 100 days or more before it is complete. It is the hydration of C_2S and C_3S to tobomorite gel, the main bonding material which occupies 70% of the structure

$$2C_2S + 4H \rightarrow \boxed{C_2S_2H_3} + CH + \text{heat} \tag{20.6}$$
$$2C_3S + 6H \rightarrow \boxed{C_3S_2H_3} + 3CH + \text{heat}. \tag{20.7}$$

$$\uparrow$$

Tobomorite gel.

Portland cement is stronger than pozzolana because gel forms in the bulk of the cement, not merely at its surface with the filler particles. The development of strength is shown in Fig. 20.2(a). The reactions give off a good deal of heat (Fig. 20.2b). It is used, in cold countries, to raise the temperature of the cement, preventing the water it contains from freezing. But in very large

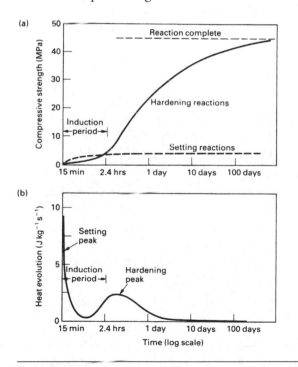

Figure 20.2   (a) The hardening of Portland cement. The setting reaction (eqn. 20.5) is followed by the hardening reactions (eqns 20.6 and 20.7). Each is associated with the evolution of heat (b).

structures such as dams, heating is a problem: then cooling pipes are embedded in the concrete to pump the heat out, and left in place afterwards as a sort of reinforcement.

High-alumina cement is fundamentally different from Portland cement. As its name suggests, it consists mainly of CA, with very little C_2S or C_3S. Its attraction is its high hardening rate: it achieves in a day what Portland cement achieves in a month. The hardening reaction is

$$CA + 10H \rightarrow CAH_{10} + heat. \tag{20.8}$$

But its long-term strength can be a problem. Depending on temperature and environment, the cement may deteriorate suddenly and without warning by "conversion" of the metastable CAH_{10} to the more stable C_3AH_6 (which formed in Portland cement). There is a substantial decrease in volume, creating porosity and causing drastic loss of strength. In cold, dry environments the changes are slow, and the effects may not be evident for years. But warm, wet conditions are disastrous, and strength may be lost in a few weeks.

The structure of Portland cement

The structure of cement, and the way in which it forms, are really remarkable. The angular cement powder is mixed with water (Fig. 20.3). Within 15 minutes

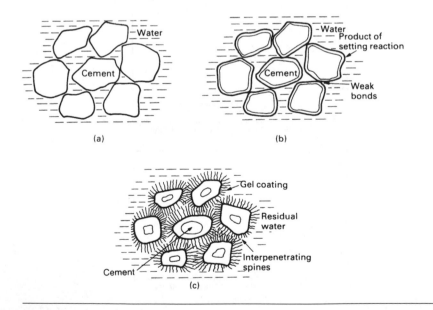

Figure 20.3   The setting and hardening of Portland cement. At the start (**a**) cement grains are mixed with water, H. After 15 minutes (**b**) the setting reaction gives a weak bond. Real strength comes with the hardening reaction (**c**), which takes some days.

the setting reaction (eqn. 20.5) coats the grains with a gelatinous envelope of hydrate (C_3AH_6). The grains are bridged at their point of contact by these coatings, giving a network of weak bonds which cause a loss of plasticity. The bonds are easily broken by stirring, but they quickly form again.

Hardening (eqns 20.6 and 20.7) starts after about 3 hours. The gel coating develops protuberances which grow into thin, densely packed rods radiating like the spines of a sea urchin from the individual cement grains. These spines are the $C_3S_2H_3$ of the second set of reactions. As hydration continues, the spines grow, gradually penetrating the region between the cement grains. The interlocked network of needles eventually consolidates into a rigid mass, and has the further property that it grows into, and binds to, the porous surface of brick, stone or pre-cast concrete.

The mechanism by which the spines grow is fascinating (Fig. 20.4). The initial envelope of hydrate on the cement grains, which gave setting, also acts as a semipermeable membrane for water. Water is drawn through the coating because of the high concentration of calcium inside, and a pressure builds up within the envelope (the induction period, shown in Fig. 20.2). This pressure bursts through the envelope, squirting little jets of a very concentrated solution of C_3S and C_2S into the surrounding water. The outer surface of the jet hydrates further to give a tube of $C_3S_2H_3$. The liquid within the tube, protected from the surrounding water, is pumped to the end by the osmotic pressure where it reacts, extending the tube. This osmotic pump continues to operate, steadily supplying reactants to the tube ends, which continue to grow until all the water or all the unreacted cement are used up.

Hardening is just another (rather complicated) example of nucleation and growth. Nucleation requires the formation, and then breakdown, of the hydrate

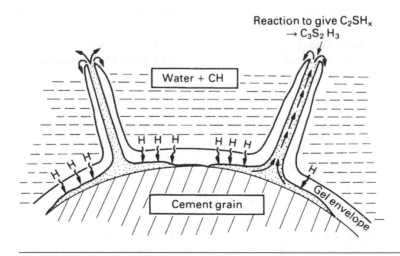

Figure 20.4   The mechanism by which the spiney structure of $C_3S_2H_3$ grows.

coating; the "induction period" shown in Fig. 20.2 is the nucleation time. Growth involves the passage of water by osmosis through the hydrate film and its reaction with the cement grain inside. The *driving force* for the transformation is the energy released when C_2S and C_3S react to give tobomorite gel $C_3S_2H_3$. The rate of the reaction is controlled by the rate at which water molecules diffuse through the film, and thus depends on temperature as

$$\text{rate} \propto \exp\left(-Q/RT\right). \tag{20.9}$$

Obviously, too, the rate will depend on the total surface area of cement grains available for reaction, and thus on the fineness of the powder. So hardening is accelerated by raising the temperature, and by grinding the powder more finely.

Concrete

Concrete is a mixture of stone and sand (the *aggregate*), glued together by cement (Fig. 20.5). The aggregate is dense and strong, so the weak phase is the hardened cement paste, and this largely determines the strength. Compared with other materials, cement is cheap; but aggregate is cheaper, so it is normal to pack as much aggregate into the concrete as possible whilst still retaining workability.

The best way to do this is to grade the aggregate so that it packs well. If particles of equal size are shaken down, they pack to a relative density of about 60%. The density is increased if smaller particles are mixed in: they fill the spaces between the bigger ones. A good combination is a 60–40 mixture of sand and gravel. The denser packing helps to fill the voids in the concrete, which are bad for obvious reasons: they weaken it, and they allow water to penetrate (which, if it freezes, will cause cracking).

When concrete hardens, the cement paste shrinks. The gravel, of course, is rigid, so that small shrinkage cracks are created. It is found that *air entrainment*

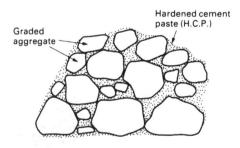

Figure 20.5 Concrete is a particulate composite of aggregate (60% by volume) in a matrix of hardened cement paste.

(mixing small bubbles of air into the concrete before pouring) helps prevent the cracks spreading.

The strength of cement and concrete

The strength of Portland cement largely depends on its *age* and its *density*. The development of strength with time was shown in Fig. 20.2(a): it still increases slowly after a year. Too much water in the original mixture gives a weak low-density cement (because of the space occupied by the excess water). Too little water is bad too because the workability is low and large voids of air get trapped during mixing. A water/cement ratio of 0.5 is a good compromise, though a ratio of 0.38 actually gives enough water to allow the reactions to go to completion.

The Young's modulus of cement paste varies with density as

$$\frac{E}{E_s} = \left(\frac{\rho}{\rho_s}\right)^3 \tag{20.10}$$

where E_s and ρ_s are the modulus and the density of solid tobomorite gel (32 GPa and 2.5 Mg m^{-3}). Concrete, of course, contains a great deal of gravel with a modulus three or so times greater than that of the paste. Its modulus can be calculated by the methods used for composite materials, giving

$$E_{\text{concrete}} = \left\{\frac{V_a}{E_a} + \frac{V_p}{E_p}\right\}^{-1}. \tag{20.11}$$

Here, V_a and V_p are the volume fractions of aggregate and cement paste, and E_a and E_p are their moduli. As Fig. 20.6 shows, experimental data for typical concretes fit this equation well.

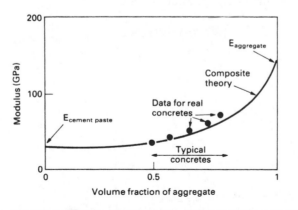

Figure 20.6   The modulus of concrete is very close to that given by simple composite theory (eqn. 20.11).

When cement is made, it inevitably contains flaws and cracks. The gel (like all ceramics) has a low fracture toughness: K_c is about 0.3 MPa m$^{1/2}$. In tension it is the longest crack which propagates, causing failure. The tensile strength of cement and concrete is around 4 MPa, implying a flaw size of 1 mm or so. The fracture toughness of concrete is a little higher than that of cement, typically 0.5 MPa m$^{1/2}$. This is because the crack must move round the aggregate, so the total surface area of the crack is greater. But this does not mean that the tensile strength is greater. It is difficult to make the cement penetrate evenly throughout the aggregate, and if it does not, larger cracks or flaws are left. And shrinkage, mentioned earlier, creates cracks on the same scale as the largest aggregate particles. The result is that the tensile strength is usually a little lower than that of carefully prepared cement. These strengths are so low that engineers, when designing with concrete and cement, arrange that it is always loaded in compression.

In compression, a single large flaw is not fatal (as it is tension). As explained in Chapter 17, cracks at an angle to the compression axis propagate in a *stable* way (requiring a progressive increase in load to make them propagate further). And they bend so that they run parallel to the compression axis (Fig. 20.7). The stress–strain curve therefore rises (Fig. 20.8), and finally reaches a maximum when the density of cracks is so large that they link to give a general crumbling of the material. In slightly more detail:

(a) Before loading, the cement or concrete contains cracks due to porosity, incomplete consolidation, and shrinkage stresses.
(b) At low stresses the material is linear elastic, with modulus given in Table 15.7. But even at low stresses, new small cracks nucleate at the surfaces between aggregate and cement.
(c) Above 50% of the ultimate crushing stress, cracks propagate stably, giving a stress–strain curve that continues to rise.

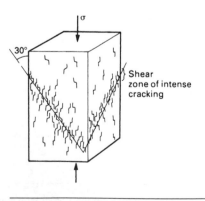

Figure 20.7   The compressive crushing of a cement or concrete block.

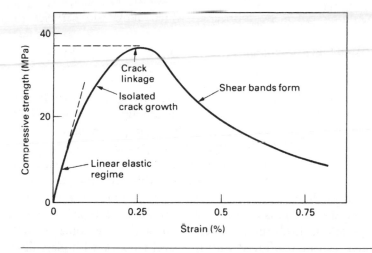

Figure 20.8   The stress–strain curve for cement or concrete in compression. Cracking starts at about half the ultimate strength.

(d) Above 90% of the maximum stress, some of the cracks become unstable, and continue to grow at constant load, linking with their neighbours. A failure surface develops at an angle of 30° to the compression axis. The load passes through a maximum and then drops – sometimes suddenly, but more usually rather slowly.

A material as complicated as cement shows considerable variation in strength. The mean crushing strength of 100 mm cubes of concrete is (typically) 50 MPa; but a few of the cubes fail at 40 MPa and a few survive to 60 MPa. There is a size effect too: 150 mm cubes have a strength which is lower, by about 10%, than that of 100 mm cubes. This is exactly what we would expect from Weibull's treatment of the strength of brittle solids (Chapter 18). There are, for concrete, additional complexities. But to a first approximation, design can be based on a median strength of 30 MPa and a Weibull exponent of 12, provided the mixing and pouring are good. When these are poor, the exponent falls to about 8.

High-strength cements

The low tensile strength of cement paste is, as we have seen, a result of low fracture toughness (0.3 MPa m$^{1/2}$) and a distribution of large inherent flaws. The scale of the flaws can be greatly reduced by four steps:

(a) Milling the cement to finer powder.
(b) Using the "ideal" water/cement ratio (0.38).

(c) Adding polymeric lubricants (which allow the particles to pack more densely).

(d) Applying pressure during hardening (which squeezes out residual porosity).

The result of doing all four things together is a remarkable material with a porosity of less than 2% and a tensile strength of up to 90 MPa. It is light (density 2.5 Mg m^{-3}) and, potentially, a cheap competitor in many low-stress applications now filled by polymers.

There are less exotic ways of increasing the strength of cement and concrete. One is to impregnate it with a polymer, which fills the pores and increases the fracture toughness a little. Another is by fibre reinforcement (Chapter 25). Steel-reinforced concrete is a sort of fibre-reinforced composite: the reinforcement carries tensile loads and, if prestressed, keeps the concrete in compression. Cement can be reinforced with fine steel wire, or with glass fibres. But these refinements, though simple, greatly increase the cost and mean that they are only viable in special applications. Plain Portland cement is probably the world's cheapest and most successful material.

Examples

20.1 In what way would you expect the setting and hardening reactions in cement paste to change with temperature? Indicate the practical significance of your result.

20.2 A concrete consists of 60% by volume of limestone aggregate plus 40% by volume of cement paste. Estimate the Young's modulus of the concrete, given that E for limestone is 63 GPa and E for cement paste is 25 GPa.

Answer

39 GPa.

20.3 Why is the tensile strength of conventional cement only about 4 MPa? How can the tensile strength of cement be increased by improvements in processing? What is the maximum value of tensile strength which can be achieved by processing improvements?

Answer

90 MPa approximately.

20.4 Make a list, based on your own observations, of selected examples of components and structures made from cement and concrete. Discuss how the way

in which the materials are used in each example is influenced by the low
(and highly variable) tensile strength of cement and concrete.

20.5   The concrete roof of the Sydney Opera House is not left bare, but is covered
with a layer of glazed ceramic tiles (see photograph below). What are the
advantages of this form of construction? How would you fix the tiles in place?

The concrete roof of the Sydney Opera House.

C. Polymers and composites

Chapter 21
Polymers

Introduction

Where people have, since the industrial revolution, used metals, nature uses polymers. Almost all biological systems are built of polymers which not only perform mechanical functions (like wood, bone, cartilage, leather) but also contain and regulate chemical reactions (leaf, veins, cells). People use these natural polymers, of course, and have done so for thousands of years. But it is only in this century that they have learned how to make polymers of their own. Early efforts (bakelite, celluloid, formaldehyde plastics) were floppy and not very strong; it is still a characteristic of most simple synthetic polymers that their stiffness (for a given section) is much less than that of metal or, indeed, of wood or bone. That is because wood and bone are composites: they are really made up of stiff fibres or particles, embedded in a matrix of simple polymer. People have learned how to make composites too: the industries which make high-performance glass, carbon, or Kevlar-fibre reinforced polymers (GFRP, CFRP, KFRP) enjoy a faster growth rate (over 10% per year) than almost any other branch of materials production. These new materials are stiff, strong and light. Though expensive, they are finding increasing use in aerospace, transport and sporting goods. And there are many opportunities for their wider application in other fields like hiking equipment, medical goods and even apparently insignificant things like spectacle frames: world-wide, at least 1,000,000,000 people wear spectacles.

And the new polymers are as exciting as the new composites. By crystallising, or by cross-linking, or by orienting the chains, new polymers are being made which are as stiff as aluminium; they will quickly find their way into production. The new processing methods can impart resistance to heat as well as to mechanical deformation, opening up new ranges of application for polymers which have already penetrated heavily into a market which used to be dominated by metals. No designer can afford to neglect the opportunities now offered by polymers and composites.

But it is a mistake to imagine that metal components can simply be replaced by components of these newer materials without rethinking the design. Polymers are less stiff, less strong and less tough than most metals, so the new component requires careful redesign. Composites, it is true, are stiff and strong. But they are often very anisotropic, and because they are bound by polymers, their properties can change radically with a small change in temperature. Proper

241

design with polymers requires a good understanding of their properties and where they come from. That is the function of the next four chapters.

In this chapter we introduce the main engineering polymers. They form the basis of a number of major industries, among them paints, rubbers, plastics, synthetic fibres and paper. As with metals and ceramics, there is a bewilderingly large number of polymers and the number increases every year. So we shall select a number of "generic" polymers which typify their class; others can be understood in terms of these. The classes of interest to us here are:

(a) *Thermoplastics* such as polyethylene, which soften on heating.
(b) *Thermosets* or *resins* such as epoxy which harden when two components (a resin and a hardener) are heated together.
(c) *Elastomers* or *rubbers*.
(d) *Natural polymers* such as cellulose, lignin and protein, which provide the mechanical basis of most plant and animal life.

Although their properties differ widely, all polymers are made up of long molecules with a covalently bonded backbone of carbon atoms. These long molecules are bonded together by weak Van der Waals and hydrogen ("secondary") bonds, or by these plus covalent cross-links. The melting point of the weak bonds is low, not far from room temperature. So we use these materials at a high fraction of the melting point of the weak bonds (though not of the much stronger covalent backbone). Not surprisingly, they show some of the features of a material near its melting point: they *creep*, and the elastic deflection which appears on loading increases with time. This is just one important way in which polymers differ from metals and ceramics, and it necessitates a different design approach (Chapter 27).

Most polymers are made from oil; the technology needed to make them from coal is still poorly developed. But one should not assume that dependence on oil makes the polymer industry specially vulnerable to oil price or availability. The value-added when polymers are made from crude oil is large. At 1998 prices, one tonne of oil is about $150; 1 tonne of polyethylene is about $800. So doubling the price of oil does not double the price of the polymer. And the energy content of metals is large too: that of aluminium is nearly twice as great as that of most polymers. So polymers are no more sensitive to energy prices than are most other commodities, and they are likely to be with us for a very long time to come.

The generic polymers

Thermoplastics

Polyethylene is the commonest of the thermoplastics. They are often described as *linear* polymers, that is the chains are not cross-linked (though they may branch occasionally). That is why they soften if the polymer is heated: the secondary bonds which bind the molecules to each other melt so that it flows like a viscous liquid, allowing it to be formed. The molecules in linear polymers have a range of molecular weights, and they pack together in a variety of configurations. Some, like polystyrene, are amorphous; others, like polyethylene, are partly crystalline. This range of molecular weights and packing geometries means that thermoplastics do not have a sharp melting point. Instead, their viscosity falls over a range of temperature, like that of an inorganic glass.

Thermoplastics are made by adding together ("polymerising") sub-units ("monomers") to form long chains. Many of them are made of the unit

repeated many times. The radical $\boxed{R}$ may simply be hydrogen (as in polyethylene), or $-CH_3$ (polypropylene) or $-Cl$ (polyvinylchloride). A few, like nylon, are more complicated. The generic thermoplastics are listed in Table 21.1. The fibre and film-forming polymers polyacrylonitrile (ACN) and polyethylene teraphthalate (PET, Terylene, Dacron, Mylar) are also thermoplastics.

Thermosets or resins

Epoxy, familiar as an adhesive and as the matrix of fibre-glass, is a thermoset (Table 21.2). Thermosets are made by mixing two components (a *resin* and a *hardener*) which react and harden, either at room temperature or on heating. The resulting polymer is usually heavily cross-linked, so thermosets are sometimes described as *network* polymers. The cross-links form during the polymerisation of the liquid resin and hardener, so the structure is almost always amorphous. On reheating, the additional secondary bonds melt, and the modulus of the polymer drops; but the cross-links prevent true melting or viscous flow so the polymer cannot be hot-worked (it turns into a rubber). Further heating just causes it to decompose.

The generic thermosets are the epoxies and the polyesters (both widely used as matrix materials for fibre-reinforced polymers) and the formaldehyde-based plastics (widely used for moulding and hard surfacing). Other formaldehyde plastics, which now replace bakelite, are ureaformaldehyde (used for electrical fittings) and melamine-formaldehyde (used for tableware).

Table 21.1   Generic thermoplastics

Thermoplastic	Composition	Uses
Polyethylene, PE	$\left(\begin{array}{c} H \\ \| \\ -C- \\ \| \\ H \end{array}\right)_n$ Partly crystalline.	Tubing, film, bottles, cups, electrical insulation, packaging.
Polypropylene, PP	$\left(\begin{array}{cc} H & H \\ \| & \| \\ -C- & C- \\ \| & \| \\ H & CH_3 \end{array}\right)_n$ Partly crystalline.	Same uses as PE, but lighter, stiffer, more resistant to sunlight.
Polytetrafluoroethylene, PTFE	$\left(\begin{array}{c} F \\ \| \\ -C- \\ \| \\ F \end{array}\right)_n$ Partly crystalline.	Teflon. Good, high-temperature polymer with very low friction and adhesion characteristics. Non-stick saucepans, bearings, seals.
Polystyrene, PS	$\left(\begin{array}{cc} H & H \\ \| & \| \\ -C- & C- \\ \| & \| \\ H & C_6H_5 \end{array}\right)_n$ Amorphous.	Cheap moulded objects. Toughened with butadiene to make high-impact polystyrene (HIPS). Foamed with CO_2 to make common packaging.
Polyvinylchloride, PVC	$\left(\begin{array}{cc} H & H \\ \| & \| \\ -C- & C- \\ \| & \| \\ H & Cl \end{array}\right)_n$ Amorphous.	Architectural uses (window frames, etc.). Plasticised to make artificial leather, hoses, clothing.
Polymethylmethacrylate, PMMA	$\left(\begin{array}{cc} H & CH_3 \\ \| & \| \\ -C- & C- \\ \| & \| \\ H & COOCH_3 \end{array}\right)_n$ Amorphous.	Perspex, lucite. Transparent sheet and mouldings. Aircraft windows, laminated windscreens.
Nylon 66	$(-C_6H_{11}NO-)_n$ Partly crystalline when drawn.	Textiles, rope, mouldings.

Elastomers

Elastomers or rubbers are almost-linear polymers with occasional cross-links in which, at room temperature, the secondary bonds have already melted. The cross-links provide the "memory" of the material so that it returns to its

Table 21.2 Generic thermosets or resins

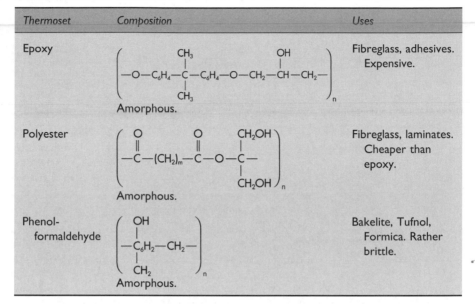

Thermoset	Composition	Uses
Epoxy	Amorphous.	Fibreglass, adhesives. Expensive.
Polyester	Amorphous.	Fibreglass, laminates. Cheaper than epoxy.
Phenol-formaldehyde	Amorphous.	Bakelite, Tufnol, Formica. Rather brittle.

original shape on unloading. The common rubbers are all based on the single structure

$$\left(\begin{array}{cccc} & H & & H \\ & | & & | \\ -C & -C=C & -C- \\ & | & | & | & | \\ & H & H & \boxed{R} & H \end{array}\right)_n$$

with the position $\boxed{R}$ occupied by H, CH_3 or Cl. They are listed in Table 21.3.

Natural polymers

The rubber polyisoprene is a natural polymer. So, too, are cellulose and lignin, the main components of wood and straw, and so are proteins like wool or silk. We use cellulose in vast quantities as paper and (by treating it with nitric acid) we make celluloid and cellophane out of it. But the vast surplus of lignin left from wood processing, or available in straw, cannot be processed to give a useful polymer. If it could, it would form the base for a vast new industry. The natural polymers are not as complicated as you might expect. They are listed in Table 21.4.

Table 21.3   Generic elastomers (rubbers)

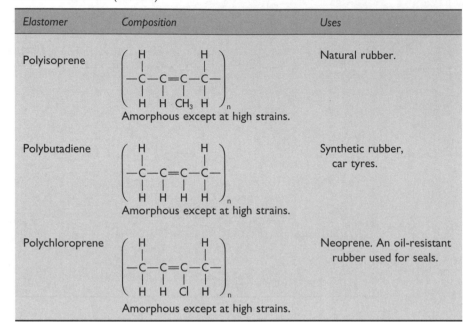

Elastomer	Composition	Uses
Polyisoprene	Amorphous except at high strains.	Natural rubber.
Polybutadiene	Amorphous except at high strains.	Synthetic rubber, car tyres.
Polychloroprene	Amorphous except at high strains.	Neoprene. An oil-resistant rubber used for seals.

Table 21.4   Generic natural polymers

Natural polymer	Composition	Uses
Cellulose	$(-C_6H_9O_6-)_n$ Crystalline.	Framework of all plant life, as the main structural component in cell walls.
Lignin	Amorphous.	The other main component in cell walls of all plant life.
Protein	R is a radical. Partly crystalline.	Gelatin, wool, silk.

Material data

Data for the properties of the generic polymers are shown in Table 21.5. But you have to be particularly careful in selecting and using data for the properties of polymers. Specifications for metals and alloys are defined fairly

Table 21.5 Properties of polymers

Polymer	Cost (UK£) ($US tonne⁻¹)	Density (Mg m⁻³)	Young's modulus (20°C 100 s) (GPa)	Tensile strength (MPa)	Fracture toughness (20°C) (MPa m$^{1/2}$)	Glass temperature T_G (K)	Softening expansion temperature T_s (K)	Specific heat (J Kg⁻¹ K⁻¹)	Thermal conductivity (W m⁻¹ K⁻¹)	Thermal coefficient (MK⁻¹)
Thermoplastics										
Polyethylene, PE (low density)	560 (780)	0.91–0.94	0.15–0.24	7–17	1–2	270	355	2250	0.35	160–190
Polyethylene, PE (high density)	510 (700)	0.95–0.98	0.55–1.0	20–37	2–5	300	390	2100	0.52	50–300
Polypropylene, PP	675 (950)	0.91	1.2–1.7	50–70	3.5	253	310	1900	0.2	100–300
Polytetrafluoroethylene, PTFE	–	2.2	0.35	17–28	–	–	395	1050	0.25	70–100
Polystyrene, PS	650 (910)	1.1	3.0–3.3	35–68	2	370	370	1350–1500	0.1–0.15	70–100
Polyvinyl chloride, PVC (unplasticised)	425 (595)	1.4	2.4–3.0	40–60	2.4	350	370	–	0.15	50–70
Polymethylmethacrylate, PMMA	1070 (1550)	1.2	3.3	80–90	1	378	400	1500	0.2	5–72
Nylons	2350 (3300)	1.15	2–3.5	60–110	3–5	340	350–420	1900	0.2–0.25	80–95
Resins or thermosets										
Epoxies	1150 (1600)	1.2–1.4	2.1–5.5	40–85	0.6–1.0	380	400–440	1700–2000	0.2–0.5	55–90
Polyesters	930 (1300)	1.1–1.4	1.3–4.5	45–85	0.5	340	420–440	1200–2400	0.2–0.24	50–100
Phenolformaldehyde	750 (1050)	1.27	8	35–55	–	–	370–550	1500–1700	0.12–0.24	26–60
Elastomers (rubbers)										
Polyisoprene	610 (850)	0.91	0.002–0.1	≈10	–	220	≈350	≈2500	≈0.15	≈600
Polybutadiene	610 (850)	1.5	0.004–0.1	–	–	171	≈350	≈2500	≈0.15	≈600
Polychloroprene	1460 (2050)	0.94	≈0.01	–	–	200	≈350	≈2500	≈0.15	≈600
Natural polymers										
Cellulose fibres		1.5	25–40	≈1000	–	–	–	–	–	–
Lignin		1.4	2.0	–	–	–	–	–	–	–
Protein		1.2–1.4	–	–	–	–	–	–	–	–

tightly; two pieces of Type 316L stainless steel from two different manufacturers will not differ much. Not so with polymers: polyethylene made by one manufacturer may be very different from polyethylene made by another. It is partly because all polymers contain a spectrum of molecular lengths; slight changes in processing change this spectrum. But it is also because details of the polymerisation change the extent of molecular branching and the degree of crystallinity in the final product; and the properties can be further changed by mechanical processing (which can, in varying degrees, align the molecules) and by proprietary additives. For all these reasons, data from compilations (like Table 21.5), or data books, are at best *approximate*. For accurate data you *must* use the manufacturers' data sheets, or conduct your own tests.

There are other ways in which polymer data differ from those for metals or ceramics. Polymers are held together by two sorts of bonds: strong covalent bonds which form the long chain backbone, and weak secondary bonds which stick the long chains together. At the glass temperature T_G, which is always near room temperature, the secondary bonds melt, leaving only the covalent bonds. The moduli of polymers reflect this. Below T_G most polymers have a modulus of around 3 GPa. (If the polymer is *drawn* to fibres or sheet, the molecules are aligned by the drawing process, and the modulus in the draw-direction can be larger.) But even if T_{room} is below T_G, T_{room} will still be a large fraction of T_G. Under load, the secondary bonds creep, and the modulus falls.* The table lists moduli for a loading time of 100 s at room temperature (20°C); for loading times of 1000 hours, the modulus can fall to one-third of that for the short (100 s) test. And above T_G, the secondary bonds melt completely: linear polymers become very viscous liquids, and cross-linked polymers become rubbers. Then the modulus can fall dramatically, from 3 GPa to 3 MPa or less.

You can see that design with polymers involves considerations which may differ from those for design with metals or ceramics. And there are other differences. One of the most important is that the yield or tensile strength of a polymer is a large fraction of its modulus; typically, $\sigma_y = E/20$. This means that design based on general yield (plastic design) gives large elastic deflections, much larger than in metals and ceramics. The excessive "give" of a poorly designed polymer component is a common experience, although it is often an advantage to have deflections without damage – as in polyethylene bottles, tough plastic luggage, or car bumpers.

The nearness of T_G to room temperature has other consequences. Near T_G most polymers are fairly tough, but K_c can drop steeply as the temperature is reduced. (The early use of polymers for shelving in refrigerators resulted in frequent fractures at +4°C. These were not anticipated because the polymer was ductile and tough at room temperature.)

---

* Remember that the modulus $E = \sigma/\epsilon$. ϵ will increase during creep at constant σ. This will give a lower apparent value of E. Long tests give large creep strains and even lower apparent moduli.

The specific heats of polymers are large – typically 5 times more than those of metals when measured per kg. When measured per m³, however, they are about the same because of the large differences in density. The coefficients of thermal expansion of polymers are enormous, 10 to 100 times larger than those of metals. This can lead to problems of thermal stress when polymers and metals are joined. And the thermal conductivities are small, 100 to 1000 times smaller than those of metals. This makes polymers attractive for thermal insulation, particularly when foamed.

In summary, then, design with polymers requires special attention to time-dependent effects, large elastic deformation and the effects of temperature, even close to room temperature. Room temperature data for the generic polymers are presented in Table 21.5. As emphasised already, they are approximate, suitable only for the first step of the design project. For the next step you should consult books (see Appendix 3), and when the choice has narrowed to one or a few candidates, data for them should be sought from manufacturers' data sheets, or from your own tests. Many polymers contain additives – plasticisers, fillers, colourants – which change the mechanical properties. Manufacturers will identify the polymers they sell, but will rarely disclose their additives. So it is essential, in making a final choice of material, that both the polymer and its source are identified and data *for that polymer, from that source*, are used in the design calculations.

Examples

21.1 What are the four main generic classes of polymers? For each generic class:

(a) give one example of a specific component made from that class;
(b) indicate why that class was selected for the component.

21.2 How do the unique characteristics of polymers influence the way in which these materials are used?

21.3 The photograph shows a flanged connector in an internally pressurised piping system. The material is a polymethylmethacrylate (PMMA). The connector failed during a hydrostatic pressure test, at a pressure of only 70% of the working pressure. At this point, the maximum hoop stress had reached 10 MPa. Table 21.5 gives a tensile strength of 80 MPa minimum for PMMA. Why do you think the connector failed at only 1/8 of this value? Justify your explanation with a short calculation. Why is PMMA a poor choice of material for the connector? What materials property in Table 21.5 is the critical design parameter in this application? The wall thickness of the tubular portion of the connector is 50 mm.

Flanged connector in an internally pressurised piping system.

Chapter 22
The structure of polymers

Introduction

If the architecture of metal crystals is thought of as classical, then that of polymers is baroque. The metal crystal is infused with order, as regular and symmetrical as the Parthenon; polymer structures are as exotic and convoluted as an Austrian altarpiece. Some polymers, it is true, form crystals, but the molecular packing in these crystals is more like that of the woven threads in a horse blanket than like the neat stacking of spheres in a metal crystal. Most are amorphous, and then the long molecules twine around each other like a bag full of tangled rope. And even the polymers which can crystallise are, in the bulk form in which engineers use them, only partly crystalline: segments of the molecules are woven into little crystallites, but other segments form a hopeless amorphous tangle in between.

The simpler polymers (like polyethylene, PMMA and polystyrene) are linear: the chains, if straightened out, would look like a piece of string. These are the thermoplastics: if heated, the strings slither past each other and the polymer softens and melts. And, at least in principle, these polymers can be drawn in such a way that the flow orients the strings, converting the amorphous tangle into sheet or fibre in which the molecules are more or less aligned. Then the properties are much changed: if you pull on the fibre (for example) you now stretch the molecular strings instead of merely unravelling them, and the stiffness and strength you measure are much larger than before.

The less simple polymers (like the epoxies, the polyesters and the formaldehyde-based resins) are networks: each chain is cross-linked in many places to other chains, so that, if stretched out, the array would look like a piece of Belgian lace, somehow woven in three dimensions. These are the thermosets: if heated, the structure softens but it does not melt; the cross-links prevent viscous flow. Thermosets are usually a bit stiffer than amorphous thermoplastics because of the cross-links, but they cannot easily be crystallised or oriented, so there is less scope for changing their properties by processing.

In this chapter we review, briefly, the essential features of polymer structures. They are more complicated than those of metal crystals, and there is no formal framework (like that of crystallography) in which to describe them exactly. But a looser, less precise description is possible, and is of enormous value in understanding the properties that polymers exhibit.

Molecular length and degree of polymerisation

Ethylene, C_2H_4, is a molecule. We can represent it as shown in Fig. 22.1(a), where the square box is a carbon atom, and the small circles are hydrogen. Polymerisation breaks the double bond, activating the ethylene monomer (Fig. 22.1b), and allowing it to link to others, forming a long chain or *macromolecule* (Fig. 22.1c). The ends of the chain are a problem: they either link to other macromolecules, or end with a *terminator* (such as an —OH group), shown as a round blob.

If only two or three molecules link, we have created a polymer. But to create a solid with useful mechanical properties, the chains must be longer – at least 500 monomers long. They are called *high polymers* (to distinguish them from the short ones) and, obviously, their length, or total molecular weight, is an important feature of their structure. It is usual to speak of the *degree of polymerisation* or DP: the number of monomer units in a molecule. Commercial polymers have a DP in the range 10^3 to 10^5.

The molecular weight of a polymer is simply the DP times the molecular weight of the monomer. Ethylene, C_2H_4, for example, has a molecular weight of 28. If the DP for a batch of polyethylene is 10^4, then the molecules have an *average* molecular weight of 280,000. The word "average" is significant. In

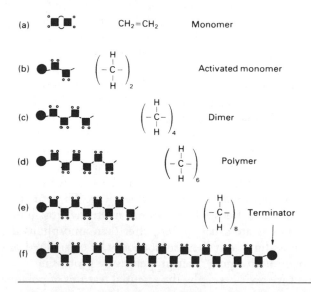

Figure 22.1   (a) The ethylene molecule or monomer; (b) the monomer in the activated state, ready to polymerise with others; (c)–(f) the ethylene polymer ("polyethylene"); the chain length is limited by the addition of terminators like —OH. The DP is the number of monomer units in the chain.

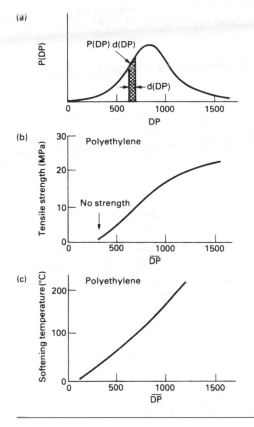

Figure 22.2　(a) Linear polymers are made of chains with a spectrum of lengths, or DPs. The probability of a given DP is P(DP); (b) and (c) the strength, the softening temperature and many other properties depend on the average DP.

all commercial polymers there is a range of DP, and thus of molecular lengths (Fig. 22.2a). Then the average is simply

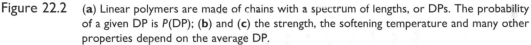

$$\overline{\mathrm{DP}} = \int_0^\infty \mathrm{DP}P(\mathrm{DP})\mathrm{d}(\mathrm{DP}) \qquad (22.1)$$

where P(DP)d(DP) is the fraction of molecules with DP values between DP and $\mathrm{DP} + \mathrm{d}(\mathrm{DP})$. The molecular weight is just $m\mathrm{DP}$ where m is the molecular weight of the monomer.

Most polymer properties depend on the average DP. Figure 22.2(b,c), for polyethylene, shows two: the tensile strength, and the softening temperature. DPs of less than 300 give no strength because the short molecules slide apart too easily. The strength rises with $\overline{\mathrm{DP}}$, but so does the viscosity; it is hard to mould polyethylene if the $\overline{\mathrm{DP}}$ is much above 10^3. The important point is that a material like polyethylene does not have a unique set of properties. There are many polyethylenes; the properties of a given batch depend on (among other things) the molecular length or $\overline{DP}$.

The molecular architecture

Thermoplastics are the largest class of engineering polymer. They have linear molecules: they are not cross-linked, and for that reason they soften when heated, allowing them to be formed (ways of doing this are described in Chapter 24). Monomers which form linear chains have two active bonds (they are *bifunctional*). A molecule with only one active bond can act as a chain terminator, but it cannot form a link in a chain. Monomers with three or more active sites (*polyfunctional* monomers) form networks: they are the basis of thermosetting polymers, or resins.

The simplest linear-chain polymer is polyethylene (Fig. 22.3a). By replacing one H atom of the monomer by a *side-group* or *radical* R (sausages on Fig. 22.3b, c, d) we obtain the *vinyl* group of polymers: R = Cl gives polyvinyl chloride; R = CH_3 gives polypropylene; R = C_6H_5 gives polystyrene. The radical gives asymmetry to the monomer unit, and there is then more than one way in which the unit can be attached to form a chain. Three arrangements are shown in Fig. 22.3. If all the side-groups are on the same side, the molecule is called *isotactic*. If they alternate in some regular way round

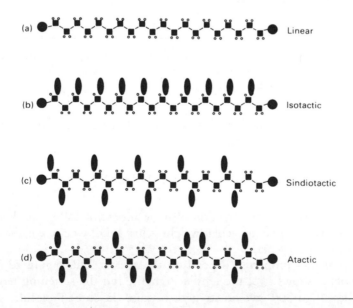

Figure 22.3   (a) Linear polyethylene; (b) an isotactic linear polymer: the side-groups are all on the same side; (c) a sindiotactic linear polymer: the side-groups alternate regularly; (d) an atactic linear polymer: the side-groups alternate irregularly.

the chain it is called *sindiotactic*. If they alternate randomly it is called *atactic*. These distinctions may seem like splitting hairs (protein, another linear polymer), but they are important: the tacticity influences properties. The regular molecules (Figs 22.3a,b,c) can stack side-by-side to form crystals: the regularly spaced side-groups nestle into the regular concavities of the next molecule. The irregular, atactic, molecules cannot: their side-groups clash, and the molecules are forced into lower-density, non-crystalline arrangements. Even the type of symmetry of the regular molecules matters: the isotactic (one-sided) molecules carry a net electric dipole and can be electroactive (showing piezoelectric effects, for instance), and others cannot.

Some polymerisation processes (such as the Ziegler process for making polyethylene) are delicate and precise in their operation: they produce only linear chains, and with a narrow spread of lengths. Others (like the older, high-pressure, ICI process) are crude and violent: side-groups may be torn from a part-formed molecule, and other growing molecules may attach themselves there, giving *branching*. Branching hinders crystallisation, just as atacticity does. Low-density polyethylene is branched, and for that reason has a low fraction of crystal ($\approx$50%), a low density, and low softening temperature (75°C). High-density PE is not branched: it is largely crystalline ($\approx$80%), it is 5% denser, and it softens at a temperature which is 30°C higher.

The next simplest group of linear polymers is the *vinylidene* group. Now two of the hydrogens of ethylene are replaced by radicals. Polymethylmethacrylate (alias PMMA, Perspex, Plexiglas or lucite) is one of these: the two radicals are $-CH_3$ and $-COOCH_3$. Now the difficulties of getting regular arrangements increases, and most of these polymers are amorphous.

Linear-chain thermoplastics are the most widely used of polymers, partly because of the ease with which they can be formed. Their plasticity allows them to be drawn into sheet, and in so doing, the molecules become aligned in the plane of the sheet, increasing the modulus and strength in this plane. Alignment is even more dramatic when linear polymers are drawn to fibres: the high strength of nylon, Dacron and Kevlar fibres reflects the near-perfect lining up of the macromolecules along the fibre axis.

Most *thermosets* start from large polyfunctional monomers. They react with each other or with small, linking molecules (like formaldehyde) in a condensation reaction – one which plucks an $-OH$ from one molecule and an $-H$ from the other to give H_2O (a by-product), welding the two molecules together at the severed bonds. Since one of the two molecules is polyfunctional, random three-dimensional networks are possible. Because of the cross-linking, thermosets do not melt when heated (though they ultimately decompose), they do not dissolve in solvents (as linear polymers do), and they cannot be formed after polymerisation (as linear polymers can). But for the same reason they are chemically more stable, are useful to a higher temperature, and are generally stiffer than thermoplastics. The irreversible setting reaction makes thermosets particularly good as adhesives, as coatings, and as the matrix for composites.

Elastomers are a special sort of cross-linked polymer. First, they are really linear polymers with just a few cross-links – one every hundred or more monomer units – so that a molecule with a DP of 500 might have fewer than five cross-link points along its length. And second, the polymer has a glass temperature which is well below room temperature, so that (at room temperature) the secondary bonds have melted. Why these two features give an elastomer is explained later (Chapter 23).

Packing of polymer molecules and the glass transition

Although we have drawn them as straight, a free polymer molecule is never so. Each C—C joint in its backbone has rotational freedom, so that the direction of the molecule changes at each step along the chain, allowing it to spiral, twist and tangle in the most extravagant way. When a linear polymer melts, its structure is that of a dense spaghetti-like tangle of these meandering molecules. Each is free to slither past the others in the melt, so the chain-links bend in a random way (Fig. 22.4). The average distance between the start of the chain and its end is then calculated in the same way that you calculate the distance a drunk staggers from the pub: if steps (of length λ) are equally likely in all directions (a "random walk"), the distance from the pub after n steps is $(\sqrt{n})\lambda$. So, if the polymer has n units of length λ, the distance from its head to its tail is, on average, $(\sqrt{n})\lambda$, not $n\lambda$ as you might at first think.

When the melt is cooled, the spaghetti tangle may simply freeze without rearranging; the resulting solid polymer then has an *amorphous* structure. But during cooling molecules can move, and (depending on their architecture) they may partly line up to form *crystallites*. We now consider each of the structures, starting with the crystallites.

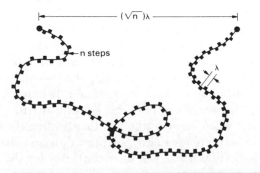

Figure 22.4   The random walk of a chain in a polymer melt, or in a solid, glassy polymer means that, on average, one end of the molecule is $(\sqrt{n})\lambda$ away from the other end. Very large strains (≈ 4) are needed to straighten the molecule out.

Polymer crystals

Linear-chain molecules can crystallise. High-density polyethylene is an example. The molecules have no side-groups or branches. On cooling, secondary bonds tend to pull the molecules together into parallel bundles, not perfectly crystalline, but not amorphous (that is, devoid of all order) either. Under some circumstances, well-defined *chain-folded* crystals form (Fig. 22.5): the long molecules fold like computer paper into a stack with a width much less than the length of the molecule. Actually, the crystals are rarely as neatly folded as computer paper. The folds are not perfectly even, and the tails of the molecules may not tuck in properly; it is more like a badly woven carpet. Nonetheless, the crystallinity is good enough for the polymer to diffract X-rays like a metal crystal, and a unit cell can be defined (Fig. 22.5). Note that the cell is much smaller than the molecule itself.

But even the most crystalline of polymers (e.g. high-density PE) is only 80% crystal. The structure probably looks something like Fig. 22.6: bundles, and chain-folded segments, make it largely crystalline, but the crystalline parts are separated by regions of disorder – amorphous, or glassy regions. Often the crystalline platelets organise themselves into *spherulites*: bundles of crystallites that, at first sight, seem to grow radially outward from a central point, giving

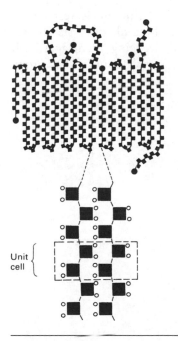

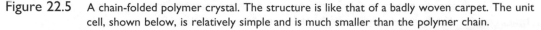

Figure 22.5   A chain-folded polymer crystal. The structure is like that of a badly woven carpet. The unit cell, shown below, is relatively simple and is much smaller than the polymer chain.

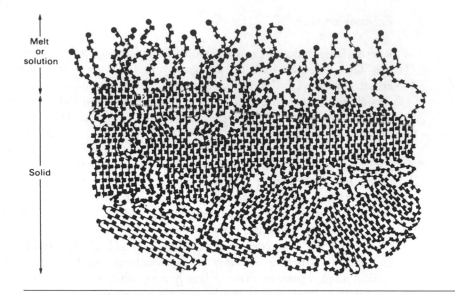

Figure 22.6 A schematic drawing of a largely crystalline polymer like high-density polyethylene. At the top the polymer has melted and the chain-folded segments have unwound.

crystals with spherical symmetry. The structure is really more complicated than that. The growing ends of a small bundle of crystallites (Fig. 22.7a) trap amorphous materials between them, wedging them apart. More crystallites nucleate on the bundle, and they, too, splay out as they grow. The splaying

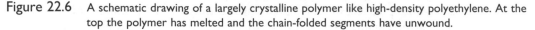

Figure 22.7 The formation and structure of a spherulite.

continues until the crystallites bend back on themselves and touch; then it can go no further (Fig. 22.7b). The spherulite then grows as a sphere until it impinges on others, to form a grain-like structure. Polythene is, in fact, like this, and polystyrene, nylon and many other linear polymers do the same thing.

When a liquid crystallises to a solid, there is a sharp, sudden decrease of volume at the melting point (Fig. 22.8a). The random arrangement of the atoms or molecules in the liquid changes discontinuously to the ordered, neatly

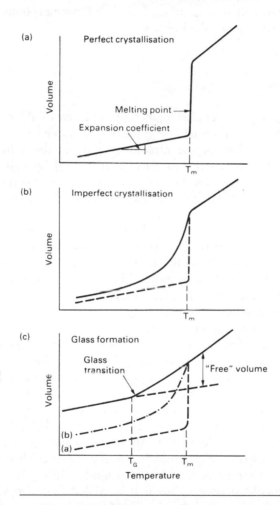

Figure 22.8   (a) The volume change when a simple melt (like a liquid metal) crystallises defines the melting point, T_m; (b) the spread of molecular weights blurs the melting point when polymers crystallise; (c) when a polymer solidifies to a glass the melting point disappears completely, but a new temperature at which the free volume disappears (the glass temperature, T_G) can be defined and measured.

packed, arrangement of the crystal. Other properties change discontinuously at the melting point also: the viscosity, for example, changes sharply by an enormous factor (10^{10} or more for a metal). Broadly speaking, polymers behave in the same way: a crystalline polymer has a fairly well-defined melting point at which the volume changes rapidly, though the sharpness found when metals crystallise is blurred by the range of molecular weights (and thus melting points) as shown in Fig. 22.8(b). For the same reason, other polymer properties (like the viscosity) change rapidly at the melting point, but the true discontinuity of properties found in simple crystals is lost.

When, instead, the polymer solidifies to a glass (an amorphous solid) the blurring is much greater, as we shall now see.

Amorphous polymers

Cumbersome side-groups, atacticity, branching and cross-linking all hinder crystallisation. In the melt, thermal energy causes the molecules to rearrange continuously. This wriggling of the molecules increases the volume of the polymer. The extra volume (over and above that needed by tightly packed, motionless molecules) is called the *free-volume*. It is the free-volume, aided by the thermal energy, that allows the molecules to move relative to each other, giving viscous flow.

As the temperature is decreased, free-volume is lost. If the molecular shape or cross-linking prevent crystallisation, then the liquid structure is retained, and free-volume is not all lost immediately (Fig. 22.8c). As with the melt, flow can still occur, though naturally it is more difficult, so the viscosity increases. As the polymer is cooled further, more free volume is lost. There comes a point at which the volume, though sufficient to contain the molecules, is too small to allow them to move and rearrange. All the free volume is gone, and the curve of specific volume flattens out (Fig. 22.8c). This is the *glass transition temperature*, T_G. Below this temperature the polymer is a *glass*.

The glass transition temperature is as important for polymers as the melting point is for metals (data for T_G are given in Table 21.5). Below T_G, secondary bonds bind the molecules into an amorphous solid; above, they start to melt, allowing molecular motion. The glass temperature of PMMA is 100°C, so at room temperature it is a brittle solid. Above T_G, a polymer becomes first *leathery*, then *rubbery*, capable of large elastic extensions without brittle fracture. The glass temperature for natural rubber is around −70°C, and it remains flexible even in the coldest winter; but if it is cooled to −196°C in liquid nitrogen, it becomes hard and brittle, like PMMA at room temperature.

That is all we need to know about structure for the moment, though more information can be found in the books listed under References. We now examine the origins of the strength of polymers in more detail, seeking the criteria which must be satisfied for good mechanical design.

Examples

22.1 Describe, in a few words, with an example or sketch where appropriate, what is meant by each of the following:

(a) a linear polymer;
(b) an isotactic polymer;
(c) a sindiotactic polymer;
(d) an atactic polymer;
(e) degree of polymerization;
(f) tangling;
(g) branching;
(h) cross-linking;
(i) an amorphous polymer;
(j) a crystalline polymer;
(k) a network polymer;
(l) a thermoplastic;
(m) a thermoset;
(n) an elastomer, or rubber;
(o) the glass transition temperature.

22.2 The density of a polyethylene crystal is $1.014 \, Mg \, m^{-3}$ at 20°C. The density of amorphous polyethylene at 20°C is $0.84 \, Mg \, m^{-3}$. Estimate the percentage crystallinity in:

(a) a low-density polyethylene with a density of $0.92 \, Mg \, m^{-3}$ at 20°C;
(b) a high-density polyethylene with a density of $0.97 \, Mg \, m^{-3}$ at 20°C.

Answers

(a) 46%, (b) 75%.

Chapter 23
Mechanical behaviour of polymers

Introduction

All polymers have a spectrum of mechanical behaviour, from *brittle-elastic* at low temperatures, through *plastic* to *viscoelastic* or *leathery*, to *rubbery* and finally to *viscous* at high temperatures. Metals and ceramics, too, have a range of mechanical behaviour, but, because their melting points are high, the variation near room temperature is unimportant. With polymers it is different: between $-20°C$ and $+200°C$ a polymer can pass through all of the mechanical states listed above, and in doing so its modulus and strength can change by a factor of 10^3 or more. So while we could treat metals and ceramics as having a constant stiffness and strength for design near ambient temperatures, we cannot do so for polymers.

The mechanical state of a polymer depends on its molecular weight and on the temperature; or, more precisely, on how close the temperature is to its glass temperature T_G. Each mechanical state covers a certain range of *normalised temperature* T/T_G (Fig. 23.1). Some polymers, like PMMA, and many epoxies, are brittle at room temperature because their glass temperatures are high and room temperature is only $0.75\ T_G$. Others, like the polyethylenes, are leathery; for these, room temperature is about $1.0\ T_G$. Still others, like polyisoprene, are elastomers; for these, room temperature is well above T_G (roughly $1.5\ T_G$). So it makes sense to plot polymer properties not against temperature T, but against T/T_G since that is what really determines the mechanical state. The modulus diagrams and strength diagrams described in this chapter are plotted in this way.

It is important to distinguish between the *stiffness* and the *strength* of a polymer. The stiffness describes the resistance to elastic deformation, the strength describes the resistance to collapse by plastic yielding or by fracture. Depending on the application, one or the other may be design-limiting. And both, in polymers, have complicated origins, which we will now explain.

Stiffness: the time- and temperature-dependent modulus

Much engineering design – particularly with polymers – is based on *stiffness*: the designer aims to keep the elastic deflections below some critical limit. Then the material property which is most important is Young's modulus, E. Metals and ceramics have Young's moduli which, near room temperature, can be

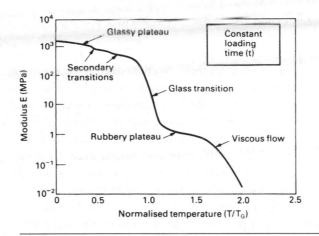

Figure 23.1 Schematic showing the way in which Young's modulus E for a linear polymer changes with temperature for a fixed loading time.

thought of as constant. Those of polymers cannot. When a polymer is loaded, it deflects by an amount which increases with the loading time t and with the temperature T. The deflection is elastic – on unloading, the strain disappears again (though that, too, may take time). So it is usual to speak of the *time- and temperature-dependent modulus, E(t, T)* (from now on simply called E). It is defined, just like any other Young's modulus, as the stress σ divided by the elastic strain ε

$$E = \frac{\sigma}{\varepsilon(t, T)}. \tag{23.1}$$

The difference is that the strain now depends on time and temperature.

The modulus E of a polymer can change enormously – by as much as a factor of 1000 – when the temperature is changed. We will focus first on the behaviour of linear-amorphous polymers, examining the reasons for the enormous range of modulus, and digressing occasionally to explain how cross-linking, or crystallisation, change things.

Linear-amorphous polymers (like PMMA or PS) show five regimes of deformation in each of which the modulus has certain characteristics, illustrated by Fig. 23.1. They are:

(a) the glassy regime, with a large modulus, around 3 GPa;
(b) the glass-transition regime, in which the modulus drops steeply from 3 GPa to around 3 MPa;
(c) the rubbery regime, with a low modulus, around 3 MPa;
(d) the viscous regime, when the polymer starts to flow;
(e) the regime of decomposition in which chemical breakdown starts.

We now examine each regime in a little more detail.

The glassy regime and the secondary relaxations

The glass temperature, T_G, you will remember, is the temperature at which the secondary bonds start to melt. Well below T_G the polymer molecules pack tightly together, either in an amorphous tangle, or in poorly organised crystallites with amorphous material in between. Load stretches the bonds, giving elastic deformation which is recovered on unloading. But there are two sorts of bonds: the taut, muscular, covalent bonds that form the backbone of the chains; and the flabby, soft, secondary bonds between them. Figure 23.2 illustrates this: the covalent chain is shown as a solid line and the side groups or radicals as full circles; they bond to each other by secondary bonds shown as dotted lines (this scheme is helpful later in understanding elastic deformation).

The modulus of the polymer is an average of the stiffnesses of its bonds. But it obviously is not an arithmetic mean: even if the stiff bonds were completely rigid, the polymer would deform because the weak bonds would stretch. Instead, we calculate the modulus by summing the deformation in each type of bond using the methods of composite theory (Chapter 25). A stress σ produces a strain which is the weighted sum of the strains in each sort of bond

$$\varepsilon = f\frac{\sigma}{E_1} + (1-f)\frac{\sigma}{E_2} = \sigma\left\{\frac{f}{E_1} + \frac{(1-f)}{E_2}\right\}. \qquad (23.2)$$

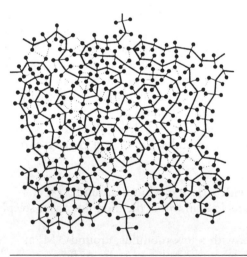

Figure 23.2   A schematic of a linear-amorphous polymer, showing the strong covalent bonds (full lines) and the weak secondary bonds (dotted lines). When the polymer is loaded below T_G, it is the secondary bonds which stretch.

Here f is the fraction of stiff, covalent bonds (modulus E_1) and $1 - f$ is the fraction of weak, secondary bonds (modulus E_2). The polymer modulus is

$$E = \frac{\sigma}{\varepsilon} = \left\{ \frac{f}{E_1} + \frac{(1-f)}{E_2} \right\}^{-1}. \tag{23.3}$$

If the polymer is completely cross-linked ($f = 1$) then the modulus (E_1) is known: it is that of diamond, 10^3 GPa. If it has no covalent bonds at all, then the modulus (E_2) is that of a simple hydrocarbon like paraffin wax, and that, too, is known: it is 1 GPa.

Substituting this information into the last equation gives an equation for the glassy modulus as a function of the fraction of covalent bonding

$$E = \left\{ \frac{f}{10^3} + \frac{(1-f)}{1} \right\}^{-1} \text{GPa}. \tag{23.4}$$

This function is plotted in Fig. 23.3. The glassy modulus of random, linear polymers ($f = \frac{1}{2}$) is always around 3 GPa. Heavily cross-linked polymers have a higher modulus because f is larger – as high as 0.75 – giving $E = 8$ GPa. Drawn polymers are different: they are anisotropic, having the chains lined up along the draw direction. Then the fraction of covalent bonds in the loading direction is increased dramatically. In extreme drawing of fibres like nylon or Kevlar this fraction reaches 98%, and the modulus rises to 100 GPa, about

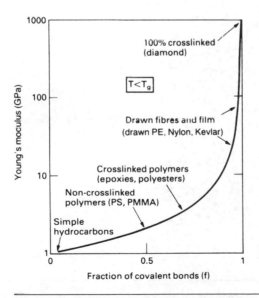

Figure 23.3   The way in which the modulus of polymers changes with the fraction of covalent bonds in the loading direction. Cross-linking increases this fraction a little; drawing increases it much more.

the same as that of aluminium. This *orientation strengthening* is a potent way of increasing the modulus of polymers. The stiffness normal to the drawing direction, of course, decreases because f falls towards zero in that direction.

You might expect that the glassy modulus (which, like that of metals and ceramics, is just due to bond-stretching) should not depend much on temperature. At very low temperatures this is correct. But the tangled packing of polymer molecules leaves some "loose sites" in the structure: side groups or chain segments, with a little help from thermal energy, readjust their positions to give a little extra strain. These secondary relaxations (Fig. 23.1) can lower the modulus by a factor of 2 or more, so they cannot be ignored. But their effect is small compared with that of the visco-elastic, or glass transition, which we come to next.

The glass, or visco-elastic transition

As the temperature is raised, the secondary bonds start to melt. Then segments of the chains can slip relative to each other like bits of greasy string, and the modulus falls steeply (Fig. 23.1). It is helpful to think of each polymer chain as contained within a tube made up by the surrounding nest of molecules (Fig. 23.4). When the polymer is loaded, bits of the molecules slide slightly in the tubes in a snake-like way (called "reptation") giving extra strain and dissipating energy. As the temperature rises past T_G, the polymer expands and

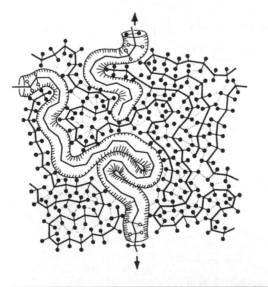

Figure 23.4   Each molecule in a linear polymer can be thought of as being contained in a tube made up by its surroundings. When the polymer is loaded at or above T_G, each molecule can move (reptate) in its tube, giving strain.

the extra free volume (Chapter 22) lowers the packing density, allowing more regions to slide, and giving a lower apparent modulus. But there are still non-sliding (i.e. elastic) parts. On unloading, these elastic regions pull the polymer back to its original shape, though they must do so against the reverse viscous sliding of the molecules, and that takes time. The result is that the polymer has *leathery* properties, as do low-density polyethylene and plasticised PVC at room temperature.

Within this regime it is found that the modulus E at one temperature can be related to that at another by a change in the time scale only, that is, there is an *equivalence between time and temperature*. This means that the curve describing the modulus at one temperature can be superimposed on that for another by a constant horizontal displacement $\log(a_T)$ along the $\log(t)$ axis, as shown in Fig. 23.5.

A well-known example of this time–temperature equivalence is the steady-state creep of a crystalline metal or ceramic, where it follows immediately from the kinetics of thermal activation (Chapter 6). At a constant stress σ the creep rate varies with temperature as

$$\dot{\varepsilon}_{ss} = \frac{\varepsilon}{t} = A\exp(-Q/RT) \tag{23.5}$$

giving

$$\varepsilon(t, T) = tA\exp(-Q/RT). \tag{23.6}$$

From eqn. (23.1) the apparent modulus E is given by

$$E = \frac{\sigma}{\varepsilon(t, T)} = \frac{\sigma}{tA}\exp(Q/RT) - \frac{B}{t}\exp(Q/RT). \tag{23.7}$$

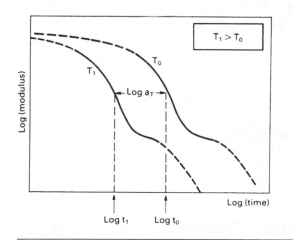

Figure 23.5   Schematic of the time–temperature equivalence for the modulus. Every point on the curve for temperature T_1 lies at the same distance, $\log(a_T)$, to the left of that for temperature T_0.

If we want to match the modulus at temperature T_1 to that at temperature T_0 (see Fig. 23.5) then we need

$$\frac{\exp(Q/RT_1)}{t_1} = \frac{\exp(Q/RT_0)}{t_0} \tag{23.8}$$

or

$$\frac{t_1}{t_0} = \frac{\exp(Q/RT_1)}{\exp(Q/RT_0)} = \exp\frac{Q}{R}\left\{\frac{1}{T_1} - \frac{1}{T_0}\right\}. \tag{23.9}$$

Thus

$$\ln\left(\frac{t_0}{t_1}\right) = -\frac{Q}{R}\left\{\frac{1}{T_1} - \frac{1}{T_0}\right\}, \tag{23.10}$$

and

$$\log(a_T) = \log(t_0/t_1) = \log\,t_0 - \log\,t_1$$
$$= \frac{-Q}{2.3R}\left\{\frac{1}{T_1} - \frac{1}{T_0}\right\}. \tag{23.11}$$

This result says that a simple shift along the time axis by $\log(a_T)$ will bring the response at T_1 into coincidence with that at T_0 (see Fig. 23.5).

Polymers are a little more complicated. The drop in modulus (like the increase in creep rate) is caused by the increased ease with which molecules can slip past each other. In metals, which have a crystal structure, this reflects the increasing number of vacancies and the increased rate at which atoms jump into them. In polymers, which are amorphous, it reflects the increase in free volume which gives an increase in the rate of reptation. Then the shift factor is given, not by eqn. (23.11) but by

$$\log(a_T) = \frac{C_1(T_1 - T_0)}{C_2 + T_1 - T_0} \tag{23.12}$$

where C_1 and C_2 are constants. This is called the "WLF equation" after its discoverers, Williams, Landel and Ferry, and (like the Arrhenius law for crystals) is widely used to predict the effect of temperature on polymer behaviour. If T_0 is taken to be the glass temperature, then C_1 and C_2 are roughly constant for all amorphous polymers (and inorganic glasses too); their values are $C_1 = 17.5$ and $C_2 = 52\,\mathrm{K}$.

Rubbery behaviour and elastomers

As the temperature is raised above T_G, one might expect that flow in the polymer should become easier and easier, until it becomes a rather sticky liquid.

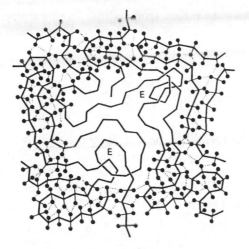

Figure 23.6   A schematic of a linear-amorphous polymer, showing entanglement points (marked "E") which act like chemical cross-links.

Linear polymers with fairly short chains ($\overline{DP} < 10^3$) do just this. But polymers with longer chains ($\overline{DP} > 10^4$) pass through a *rubbery state*.

The origin of rubber elasticity is more difficult to picture than that of a crystal or glass. The long molecules, intertwined like a jar of exceptionally long worms, form *entanglements* – points where molecules, because of their length and flexibility, become knotted together (Fig. 23.6). On loading, the molecules reptate (slide) except at entanglement points. The entanglements give the material a shape-memory: load it, and the segments between entanglements straighten out; remove the load and the wriggling of the molecules (being above T_G) draws them back to their original configuration, and thus shape. Stress tends to *order* the molecules of the material; removal of stress allows it to *disorder* again. The rubbery modulus is small, about one-thousandth of the glassy modulus, T_G, but it is there nonetheless, and gives the plateau in the modulus shown in Fig. 23.1.

Much more pronounced rubbery behaviour is obtained if the chance entanglements are replaced by deliberate cross-links. The number of cross-links must be small – about 1 in every few hundred monomer units. But, being strong, the covalent cross-links do not melt, and this makes the polymer above T_G into a true *elastomer*, capable of elastic extensions of 300% or more (the same as the draw ratio of the polymer in the plastic state – see the next section) which are recovered completely on unloading. Over-frequent cross-links destroy the rubbery behaviour. If every unit on the polymer chain has one (or more) cross-links to other chains, then the covalent bonds form a three-dimensional network, and melting of the secondary bonds does not leave long molecular spans which can straighten out under stress. So good elastomers, like polyisoprene (natural rubber) are linear polymers with just a few cross-links, well above their glass temperatures (room temperature is 1.4 T_G for polyisoprene). If they are cooled below T_G, the modulus rises steeply and the rubber becomes hard and brittle, with properties like those of PMMA at room temperature.

Viscous flow

At yet higher temperatures ($>1.4T_G$) the secondary bonds melt completely and even the entanglement points slip. This is the regime in which thermoplastics are moulded: linear polymers become viscous liquids. The viscosity is always defined (and usually measured) in shear: if a shear stress σ_s produces a rate of shear $\dot{\gamma}$ then the viscosity (Chapter 19) is

$$\eta = \frac{\sigma_s}{10\dot{\gamma}}. \tag{23.13}$$

Its units are poise (P) or 10^{-1} Pa s.

Polymers, like inorganic glasses, are formed at a viscosity in the range 10^4 to 10^6 poise, when they can be blown or moulded. (When a metal melts, its viscosity drops discontinuously to a value near 10^{-3} poise – about the same as that of water; that is why metals are formed by casting, not by the more convenient methods of blowing or moulding.) The viscosity depends on temperature, of course; and at very high temperatures the dependence is well described by an Arrhenius law, like inorganic glasses (Chapter 19). But in the temperature range 1.3–1.5 T_G, where most thermoplastics are formed, the flow has the same time–temperature equivalence as that of the viscoelastic regime (eqn. 23.12) and is called "rubbery flow" to distinguish it from the higher-temperature Arrhenius flow. Then, if the viscosity at one temperature T_0 is η_0, the viscosity at a higher temperature T_1 is

$$\eta_1 = \eta_0 \exp\left\{-\frac{C_1(T_1 - T_0)}{C_2 + T_1 - T_0}\right\}. \tag{23.14}$$

When you have to estimate how a change of temperature changes the viscosity of a polymer (in calculating forces for injection moulding, for instance), this is the equation to use.

Cross-linked polymers do not melt. But if they are made hot enough, they, like linear polymers, decompose.

Decomposition

If a polymer gets too hot, the thermal energy exceeds the cohesive energy of some part of the molecular chain, causing depolymerisation or degradation. Some (like PMMA) decompose into monomer units; others (PE, for instance) randomly degrade into many products. It is commercially important that no decomposition takes place during high-temperature moulding, so a maximum safe working temperature is specified for each polymer; typically, it is about 1.5 T_G.

Modulus diagrams for polymers

The above information is conveniently summarised in the *modulus diagram* for a polymer. Figure 23.7 shows an example: it is a modulus diagram for PMMA, and is typical of linear-amorphous polymers (PS, for example, has a very similar diagram). The modulus E is plotted, on a log scale, on the vertical axis: it runs from 0.01 MPa to 10,000 MPa. The temperature, normalised by the glass temperature T_G, is plotted linearly on the horizontal axis: it runs from 0 (absolute zero) to 1.6 T_G (below which the polymer decomposes).

The diagram is divided into five *fields*, corresponding to the five regimes described earlier. In the glassy field the modulus is large – typically 3 GPa – but it drops slightly as the secondary transitions cause local relaxations. In the

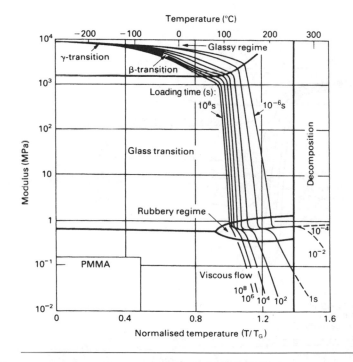

Figure 23.7   A modulus diagram for PMMA. It shows the glassy regime, the glass–rubber transition, the rubbery regime and the regime of viscous flow. The diagram is typical of linear-amorphous polymers.

glassy or viscoelastic–transition regime, the modulus drops steeply, flattening out again in the rubbery regime. Finally, true melting or decomposition causes a further drop in modulus.

Time, as well as temperature, affects the modulus. This is shown by the *contours of loading time*, ranging from very short (10^{-6} s) to very long (10^8 s). The diagram shows how, even in the glassy regime, the modulus at long loading times can be a factor of 2 or more less than that for short times; and in the glass transition region the factor increases to 100 or more. The diagrams give a compact summary of the small-strain behaviour of polymers, and are helpful in seeing how a given polymer will behave in a given application.

Cross-linking raises and extends the rubbery plateau, increasing the rubber-modulus, and suppressing melting. Figure 23.8 shows how, for a single loading time, the contours of the modulus diagram are pushed up as the cross-link density is increased. Crystallisation increases the modulus too (the crystal is stiffer than the amorphous polymer because the molecules are more densely packed) but it does not suppress melting, so crystalline linear-polymers (like high-density PE) can be formed by heating and moulding them, just like linear-amorphous polymers; cross-linked polymers cannot.

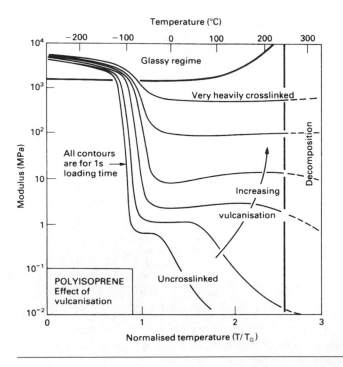

Figure 23.8   The influence of cross-linking on a contour of the modulus diagram for polyisoprene.

Strength: cold drawing and crazing

Engineering design with polymers starts with stiffness. But strength is also important, sometimes overridingly so. A plastic chair need not be very stiff – it may be more comfortable if it is a bit flexible – but it must not collapse plastically, or fail in a brittle manner, when sat upon. There are numerous examples of the use of polymers (luggage, casings of appliances, interior components for automobiles) where strength, not stiffness, is the major consideration.

The "strength" of a solid is the stress at which something starts to happen which gives a permanent shape change: plastic flow, or the propagation of a brittle crack, for example. At least five strength-limiting processes are known in polymers. Roughly in order of increasing temperature, they are:

(a) brittle fracture, like that in ordinary glass;
(b) cold drawing, the drawing-out of the molecules in the solid state, giving a large shape change;
(c) shear banding, giving slip bands rather like those in a metal crystal;
(d) crazing, a kind of microcracking, associated with local cold-drawing;
(e) viscous flow, when the secondary bonds in the polymer have melted.

We now examine each of these in a little more detail.

Brittle fracture

Below about 0.75 T_G, polymers are brittle (Fig. 23.9). Unless special care is taken to avoid it, a polymer sample has small surface cracks (depth c) left by

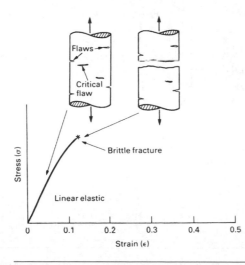

Figure 23.9   Brittle fracture: the largest crack propagates when the fast-fracture criterion is satisfied.

machining or abrasion, or caused by environmental attack. Then a tensile stress σ will cause brittle failure if

$$\sigma = \frac{K_c}{\sqrt{\pi c}} \qquad (23.15)$$

where K_c is the fracture toughness of the polymer. The fracture toughness of most polymers (Table 21.5) is, very roughly, 1 MPa m$^{1/2}$, and the incipient crack size is, typically, a few micrometres. Then the fracture strength in the brittle regime is about 100 MPa. But if deeper cracks or stress concentrations are cut into the polymer, the stress needed to make them propagate is, of course, lower. When designing with polymers you must remember that below 0.75 T_G they are low-toughness materials, and that anything that concentrates stress (like cracks, notches, or sharp changes of section) is dangerous.

Cold drawing

At temperatures 50°C or so below T_G, thermoplastics become plastic (hence the name). The stress–strain curve typical of polyethylene or nylon, for example, is shown in Fig. 23.10. It shows three regions.

At low strains the polymer is *linear elastic*, which is the modulus we have just discussed. At a strain of about 0.1 the polymer *yields* and then *draws*. The chains unfold (if chain-folded) or draw out of the amorphous tangle (if glassy),

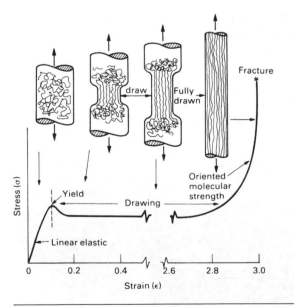

Figure 23.10   Cold-drawing of a linear polymer: the molecules are drawn out and aligned giving, after a draw ratio of about 4, a material which is much stronger in the draw direction than it was before.

and straighten and align. The process starts at a point of weakness or of stress concentration, and a segment of the gauge length draws down, like a neck in a metal specimen, until the *draw ratio* (l/l_0) is sufficient to cause alignment of the molecules (like pulling cotton wool). The draw ratio for alignment is between 2 and 4 (nominal strains of 100 to 300%). The neck propagates along the sample until it is all drawn (Fig. 23.10).

The drawn material is stronger in the draw direction than before; that is why the neck spreads instead of simply causing failure. When drawing is complete, the stress–strain curve rises steeply to final fracture. This draw-strengthening is widely used to produce high-strength fibres and film (Chapter 24). An example is nylon made by melt spinning: the molten polymer is squeezed through a fine nozzle and then pulled (draw ratio ≈ 4), aligning the molecules along the fibre axis; if it is then cooled to room temperature, the reorientated molecules are frozen into position. The drawn fibre has a modulus and strength some 8 times larger than that of the bulk, unoriented, polymer.

Crazing

Many polymers, among them PE, PP and nylon, draw at room temperature. Others with a higher T_G, such as PS, do not – although they draw well at higher temperatures. If PS is loaded in tension at room temperature it *crazes*. Small crack-shaped regions within the polymer draw down, but being constrained by the surrounding undeformed solid, the drawn material ends up as ligaments which link the craze surfaces (Fig. 23.11). The crazes are easily

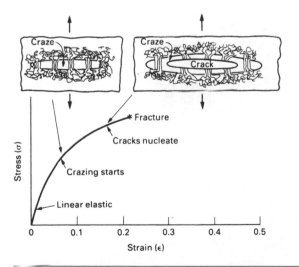

Figure 23.11   Crazing in a linear polymer: molecules are drawn out as in Fig. 23.10, but on a much smaller scale, giving strong strands which bridge the microcracks.

visible as white streaks or as general whitening when cheap injection-moulded articles are bent (plastic pen tops, appliance casings, plastic caps). The crazes are a precursor to fracture. Before drawing becomes general, a crack forms at the centre of a craze and propagates – often with a crazed zone at its tip – to give final fracture (Fig. 23.11).

Shear banding

When crazing limits the ductility in tension, large plastic strains may still be possible in compression *shear banding* (Fig. 23.12). Within each band a finite shear has taken place. As the number of bands increases, the total overall strain accumulates.

Viscous flow

Well above T_G polymers flow in the viscous manner we have described already. When this happens the strength falls steeply.

Strength diagrams for polymers

Most of this information can be summarised as a *strength diagram* for a polymer. Figure 23.13 is an example, again for PMMA. Strength is less well understood than stiffness but the diagram is broadly typical of other linear

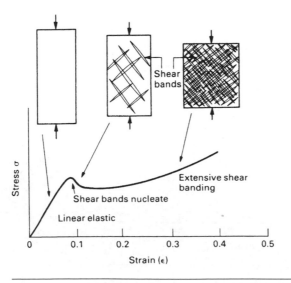

Figure 23.12   Shear banding, an alternative form of polymer plasticity which appears in compression.

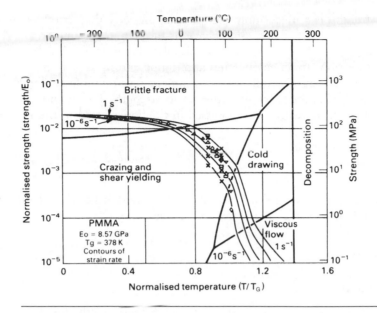

Figure 23.13   A strength diagram for PMMA. The diagram is broadly typical of linear polymers.

polymers. The diagram is helpful in giving a broad, approximate, picture of polymer strength. The vertical axis is the strength of the polymer: the stress at which inelastic behaviour becomes pronounced. The right-hand scale gives the strength in MPa; the left-hand scale gives the strength normalised by Young's modulus at 0 K. The horizontal scale is the temperature, unnormalised across the top and normalised by T_G along the bottom. (The normalisations make the diagrams more general: similar polymers should have similar normalised diagrams.)

The diagram is divided, like the modulus diagram, into *fields* corresponding to the five strength-limiting processes described earlier. At low temperatures there is a brittle field; here the strength is calculated by linear-elastic fracture mechanics. Below this lies the crazing field: the stresses are too low to make a single crack propagate unstably, but they can still cause the slow growth of microcracks, limited and stabilised by the strands of drawn material which span them. At higher temperatures true plasticity begins: cold drawing and, in compression, shear banding. And at high temperature lies the field of viscous flow.

The strength of a polymer depends on the strain rate as well as the temperature. The diagram shows *contours of constant strain rate*, ranging from very slow (10^{-6} s^{-1}) to very fast (1 s^{-1}). The diagram shows how the strength varies with temperature and strain rate, and helps identify the dominant strength-limiting mechanism. This is important because the ductility depends on mechanism: in the cold-drawing regime it is large, but in the brittle fracture regime it is zero.

Strength is a much more complicated property than stiffness. Strength diagrams summarise nicely the behaviour of laboratory samples tested in simple tension. But they (or equivalent compilations of data) must be used with circumspection. In an engineering application the stress-state may be multiaxial, not simple tension; and the environment (even simple sunlight) may attack and embrittle the polymer, reducing its strength. These, and other, aspects of design with polymers, are discussed in the books listed in the References.

Examples

23.1 Estimate the loading time needed to give a modulus of 0.2 GPa in low-density polyethylene at the glass transition temperature.

Answer

270 days.

23.2 Explain how the modulus of a polymer depends on the following factors:

(a) temperature;
(b) loading time;
(c) fraction of covalent cross-links;
(d) molecular orientation;
(e) crystallinity;
(f) degree of polymerisation.

23.3 Explain how the tensile strength of a polymer depends on the following factors:

(a) temperature;
(b) strain rate;
(c) molecular orientation;
(d) degree of polymerisation.

23.4 Explain how the toughness of a polymer is affected by:

(a) temperature;
(b) strain rate;
(c) molecular orientation.

23.5 The Young's modulus of natural rubber decreases from 3 GPa at $-200°C$ to 3 MPa at room temperature, whereas the Young's modulus of epoxy resin is approximately 10 GPa at both temperatures. Explain this difference in behaviour.

Chapter 24
Production, forming and joining of polymers

Introduction

People have used polymers for far longer than metals. From the earliest times, wood, leather, wool and cotton have been used for shelter and clothing. Many natural polymers are cheap and plentiful (not all, though; think of silk) and remarkably strong. But they evolved for specific natural purposes – to support a tree, to protect an animal – and are not always in the form best suited to meet the needs of engineering.

So people have tried to improve on nature. First, they tried to extract natural polymers, and reshape them to their purpose. Cellulose (Table 21.4), extracted from wood shavings and treated with acids, allows the replacement of the $-OH$ side group by $-COOCH_3$ to give *cellulose acetate*, familiar as rayon (used to reinforce car tyres) and as transparent acetate film. Replacement by $-NO_3$ instead gives *cellulose nitrate*, the celluloid of the film industry and a component of many lacquers. Natural latex from the rubber tree is vulcanised to give rubbers, and filled (with carbon black, for instance) to make it resistant to sunlight. But the range of polymers obtained in this way is limited.

The real breakthrough came when chemists developed processes for making large molecules from their smallest units. Instead of the ten or so natural polymers and modifications of them, the engineer was suddenly presented with hundreds of new materials with remarkable and diverse properties. The number is still increasing.

And we are still learning how best to fabricate and use them. As emphasised in the last chapter, the mechanical properties of polymers differ in certain fundamental ways from those of metals and ceramics, and the methods used to design with them (Chapter 27) differ accordingly. Their special properties also need special methods of fabrication. This chapter outlines how polymers are fabricated and joined. To understand this, we must first look, in slightly more detail, at their synthesis.

Synthesis of polymers

Plastics are made by a chemical reaction in which monomers add (with nothing left over) or condense (with H_2O left over) to give a high polymer.

Polyethylene, a linear polymer, is made by an *addition reaction*. It is started with an initiator, such as H_2O_2, which gives free, and very reactive $-OH$ radicals. One of these breaks the double-bond of an ethylene molecule, C_2H_4, when it is heated under pressure, to give

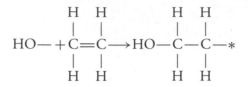

The left-hand end of the activated monomer is sealed off by the OH terminator, but the right-hand end (with the star) is aggressively reactive and now attacks another ethylene molecule, as we illustrated earlier in Fig. 22.1. The process continues, forming a longer and longer molecule by a sort of chain reaction. The $-OH$ used to start a chain will, of course, terminate one just as effectively, so excess initiator leads to short chains. As the monomer is exhausted the reaction slows down and finally stops. The DP depends not only on the amount of initiator, but on the pressure and temperature as well.

Nylon, also a linear polymer, is made by a *condensation reaction*. Two different kinds of molecule react to give a larger molecule, and a by-product (usually H_2O); the ends of large molecules are active, and react further, building a polymer chain. Note how molecules of one type condense with those of the other in this reaction of two symmetrical molecules.

$$HO-\overset{\overset{\textstyle O}{\|}}{C}-(CH_2)_8-\overset{\overset{\textstyle O}{\|}}{C}-\overline{OH+H}-\overset{\overset{\textstyle N}{|}}{N}-(CH_2)_6-\overset{\overset{\textstyle H}{|}}{N}-H.$$

The resulting chains are regular and symmetrical, and tend to crystallise easily. Condensation reactions do not rely on an initiator, so the long molecules form by the linking of shorter (but still long) segments, which in turn grow from smaller units. In this they differ from addition reactions, in which single monomer units add one by one to the end of the growing chain.

Most network polymers (the epoxies and the polyesters, for instance) are made by condensation reactions. The only difference is that one of the two reacting molecules is multifunctional (polyester is three-functional) so the reaction gives a three-dimensional lacework, not linear threads, and the resulting polymer is a thermoset.

Polymer alloys

Copolymers

If, when making an addition polymer, two monomers are mixed, the chain which forms contains both units (*copolymerisation*). Usually the units add

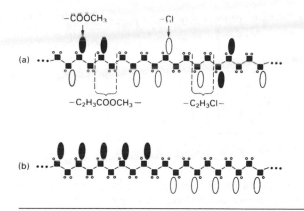

−COOCH₃

−Cl

(a)

−C₂H₃COOCH₃−

−C₂H₃Cl−

(b)

Figure 24.1   **(a)** A copolymer of vinyl chloride and vinyl acetate; the "alloy" packs less regularly, has a lower T_G, and is less brittle than simple polyvinylchloride (PVC). **(b)** A block copolymer: the two different molecules in the alloy are clustered into blocks along the chain.

randomly, and by making the molecule less regular, they favour an amorphous structure with a lower packing density, and a lower T_G. PVC, when pure, is brittle; but copolymerising it with vinyl acetate (which has a $-COOCH_3$ radical in place of the $-Cl$) gives the flexible copolymer shown in Fig. 24.1(a). Less often, the two monomers group together in blocks along the chain; the result is called a *block copolymer* (Fig. 24.1b).

Solid solutions: plasticisers

Plasticisers are organic liquids with relatively low molecular weights (100–1000) which dissolve in large quantities (up to 35%) in solid polymers. The chains are forced apart by the oily liquid, which lubricates them, making it easier for them to slide over each other. So the plasticiser does what its name suggests: it makes the polymer more flexible (and makes its surface feel slightly oily). It achieves this by lowering T_G, but this also reduces the tensile strength – so moderation must be exercised in its use. And the plasticiser must have a low vapour pressure or it will evaporate and leave the polymer brittle.

Two-phase alloys: toughened polymers

When styrene and butadiene are polymerised, the result is a mixture of distinct molecules of polystyrene and of a rubbery copolymer of styrene and butadiene. On cooling, the rubbery copolymer precipitates out, much as $CuAl_2$ precipitated out of aluminium alloys, or Fe_3C out of steels (Chapters 10 and 11). The resulting microstructure is shown in Fig. 24.2: the matrix of glassy polystyrene contains rubbery particles of the styrene–butadiene copolymer. The rubber

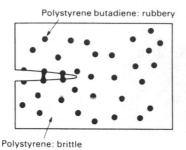

Polystyrene butadiene: rubbery

Polystyrene: brittle

Figure 24.2   A two-phase polymer alloy, made by co-polymerising styrene and butadiene in polystyrene. The precipitates are a polystyrene–butadiene copolymer.

particles stop cracks in the material, increasing its fracture toughness – for this reason the alloy is called *high-impact polystyrene*. Other polymers can be toughened in the same way.

Stabilisation and vulcanisation

Polymers are damaged by radiation, particularly by the ultraviolet in sunlight. An ultraviolet photon has enough energy to break the C—C bond in the polymer backbone, splitting it into shorter chains. Paints, especially, are exposed to this sort of radiation damage. The solution is to add a pigment or filler (like carbon) which absorbs radiation, preventing it from hitting the delicate polymer chains. Car tyres contain as much as 30 wt% of carbon to *stabilise* the polymer against attack by sunlight.

Oxygen, too, can damage polymers by creating —O— cross-links between polymer chains; it is a sort of unwanted vulcanisation. The cross-links raise T_G, and make the polymer brittle; it is a particular problem with rubbers, which lose their elasticity. Ozone (O_3) is especially damaging because it supplies oxygen in an unusually active form. Sunlight (particularly ultraviolet again) promotes oxidation, partly because it creates O_3. Polymers containing a C=C double bond are particularly vulnerable, because oxygen links to it to give —C—O—C cross-links; that is why rubbers are attacked by oxygen more than other polymers (see their structure in Table 21.3). The way to avoid oxygen attack is to avoid polymers containing double-bonds, and to protect the polymer from direct sunlight by stabilising it.

Forming of polymers

Thermoplastics soften when heated, allowing them to be formed by *injection moulding, vacuum forming, blow moulding* and *compression moulding*. Thermosets, on the other hand, are heated, formed and cured simultaneously, usually by *compression moulding*. Rubbers are formed like thermosets, by pressing and heating a mix of elastomer and vulcanising agent in a mould.

Polymers can be used as surface coatings. Linear polymers are applied as a solution; the solvent evaporates leaving a protective film of the polymer. Thermosets are applied as a fluid mixture of resin and hardener which has to be mixed just before it is used, and cures almost as soon as it is applied.

Polymer fibres are produced by forcing molten polymer or polymer in solution through fine nozzles (spinnerettes). The fibres so formed are twisted into a yarn and woven into fabric. Finally, polymers may be expanded into foams by mixing in chemicals that release CO_2 bubbles into the molten polymer or the curing resin, or by expanding a dissolved gas into bubbles by reducing the pressure.

The full technical details of these processes are beyond the scope of this book (see the References for further enlightenment), but it is worth having a slightly closer look at them to get a feel for the engineering context in which each is used.

Extrusion

A polymer extruder is like a giant cake-icer. Extrusion is a cheap continuous process for producing shapes of constant section (called "semis", meaning "semi-finished" products or stock). Granules of polymer are fed into a screw like that of an old-fashioned meat mincer, turning in a heated barrel (Fig. 24.3a). The screw compacts and mixes the polymer, which melts as it approaches the hot end of the barrel, where it is forced through a die and then cooled (so that its new shape is maintained) to give tubes, sheet, ribbon and rod. The shear-flow in the die orients the molecules in the extrusion direction and increases the strength. As the extrusion cools it recovers a bit, and this causes a significant transverse expansion. Complex die shapes lead to complex recovery patterns, so that the final section is not the same as that of the die opening. But die-makers can correct for this, and the process is so fast and cheap that it is very widely used (60% of all thermoplastics undergo some form of extrusion). So attractive is it that it has been adopted by the manufacturers of ceramic components who mix the powdered ceramic with a polymer binder, extrude the mixture, and then burn off the polymer while firing the ceramic.

Injection moulding

In *injection moulding*, polymer granules are compressed by a ram or screw, heated until molten and squirted into a cold, split-mould under pressure (Fig. 24.3b). The moulded polymer is cooled below T_G, the mould opens and the product pops out. Excess polymer is injected to compensate for contraction in the mould. The molecules are oriented parallel to the flow direction during

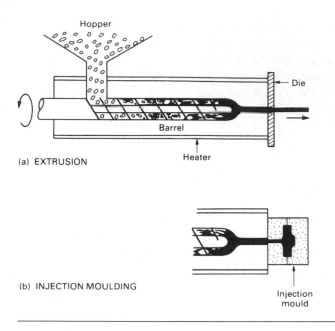

(a) EXTRUSION

(b) INJECTION MOULDING

Figure 24.3   **(a)** Extrusion: polymer granules are heated, mixed and compressed by the screw which forces the now molten polymer out through a die. **(b)** Injection moulding is extrusion into a mould. If the moulding is cooled with the pressure on, good precision and detail are obtained.

injection, giving useful strengthening, but properties that are anisotropic. The process gives high-precision mouldings, because the polymer cools with the pressure still on, but is slow (the cycle time is between 1 and 5 minutes) and the moulds are expensive. Typically, moulding temperatures for thermoplastics are between 150 and 350°C (1.3 and 1.6 T_G) and the pressures needed to give good detail are high – up to 120 MPa.

Vacuum and blow forming of sheet

In *vacuum* and *blow forming*, sheets produced by extrusion are shaped by vacuum or pressure forming. Heat-softened sheet is pressed into a mould by atmospheric pressure when a vacuum is created between the mould and the sheet, Fig. 24.4(a). Plastic bottles are made by blowing instead: heated tube is clamped in a split mould and expanded with compressed air to take up its shape (Fig. 24.4b). Both methods are cheap and quick, and can be fairly accurate.

Compression moulding

Both thermoplastics and thermosets can be formed by *compression mould-ing* (Fig. 24.5). The polymer, or mixture of resin and hardener, is heated

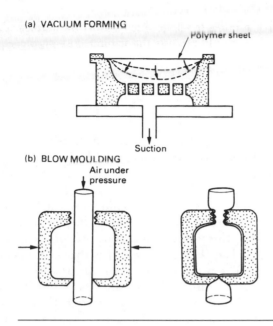

Figure 24.4   **(a)** Vacuum forming is good for making simple shapes out of sheet. **(b)** Blow moulding is used to make plastic containers.

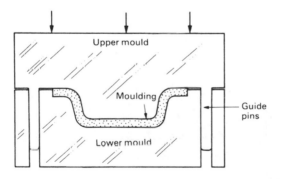

COMPRESSION MOULDING

Figure 24.5   Compression moulding: the thermoset is squeezed to shape and cured (by heating) at the same time. Once it has cured it can be taken from the mould while still hot – so the process is faster than the moulding of thermoplastics.

and compressed between dies. The method is well suited to the forming of thermosets (casings for appliances, for instance) and of composites with a thermosetting matrix (car bumpers, for example). Since a thermoset can be removed while it is still hot, the cycle time is as short as 10 seconds for small components, 10 minutes for large thick-walled mouldings. Pressures are lower than for injection mouldings, so the capital cost of the equipment is much less.

Films and fibres

Thin sheet and fine fibres of polymers are extruded, using a narrow slit, or a die with many small holes in it (a *spinnerette*). The molten polymer cools so fast when it is extruded that it solidifies in the amorphous state. A great increase in strength is possible if the film or fibres are drawn off through a tensioning device which stretches them, unravelling the tangled molecules and aligning them in the plane of the sheet or along the axis of the fibre. Film has to be stretched in two directions at once: one way of doing so is to blow an enormous thin-walled bag (Fig. 24.6) which is then cut, opened out flat, and rolled onto a drum. The properties normal to the surface of the stretched film, or across the axis of the oriented fibre, are worse than before, but they are never loaded in this direction so it does not matter. Here people emulate nature: most natural polymers (wood, wool, cotton, silk) are highly oriented in just this way.

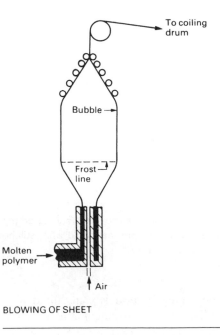

BLOWING OF SHEET

Figure 24.6   Production of plastic sheet by blowing. The bubble is split, flattened and rolled onto a drum.

Joining of polymers

Polymers are joined by cementing, by welding and by various sorts of fasteners, many themselves moulded from polymers. Joining, of course, can sometimes be avoided by integral design, in which coupled components are moulded into a single unit.

A polymer is joined to itself by *cementing* with a solution of the same polymer in a volatile solvent. The solvent softens the surfaces, and the dissolved polymer molecules bond them together. Components can be joined by *monomer-cementing*: the surfaces are coated with monomer which polymerises onto the pre-existing polymer chains, creating a bond.

Polymers can be stuck to other materials with *adhesives*, usually epoxies. They can be attached by a variety of *fasteners*, which must be designed to distribute the fastening load uniformly over a larger area than is usual for metals, to avoid fracture. Ingenious splined or split fasteners can be moulded onto polymer components, allowing the parts to be snapped together; and threads can be moulded onto parts to allow them to be screwed together. Finally, polymers can be *friction-welded* to bring the parts, rotating or oscillating, into contact; frictional heat melts the surfaces which are held under static load until they resolidify.

Examples

24.1 Describe in a few words, with an example or sketch where appropriate, what is meant by each of the following:

(a) an addition reaction;
(b) a condensation reaction;
(c) a copolymer;
(d) a block copolymer;
(e) a plasticiser;
(f) a toughened polymer;
(g) a filler.

24.2 What forming process would you use to manufacture each of the following items;

(a) a continuous rod of PTFE;
(b) thin polyethylene film;
(c) a PMMA protractor;
(d) a ureaformaldehyde electrical switch cover;
(e) a fibre for a nylon rope.

24.3 Low-density polyethylene is being extruded at 200°C under a pressure of 60 MPa. What increase in temperature would be needed to decrease the extrusion pressure to 40 MPa? The shear rate is the same in both cases. [Hint: use eqns (23.13) and (23.14) with $C_1 = 17.5$, $C_2 = 52$ K and $T_0 = T_G = 270$ K.]

Answer

32°C.

24.4   Discuss the problems involved in replacing the metal parts of an ordinary bicycle with components made from polymers. Illustrate your answer by specific reference to the frame, wheels, transmission and bearings.

Chapter 25
Composites: fibrous, particulate and foamed

Introduction

The word "composites" has a modern ring. But using the high strength of fibres to stiffen and strengthen a cheap matrix material is probably older than the wheel. The Processional Way in ancient Babylon, one of the lesser wonders of the ancient world, was made of bitumen reinforced with plaited straw. Straw and horse hair have been used to reinforce mud bricks (improving their fracture toughness) for at least 5000 years. Paper is a composite; so is concrete: both were known to the Romans. And almost all natural materials which must bear load – wood, bone, muscle – are composites.

The composite industry, however, is new. It has grown rapidly in the past 30 years with the development of *fibrous composites*: to begin with, *glass-fibre reinforced polymers* (GFRP or fibreglass) and, more recently, *carbon-fibre reinforced polymers* (CFRP). Their use in boats, and their increasing replacement of metals in aircraft and ground transport systems, is a revolution in material usage which is still accelerating.

Composites need not be made of fibres. Plywood is a *lamellar composite*, giving a material with uniform properties in the plane of the sheet (unlike the wood from which it is made). Sheets of GFRP or of CFRP are laminated together, for the same reason. And sandwich panels – composites made of stiff skins with a low-density core – achieve special properties by combining, in a sheet, the best features of two very different components.

Cheapest of all are the *particulate composites*. Aggregate plus cement gives concrete, and the composite is cheaper (per unit volume) than the cement itself. Polymers can be filled with sand, silica flour, or glass particles, increasing the stiffness and wear-resistance, and often reducing the price. And one particulate composite, tungsten-carbide particles in cobalt (known as "cemented carbide" or "hard metal"), is the basis of the heavy-duty cutting tool industry.

But high stiffness is not always what you want. Cushions, packaging and crash-padding require materials with moduli that are lower than those of any solid. This can be done with *foams* – composites of a solid and a gas – which have properties which can be tailored, with great precision, to match the engineering need.

We now examine the properties of fibrous and particulate composites and foams in a little more detail. With these materials, more than any other, properties can be designed-in; the characteristics of the material itself can be engineered.

Fibrous composites

Polymers have a low stiffness, and (in the right range of temperature) are ductile. Ceramics and glasses are stiff and strong, but are catastrophically brittle. In *fibrous composites* we exploit the great strength of the ceramic while avoiding the catastrophe: the brittle failure of fibres leads to a progressive, not a sudden, failure.

If the fibres of a composite are aligned along the loading direction, then the stiffness and the strength are, roughly speaking, an average of those of the matrix and fibres, weighted by their volume fractions. But not all composite properties are just a linear combination of those of the components. Their great attraction lies in the fact that, frequently, something extra is gained.

The toughness is an example. If a crack simply ran through a GFRP composite, one might (at first sight) expect the toughness to be a simple weighted average of that of glass and epoxy; and both are low. But that is not what happens. The strong fibres pull out of the epoxy. In pulling out, work is done and this work contributes to the toughness of the composite. The toughness is greater – often much greater – than the linear combination.

Polymer-matrix composites for aerospace and transport are made by laying up glass, carbon or Kevlar fibres (Table 25.1) in an uncured mixture of resin and hardener. The resin cures, taking up the shape of the mould and bonding to the fibres. Many composites are based on epoxies, though there is now a trend to using the cheaper polyesters.

Laying-up is a slow, labour-intensive job. It can be by-passed by using thermoplastics containing chopped fibres which can be injection moulded. The random chopped fibres are not quite as effective as laid-up continuous fibres, which can be oriented to maximise their contribution to the strength.

Table 25.1   Properties of some fibres and matrices

Material	Density ρ (Mg m^{-3})	Modulus E (GPa)	Strength σ_f (MPa)
Fibres			
Carbon, Type1	1.95	390	2200
Carbon, Type2	1.75	250	2700
Cellulose fibres	1.61	60	1200
Glass (E-glass)	2.56	76	1400–2500
Kevlar	1.45	125	2760
Matrices			
Epoxies	1.2–1.4	2.1–5.5	40–85
Polyesters	1.1–1.4	1.3–4.5	45–85

But the flow pattern in injection moulding helps to line the fibres up, so that clever mould design can give a stiff, strong product. The technique is used increasingly for sports goods (tennis racquets, for instance) and light-weight hiking gear (like back-pack frames).

Making good fibre-composites is not easy; large companies have been bankrupted by their failure to do so. The technology is better understood than it used to be; the tricks can be found in the books listed in the References. But suppose you can make them, you still have to know how to use them. That needs an understanding of their properties, which we examine next. The important properties of three common composites are listed in Table 25.2, where they are compared with a high-strength steel and a high-strength aluminium alloy of the sort used for aircraft structures.

Modulus

When two linear-elastic materials (though with different moduli) are mixed, the mixture is also linear-elastic. The modulus of a fibrous composite when loaded *along* the fibre direction (Fig. 25.1a) is a linear combination of that of the fibres, E_f, and the matrix, E_m

$$E_{c\parallel} = V_f E_f + (1 - V_f) E_m \qquad (25.1)$$

where V_f is the volume fraction of fibres (see Book 1, Chapter 6). The modulus of the same material, loaded across the fibres (Fig. 25.1b) is much less – it is only

$$E_{c\perp} = \left\{ \frac{V_f}{E_f} + \frac{1 - V_f}{E_m} \right\}^{-1} \qquad (25.2)$$

(see Book 1, Chapter 6 again).

Table 25.1 gives E_f and E_m for common composites. The moduli $E_\parallel$ and $E_\perp$ for a composite with, say, 50% of fibres, differ greatly: a uniaxial composite (one in which all the fibres are aligned in one direction) is exceedingly anisotropic. By using a cross-weave of fibres (Fig. 25.1c) the moduli in the 0 and 90° directions can be made equal, but those at 45° are still very low. Approximate isotropy can be restored by laminating sheets, rotated through 45°, to give a plywood-like *fibre laminate*.

Tensile strength and the critical fibre length

Many fibrous composites are made of strong, brittle fibres in a more ductile polymeric matrix. Then the stress–strain curve looks like the heavy line in Fig. 25.2. The figure largely explains itself. The stress–strain curve is linear,

Table 25.2 Properties, and specific properties, of composites

Material	Density ρ (Mg m^{-3})	Young's modulus E (GPa)	Strength σ_y (MPa)	Fracture toughness K_c (MPa m$^{1/2}$)	E/ρ	$E^{1/2}/\rho$	$E^{1/3}/\rho$	σ_y/ρ
Composites								
CFRP, 58% uniaxial C in epoxy	1.5	189	1050	32–45	126	9	3.8	700
GFRP, 50% uniaxial glass in polyester	2.0	48	1240	42–60	24	3.5	1.8	620
Kevlar-epoxy (KFRP), 60% uniaxial Kevlar in epoxy	1.4	76	1240	–	54	6.2	3.0	886
Metals								
High-strength steel	7.8	207	1000	100	27	1.8	0.76	128
Aluminium alloy	2.8	71	500	28	25	3.0	1.5	179

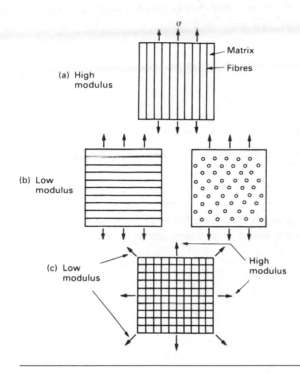

Figure 25.1   (a) When loaded along the fibre direction the fibres and matrix of a continuous-fibre composite suffer equal strains. (b) When loaded across the fibre direction, the fibres and matrix see roughly equal stress; particulate composites are the same. (c) A 0−90° laminate has high and low modulus directions; a 0−45−90−135° laminate is nearly isotropic.

with slope E (eqn. 25.1) until the matrix yields. From there on, most of the extra load is carried by the fibres which continue to stretch elastically until they fracture. When they do, the stress drops to the yield strength of the matrix (though not as sharply as the figure shows because the fibres do not all break at once). When the matrix fractures, the composite fails completely.

In any structural application it is the peak stress which matters. At the peak, the fibres are just on the point of breaking and the matrix has yielded, so the stress is given by the *yield* strength of the matrix, σ_y^m, and the *fracture* strength of the fibres, σ_f^f, combined using a rule of mixtures

$$\sigma_{TS} = V_f \sigma_f^f + (1 - V_f)\sigma_y^m. \tag{25.3}$$

This is shown as the line rising to the right in Fig. 25.3. Once the fibres have fractured, the strength rises to a second maximum determined by the fracture strength of the matrix

$$\sigma_{TS} = (1 - V_f)\sigma_f^m \tag{25.4}$$

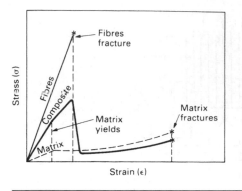

Figure 25.2   The stress–strain curve of a continuous fibre composite (heavy line), showing how it relates to those of the fibres and the matrix (thin lines). At the peak the fibres are on the point of failing.

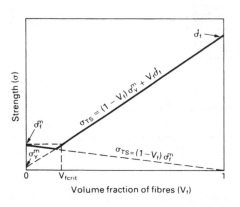

Figure 25.3   The variation of peak stress with volume fraction of fibres. A minimum volume fraction $(V_{f_{crit}})$ is needed to give any strengthening.

where σ_f^m is the fracture strength of the matrix; it is shown as the line falling to the right on Fig. 25.3. The figure shows that adding too few fibres does more harm than good: a critical volume fraction $V_{f_{crit}}$ of fibres must be exceeded to give an increase in strength. If there are too few, they fracture before the peak is reached and the ultimate strength of the material is reduced.

For many applications (e.g. body pressings), it is inconvenient to use continuous fibres. It is a remarkable feature of these materials that chopped fibre composites (convenient for moulding operations) are nearly as strong as those with continuous fibres, provided the fibre length exceeds a critical value.

Consider the peak stress that can be carried by a chopped-fibre composite which has a matrix with a yield strength in shear of $\sigma_s^m (\sigma_s^m \approx \frac{1}{2}\sigma_y^m)$. Figure 25.4

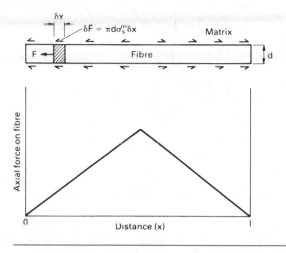

Figure 25.4   Load transfer from the matrix to the fibre causes the tensile stress in the fibre to rise to peak in the middle. If the peak exceeds the fracture strength of the fibre, it breaks.

shows that the axial force transmitted to a fibre of diameter d over a little segment δx of its length is

$$\delta F = \pi d \sigma_s^m \delta x. \tag{25.5}$$

The force on the fibre thus increases from zero at its end to the value

$$F = \int_0^x \pi d \sigma_s^m dx = \pi d \sigma_s^m x \tag{25.6}$$

at a distance x from the end. The force which will just break the fibre is

$$F_c = \frac{\pi d^2}{4} \sigma_f^f. \tag{25.7}$$

Equating these two forces, we find that the fibre will break at a distance

$$x_c = \frac{d}{4} \frac{\sigma_f^f}{\sigma_s^m} \tag{25.8}$$

from its end. If the fibre length is less than $2x_c$, the fibres do not break – but nor do they carry as much load as they could. If they are much longer than $2x_c$, then nothing is gained by the extra length. The optimum strength (and the most effective use of the fibres) is obtained by chopping them to the length $2x_c$ in the first place. The average stress carried by a fibre is then simply $\sigma_f^f/2$ and the peak strength (by the argument developed earlier) is

$$\sigma_{TS} = \frac{V_f \sigma_f^f}{2} + (1 - V_f)\sigma_y^m. \tag{25.9}$$

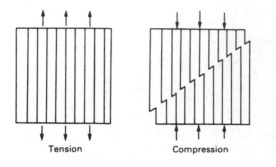

Figure 25.5   Composites fail in compression by kinking, at a load which is lower than that for failure in tension.

This is more than one-half of the strength of the continuous-fibre material (eqn. 25.3). Or it is if all the fibres are aligned along the loading direction. That, of course, will not be true in a chopped-fibre composite. In a car body, for instance, the fibres are randomly oriented in the plane of the panel. Then only a fraction of them – about $\frac{1}{4}$ – are aligned so that much tensile force is transferred to them, and the contributions of the fibres to the stiffness and strength are correspondingly reduced.

The compressive strength of composites is less than that in tension. This is because the fibres buckle or, more precisely, they *kink* – a sort of co-operative buckling, shown in Fig. 25.5. So while brittle ceramics are best in compression, composites are best in tension.

Toughness

The *toughness* G_c of a composite (like that of any other material) is a measure of the energy absorbed per unit crack area. If the crack simply propagated straight through the matrix (toughness G_c^m) and fibres (toughness G_c^f), we might expect a simple rule-of-mixtures

$$G_c = V_f G_c^f + (1 - V_f) G_c^m. \tag{25.10}$$

But it does not usually do this. We have already seen that, if the length of the fibres is less than $2x_c$, they will not fracture. And if they do not fracture they must instead pull out as the crack opens (Fig. 25.6). This gives a major new contribution to the toughness. If the matrix shear strength is σ_s^m (as before), then the work done in pulling a fibre out of the fracture surface is given approximately by

$$\int\limits_{0}^{1/2} F\mathrm{d}x = \int\limits_{0}^{1/2} \pi d\sigma_s^m x\mathrm{d}x = \pi d\sigma_s^m \frac{l^2}{8} \tag{25.11}$$

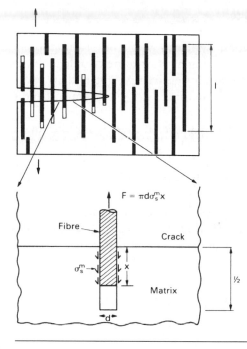

$F = \pi d \sigma_s^m x$

Fibre

Crack

σ_s^m

x

Matrix

½

d

Figure 25.6   Fibres toughen by pulling out of the fracture surface, absorbing energy as the crack opens.

The number of fibres per unit crack area is $4V_f/\pi d^2$ (because the volume fraction is the same as the *area* fraction on a plane perpendicular to the fibres). So the total work done per unit crack area is

$$G_c = \pi d \sigma_s^m \frac{l^2}{8} \times \frac{4V_f}{\pi d^2} = \frac{V_f}{2d} \sigma_s^m l^2. \qquad (25.12)$$

This assumes that l is less than the critical length $2x_c$. If l is greater than $2x_c$ the fibres will *not* pull out, but will break instead. Thus optimum toughness is given by setting $l = 2x_c$ in eqn. (25.12) to give

$$G_c = \frac{2V_f}{d} \sigma_s^m x_c^2 = \frac{2V_f}{d} \sigma_s^m \left(\frac{d}{4} \frac{\sigma_f^f}{\sigma_s^m} \right)^2 = \frac{V_f d}{8} \frac{(\sigma_f^f)^2}{\sigma_s^m}. \qquad (25.13)$$

The equation says that, to get a high toughness, you should use strong fibres in a weak matrix (though of course a weak matrix gives a low strength). This mechanism gives CFRP and GFRP a toughness ($50\,kJ\ m^{-2}$) far higher than that of either the matrix ($5\,kJ\ m^{-2}$) or the fibres ($0.1\,kJ\ m^{-2}$); without it neither would be useful as an engineering material.

Applications of composites

In designing transportation systems, weight is as important as strength. Figure 25.7 shows that, depending on the geometry of loading, the component which gives the least *deflection* for a given weight is that made of a material with a maximum E/ρ (ties in tension), $E^{1/2}/\rho$ (beam in bending) or $E^{1/3}/\rho$ (plate in bending).

When E/ρ is the important parameter, there is nothing to choose between steel, aluminium or fibre glass (Table 25.2). But when $E^{1/2}/\rho$ is controlling, aluminium is better than steel: that is why it is the principal airframe material. Fibreglass is not significantly better. Only CFRP and KFRP offer a real advantage, and one that is now exploited extensively in aircraft structures. This advantage persists when $E^{1/3}/\rho$ is the determining quantity – and for this reason both CFRP and KFRP find particular application in floor panels and large load-bearing surfaces like flaps and tail planes.

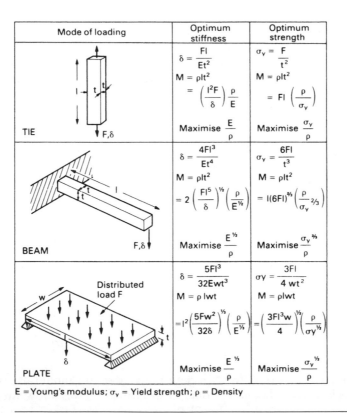

E = Young's modulus; σ_y = Yield strength; ρ = Density

Figure 25.7 The combination of properties which maximise the stiffness-to-weight ratio and the strength-to-weight ratio, for various loading geometries.

In some applications it is strength, not stiffness, that matters. Figure 25.7 shows that the component with the greatest *strength* for a given weight is that made of the material with a maximum σ_y/ρ (ties in tension), $\sigma_y^{2/3}/\rho$ (beams in bending) or $\sigma_y^{1/2}/\rho$ (plates in bending). Even when σ_y/ρ is the important parameter, composites are better than metals (Table 25.2), and the advantage grows when $\sigma_y^{2/3}/\rho$ or $\sigma_y^{1/2}/\rho$ are dominant.

Despite the high cost of composites, the weight-saving they permit is so great that their use in trains, trucks and even cars is now extensive. But, as this chapter illustrates, the engineer needs to understand the material and the way it will be loaded in order to use composites effectively.

Particulate composites

Particulate composites are made by blending silica flour, glass beads, even sand into a polymer during processing.

Particulate composites are much less efficient in the way the filler contributes to the strength. There is a small gain in stiffness, and sometimes in strength and toughness, but it is far less than in a fibrous composite. Their attraction lies more in their low cost and in the good wear resistance that a hard filler can give. Road surfaces are a good example: they are either *macadam* (a particulate composite of gravel in bitumen, a polymer) or *concrete* (a composite of gravel in cement, for which see Chapter 20).

Cellular solids, or foams

Many natural materials are cellular: wood and bone, for example; cork and coral, for instance. There are good reasons for this: cellular materials permit an optimisation of stiffness, or strength, or of energy absorption, for a given weight of material. These natural foams are widely used by people (wood for structures, cork for thermal insulation), and synthetic foams are common too: cushions, padding, packaging, insulation, are all functions filled by cellular polymers. Foams give a way of making solids which are very light and, if combined with stiff skins to make sandwich panels, they give structures which are exceptionally stiff and light. The engineering potential of foams is considerable, and, at present, incompletely realised.

Most polymers can be foamed easily. It can be done by simple mechanical stirring or by blowing a gas under pressure into the molten polymer. But by far the most useful method is to mix a chemical blowing agent with the granules of polymer before processing: it releases CO_2 during the heating cycle, generating gas bubbles in the final moulding. Similar agents can be blended into thermosets so that gas is released during curing, expanding the polymer into a foam; if it is contained in a closed mould it takes up the mould shape accurately and with a smooth, dense, surface.

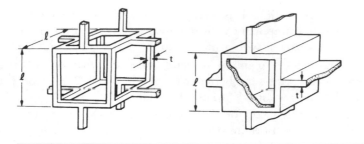

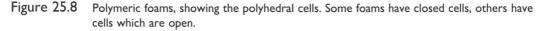

Figure 25.8   Polymeric foams, showing the polyhedral cells. Some foams have closed cells, others have cells which are open.

The properties of a foam are determined by the properties of the polymer, and by the *relative density*, ρ/ρ_s: the density of the foam (ρ) divided by that of the solid (ρ_s) of which it is made. This plays the role of the volume fraction V_f of fibres in a composite, and all the equations for foam properties contain ρ/ρ_s. It can vary widely, from 0.5 for a dense foam to 0.005 for a particularly light one.

The cells in foams are polyhedral, like grains in a metal (Fig. 25.8). The cell walls, where the solid is concentrated, can be open (like a sponge) or closed (like a flotation foam), and they can be equiaxed (like the polymer foam in the figure) or elongated (like cells in wood). But the aspect of structures which is most important in determining properties is none of these; it is the relative density. We now examine how foam properties depend on ρ/ρ_s and on the properties of the polymer of which it is made (which we covered in Chapter 23).

Mechanical properties of foams

When a foam is compressed, the stress–strain curve shows three regions (Fig. 25.9). At small strains the foam deforms in a *linear-elastic* way: there is then a *plateau* of deformation at almost constant stress; and finally there is a region of *densification* as the cell walls crush together.

At small strains the cell walls at first *bend*, like little beams of modulus E_s, built in at both ends. Figure 25.10 shows how a hexagonal array of cells is distorted by this bending. The deflection can be calculated from simple beam theory. From this we obtain the stiffness of a unit cell, and thus the modulus E of the foam, in terms of the length l and thickness t of the cell walls. But these are directly related to the relative density: $\rho/\rho_s = (t/l)^2$ for open-cell foams, the commonest kind. Using this gives the foam modulus as

$$E = E_s \left(\frac{\rho}{\rho_s} \right)^2 . \tag{25.14}$$

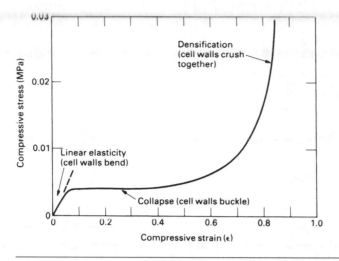

Figure 25.9   The compressive stress–strain curve for a polymeric foam. Very large compressive strains are possible, so the foam absorbs a lot of energy when it is crushed.

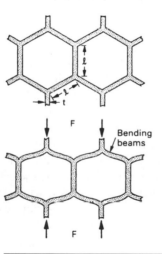

Figure 25.10   Cell wall bending gives the linear-elastic portion of the stress–strain curve.

Real foams are well described by this formula. Note that foaming offers a vast range of modulus: ρ/ρ_s can be varied from 0.5 to 0.005, a factor of 10^2, by processing, allowing E to be varied over a factor of 10^4.

Linear-elasticity, of course, is limited to small strains (5% or less). Elastomeric foams can be compressed far more than this. The deformation is still recoverable (and thus elastic) but is non-linear, giving the plateau on Fig. 25.9. It is caused by the *elastic buckling* of the columns or plates which make up the cell edges or walls, as shown in Fig. 25.11. Again using standard results of beam theory,

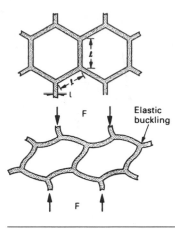

Figure 25.11   When an elastomeric foam is compressed beyond the linear region, the cell walls buckle elastically, giving the long plateau shown in Fig. 25.9.

the *elastic collapse stress* σ_{el}^* can be calculated in terms of the density ρ. The result is

$$\sigma_{el}^* = 0.05E_s \left(\frac{\rho}{\rho_s}\right)^2. \tag{25.15}$$

As before, the strength of the foam is controlled by the density, and can be varied at will through a wide range. Low-density $(\rho/\rho_s = 0.01)$ elastomeric foams collapse under tiny stresses; they are used to package small, delicate instruments. Denser foams $(\rho/\rho_s = 0.05)$ are used for seating and beds: their moduli and collapse strengths are 25 times larger. Still denser foams are used for packing heavier equipment: appliances or small machine tools, for instance.

Cellular materials can collapse by another mechanism. If the cell-wall material is *plastic* (as many polymers are) then the foam as a whole shows plastic behaviour. The stress–strain curve still looks like Fig. 25.9, but now the plateau is caused by plastic collapse. Plastic collapse occurs when the moment exerted on the cell walls exceeds its fully plastic moment, creating plastic hinges as shown in Fig. 25.12. Then the collapse stress σ_{pl}^* of the foam is related to the yield strength σ_y of the wall by

$$\sigma_{pl}^* = 0.3\sigma_y \left(\frac{\rho}{\rho_s}\right)^{3/2}. \tag{25.16}$$

Plastic foams are good for the kind of packaging which is meant to absorb the energy of a single impact: polyurethane automobile crash padding, polystyrene foam to protect a television set if it is accidentally dropped during delivery. The long plateau of the stress–strain curve absorbs energy but the foam is damaged in the process.

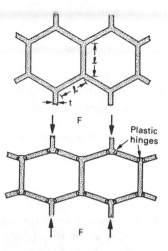

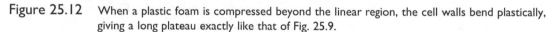

Figure 25.12 When a plastic foam is compressed beyond the linear region, the cell walls bend plastically, giving a long plateau exactly like that of Fig. 25.9.

Materials that can be engineered

The materials described in this chapter differ from most others available to the designer in that their properties can be engineered to suit, as nearly as possible, the application. The stiffness, strength and toughness of a composite are, of course, controlled by the type and volume fraction of fibres. But the materials engineering can go further than this, by orienting or laminating the fibre weave to give directional properties, or to reinforce holes or fixing points, or to give a stiffness which varies in a controlled way across a component. Foaming, too, allows new degrees of freedom to the designer. Not only can the stiffness and strength be controlled over a vast range (10^4 or more) by proper choice of matrix polymer and foam density, but gradients of foam density and thus of properties can be designed-in. Because of this direct control over properties, both sorts of composites offer special opportunities for designing *weight-optimal structures*, particularly attractive in aerospace and transport. Examples of this sort of design can be found in the books listed in the References.

Examples

25.1 A unidirectional fibre composite consists of 60% by volume of Kevlar fibres in a matrix of epoxy. Find the moduli $E_{c\parallel}$ and $E_{c\perp}$. Comment on the accuracy of your value for $E_{c\perp}$. Use the moduli given in Table 25.1, and use an average value where a range of moduli is given.

Answers

77 GPa and 9 GPa.

25.2 A unidirectional fibre composite consists of 60% by volume of continuous type-1 carbon fibres in a matrix of epoxy. Find the maximum tensile strength of the composite. You may assume that the matrix yields in tension at a stress of 40 MPa.

Answer

1336 MPa.

25.3 A composite material for a car-repair kit consists of a random mixture of short glass fibres in a polyester matrix. Estimate the maximum toughness G_c of the composite. You may assume that: the volume fraction of glass is 30%; the fibre diameter is 15 μm; the fracture strength of the fibres is 1400 MPa; and the shear strength of the matrix is 30 MPa.

Answer

$37\,\text{kJ}\,\text{m}^{-2}$.

25.4 Calculate the critical length $2x_c$ of the fibres in Example 25.3. How would you expect G_c to change if the fibres were substantially longer than $2x_c$?

Answer

0.35 mm.

25.5 The diagram shows a layer of brittle polyurethane foam which is stuck to a steel plate. The foam is part of the thermal insulation system surrounding a liquid methane tank. In service, the temperature distribution in the foam is linear, with

$$T(°C) = -100\left(\frac{x}{t}\right).$$

If the steel plate is considered to be infinitely stiff, find an expression for the tensile stress σ in the foam as a function of x. If the Young's modulus and coefficient of thermal expansion of the foam are 34 MPA and $10^{-4}\,°\text{C}^{-1}$, find the maximum value of the tensile stress. Where is this maximum stress located?

Answers

$\sigma = E\alpha 100(x/t)$; 0.34 MPa; at $x = t$.

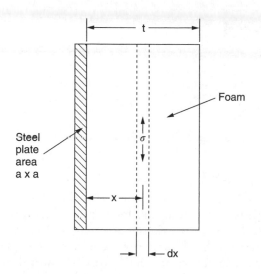

25.6   In Example 25.5, the tensile thermal stress in the foam makes it crack. The cracks form at the free surface $(x = t)$ and then run inwards through the thickness of the layer. It is important to prevent cracking in order to have a liquid-proof safety barrier next to the methane tank. How would you modify the material to prevent it cracking?

25.7   What are the requirements for a polymer foam used in (a) a seat cushion, (b) a safety helmet for climbers. How do the deformation mechanisms in the two foams differ? Why can a seat cushion last indefinitely, when a climbing helmet must be thrown away if it has been involved in a fall?

Chapter 26
Special topic: wood

Introduction

Wood is the oldest and still most widely used of structural materials. Its documented use in buildings and ships spans more than 5000 years. In the sixteenth century the demand for stout oaks for ship-building was so great that the population of suitable trees was seriously depleted, and in the seventeenth and eighteenth centuries much of Europe was deforested completely by the exponential growth in consumption of wood. Today the world production is about the same as that of iron and steel: roughly 10^9 tonnes per year. Much of this is used structurally: for beams, joists, flooring or supports which will bear load. Then the properties which interest the designer are the *moduli*, the *yield* or *crushing strength*, and the *toughness*. These properties, summarised in Table 26.1, vary considerably: oak is about 5 times stiffer, stronger and tougher than balsa, for instance. In this case study we examine the structure of wood and the way the mechanical properties depend on it, using many of the ideas developed in the preceding five chapters.

The structure of wood

It is necessary to examine the structure of wood at three levels. At the macroscopic (unmagnified) level the important features are shown in Fig. 26.1. The main cells ("fibres" or "tracheids") of the wood run axially up and down the tree: this is the direction in which the strength is greatest. The wood is divided radially by the growth rings: differences in density and cell size caused by rapid growth in the spring and summer, and sluggish growth in the autumn and winter. Most of the growing processes of the tree take place in the *cambium*, which is a thin layer just below the bark. The rest of the wood is more or less dead; its function is mechanical: to hold the tree up. With 10^8 years in which to optimise its structure, it is perhaps not surprising that wood performs this function with remarkable efficiency.

To see how it does so, one has to examine the structure at the microscopic (light microscope or scanning electron microscope) level. Figure 26.2 shows how the wood is made up of long hollow cells, squeezed together like straws, most of them parallel to the axis of the tree. The *axial* section shows roughly hexagonal cross-sections of the cells: the *radial* and the *tangential* sections show their long, thin shape. The structure appears complicated because of the fat

Table 26.1   Mechanical properties of woods

Wood	Density[1] (Mg m^{-3})	Young's modulus[1,2] (GPa)		Strength[1,3] (MPa) ∥ to grain		Fracture toughness[1] (MPa m$^{1/2}$)	
		∥ to grain	⊥ to grain	Tension	Compression	∥ to grain	⊥ to grain
Balsa	0.1–0.3	4	0.2	23	12	0.05	1.2
Mahogany	0.53	13.5	0.8	90	46	0.25	6.3
Douglas fir	0.55	16.4	1.1	70	42	0.34	6.2
Scots pine	0.55	16.3	0.8	89	47	0.35	6.1
Birch	0.62	16.3	0.9	–	–	0.56	–
Ash	0.67	15.8	1.1	116	53	0.61	9.0
Oak	0.69	16.6	1.0	97	52	0.51	4.0
Beech	0.75	16.7	1.5	–	–	0.95	8.9

[1] Densities and properties of wood vary considerably; allow ±20% on the data shown here. All properties vary with moisture content and temperature; see text.
[2] Dynamic moduli; moduli in static tests are about two-thirds of these.
[3] Anisotropy increases as the density decreases. The transverse strength is usually between 10% and 20% of the longitudinal.

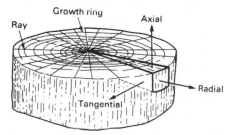

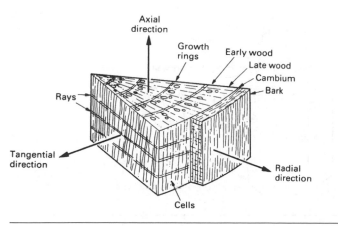

Figure 26.1   The macrostructure of wood. Note the co-ordinate system (axial, radial, tangential).

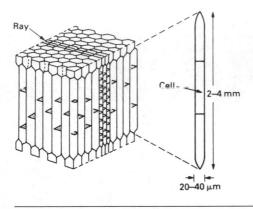

Figure 26.2   The microstructure of wood. Woods are foams of relative densities between 0.07 and 0.5, with cell walls which are fibre-reinforced. The properties are very anisotropic, partly because of the cell shape and partly because the cell-wall fibres are aligned near the axial direction.

tubular *sap channels* which run up the axis of the tree carrying fluids from the roots to the branches, and the strings of smaller cells called *rays* which run radially outwards from the centre of the tree to the bark. But neither is of much importance mechanically; it is the fibres or tracheids which give wood its stiffness, strength and toughness.

The walls of these tracheid cells have a structure at a molecular level, which (but for the scale) is like a composite – like fibreglass, for example (Fig. 26.3). The role of the strong glass fibres is taken by fibres of crystalline *cellulose*, a high polymer $(C_6H_{10}O_5)_n$, made by the tree from glucose, $C_6H_{12}O_6$, by a condensation reaction, and with a $\overline{DP}$ of about 10^4. Cellulose is a linear polymer

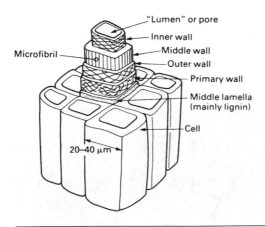

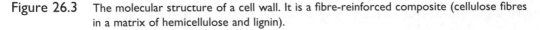

Figure 26.3   The molecular structure of a cell wall. It is a fibre-reinforced composite (cellulose fibres in a matrix of hemicellulose and lignin).

with no cumbersome side-groups, so it crystallises easily into *microfibrils* of great strength (these are the fibres you can sometimes see in coarse paper). Cellulose microfibrils account for about 45% of the cell wall. The role of the epoxy matrix is taken by the *lignin*, an amorphous polymer (like epoxy), and by *hemicellulose*, a partly crystalline polymer of glucose with a smaller DP than cellulose; between them they account for a further 40% of the weight of the wood. The remaining 10–15% is water and *extractives*: oils and salts which give wood its colour, its smell, and (in some instances) its resistance to beetles, bugs and bacteria. The chemistry of wood is summarised in Table 26.2.

It is a remarkable fact that, although woods differ enormously in appearance, the composition and structure of their cell walls do not. Woods as different as balsa and beech have cell walls with a density ρ_s of $1.5 \, \text{Mg m}^{-3}$ with the chemical make-up given in Table 26.2 and with almost the same elaborate lay-up of cellulose fibres (Fig. 26.3). The lay-up is important because it accounts, in part, for the enormous anisotropy of wood – the difference in strength along and across the grain. The cell walls are helically wound, like the handle of a CFRP golf club, with the fibre direction nearer the cell axis rather than across it. This gives the cell wall a modulus and strength which are large parallel to the axis of the cell and smaller (by a factor of about 3) across it. The properties of the cell wall are summarised in Table 26.3; it is a little less stiff, but nearly as strong as an aluminium alloy.

Wood, then, is a foamed fibrous composite. Both the foam cells and the cellulose fibres in the cell wall are aligned predominantly along the grain of the wood (i.e. parallel to the axis of the trunk). Not surprisingly, wood is mechanically very anisotropic: the properties along the grain are quite different from those across it. But if all woods are made of the same stuff, why do the

Table 26.2   Composition of cell wall of wood

Material	Structure	Approx. wt%
Fibres		
Cellulose $(C_6H_{10}O_5)_n$	Crystalline	45
Matrix		
Lignin	Amorphous	20
Hemicellulose	Semi-crystalline	20
Water	Dissolved in the matrix	10
Extractives	Dispersed in the matrix	5

Table 26.3   Properties of cell wall

Property	Axial	Transverse
Density, ρ_s (Mg m^{-3})	1.5	
Modulus, E_s (GPa)	35	10
Yield strength, σ_y (MPa)	150	50

properties range so widely from one sort of wood to another? The differences between woods are primarily due to the differences in their relative densities (see Table 26.1). This we now examine more closely.

The mechanical properties of wood

All the properties of wood depend to some extent on the amount of water it contains. Green wood can contain up to 50% water. Seasoning (for 2 to 10 years) or kiln drying (for a few days) reduces this to around 14%. The wood shrinks, and its modulus and strength increase (because the cellulose fibrils pack more closely). To prevent *movement*, wood should be dried to the value which is in equilibrium with the humidity where it will be used. In a centrally heated house (20°C, 65% humidity), for example, the equilibrium moisture content is 12%. Wood shows ordinary thermal expansion, of course, but its magnitude ($\alpha = 5\,\mathrm{MK}^{-1}$ along the grain, $50\,\mathrm{MK}^{-1}$ across the grain) is small compared to dimensional changes caused by drying.

Elasticity

Woods are visco-elastic solids: on loading they show an immediate elastic deformation followed by a further slow creep or "delayed" elasticity. In design with wood it is usually adequate to treat the material as elastic, taking a rather lower modulus for long-term loading than for short loading times (a factor of 3 is realistic) to allow for the creep. The modulus of a wood, for a given water content, then depends principally on its density, and on the angle between the loading direction and the grain.

Figure 26.4 shows how Young's modulus along the grain ("axial" loading) and across the grain ("radial" or "tangential" loading) varies with density. The axial modulus varies linearly with density and the others vary roughly as its square. This means that the anisotropy of the wood (the ratio of the modulus along the grain to that across the grain) increases as the density decreases: balsa woods are very anisotropic; oak or beech are less so. In structural applications, wood is usually loaded along the grain: then only the axial modulus is important. Occasionally it is loaded across the grain, and then it is important to know that the stiffness can be a factor of 10 or more smaller (Table 26.1).

The moduli of wood can be understood in terms of the structure. When loaded along the grain, the cell walls are extended or compressed (Fig. 26.5a). The modulus $E_{w\parallel}$ of the wood is that of the cell wall, E_s, scaled down by the fraction of the section occupied by cell wall. Doubling the density obviously doubles this section, and therefore doubles the modulus. It follows immediately that

$$E_{w\parallel} = E_s \left(\frac{\rho}{\rho_s} \right) \tag{26.1}$$

where ρ_s is the density of the solid cell wall (Table 26.3).

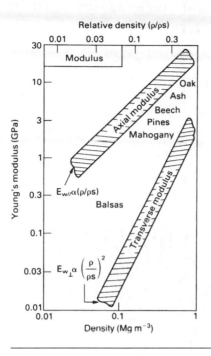

Figure 26.4   Young's modulus for wood depends mainly on the relative density ρ/ρ_s. That along the grain varies as ρ/ρ_s; that across the grain varies roughly as $(\rho/\rho_s)^2$, like polymer foams.

The transverse modulus $E_{w\perp}$ is lower partly because the cell wall is less stiff in this direction, but partly because the foam structure is intrinsically anisotropic because of the cell shape. When wood is loaded across the grain, the cell walls bend (Fig. 26.5b,c). It behaves like a foam (Chapter 25) for which

$$E_{w\perp} = E_s \left(\frac{\rho}{\rho_s} \right)^2. \tag{26.2}$$

The elastic anisotropy

$$E_{w\parallel}/E_{w\perp} = (\rho_s/\rho). \tag{26.3}$$

Clearly, the lower the density the greater is the elastic anisotropy.

The tensile and compressive strengths

The axial tensile strength of many woods is around 100 MPa – about the same as that of strong polymers like the epoxies. The ductility is low – typically 1% strain to failure.

Compression along the grain causes the kinking of cell walls, in much the same way that composites fail in compression (Chapter 25, Fig. 25.5). The

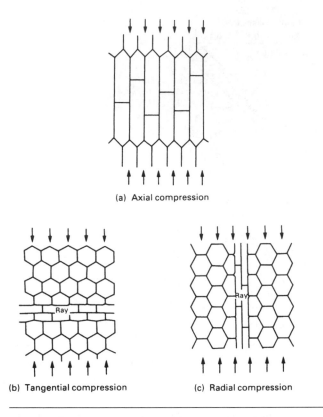

(a) Axial compression

(b) Tangential compression

(c) Radial compression

Figure 26.5   **(a)** When wood is loaded along the grain most of the cell walls are compressed axially; **(b)** when loaded across the grain, the cell walls bend like those in the foams described in Chapter 25.

kink usually initiates at points where the cells bend to make room for a ray, and the kink band forms at an angle of 45° to 60°. Because of this kinking, the compressive strength is less (by a factor of about 2 – see Table 26.1) than the tensile strength, a characteristic of composites.

Like the modulus, the tensile and compressive strengths depend mainly on the density (Fig. 26.6). The strength parallel to the grain varies linearly with density, for the same reason that the axial modulus does: it measures the strength of the cell wall, scaled by the fraction of the section it occupies, giving

$$\sigma_{\parallel} = \sigma_s \left(\frac{\rho}{\rho_s} \right), \tag{26.4}$$

where σ_s is the yield strength of the solid cell wall.

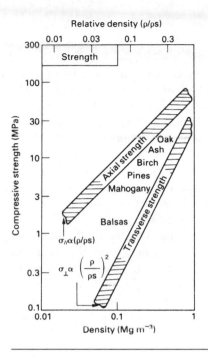

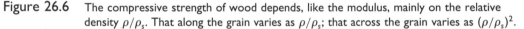

Figure 26.6   The compressive strength of wood depends, like the modulus, mainly on the relative density ρ/ρ_s. That along the grain varies as ρ/ρ_s; that across the grain varies as $(\rho/\rho_s)^2$.

Figure 26.6 shows that the transverse crushing strength $\sigma_\perp$ varies roughly as

$$\sigma_\perp = \sigma_s \left(\frac{\rho}{\rho_s}\right)^2. \tag{26.5}$$

The explanation is almost the same as that for the transverse modulus: the cell walls bend like beams, and collapse occurs when these beams reach their plastic collapse load. As with the moduli, moisture and temperature influence the crushing strength.

The toughness

The toughness of wood is important in design for exactly the same reasons that that of steel is: it determines whether a structure (a frame building, a pit prop, the mast of a yacht) will fail suddenly and unexpectedly by the propagation of a fast crack. In a steel structure the initial crack is that of a defective weld, or is formed by corrosion or fatigue; in a wooden structure the initial defect may be a knot, or a saw cut, or cell damage caused by severe mishandling.

Recognising its importance, various tests have been devised to measure wood toughness. A typical static test involves loading square section beams in three-point bending until they fail; the "toughness" is measured by the area under the load–deflection curve. A typical dynamic test involves dropping, from an increasingly great height, a weight of 1.5 kg; the height which breaks the beam is a "toughness" – of a sort. Such tests (still universally used) are good for ranking different batches or species of wood; but they do not measure a property which can be used sensibly in design.

The obvious parameter to use is the fracture toughness, K_c. Not unexpectedly, it depends on density (Fig. 26.7), varying as $(\rho/\rho_s)^{3/2}$. It is a familiar observation that wood splits easily along the grain, but with difficulty across the grain. Figure 26.7 shows why: the fracture toughness is more than a factor of 10 smaller along the grain than across it.

Loaded along the grain, timber is remarkably tough: much tougher than any simple polymer, and comparable in toughness with fibre-reinforced composites. There appear to be two contributions to the toughness. One is the very large fracture area due to cracks spreading, at right angles to the break, along the cell interfaces, giving a ragged fracture surface. The other is more important, and it is exactly what you would expect in a composite: fibre pull-out (Chapter 25,

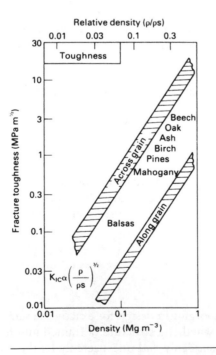

Figure 26.7   The fracture toughness of wood, like its other properties, depends primarily on relative density. That across the grain is roughly ten times larger than that along the grain. Both vary as $(\rho/\rho_s)^{3/2}$.

Table 26.4   Specific strength of structural materials

Material	$\dfrac{E}{\rho}$	$\dfrac{\sigma_y}{\rho}$	$\dfrac{K_c}{\rho}$
Woods	20–30	120–170	1–12
Al-alloy	25	179	8–16
Mild steel	26	30	18
Concrete	15	3	0.08

Fig. 25.6). As a crack passes across a cell the cellulose fibres in the cell wall are unravelled, like pulling thread off the end of a bobbin. In doing so, the fibres must be pulled out of the hemicellulose matrix, and a lot of work is done in separating them. It is this work which makes the wood tough.

For some uses, the anisotropy of timber and its variability due to knots and other defects are particularly undesirable. Greater uniformity is possible by converting the timber into board such as laminated plywood, chipboard and fibre-building board.

Summary: wood compared to other materials

The mechanical properties of wood (a structural material of first importance because of the enormous scale on which it is used) relate directly to the shape and size of its cells, and to the properties of the composite-like cell walls. Loaded along the grain, the cell walls are loaded in simple tension or compression, and the properties scale as the density. But loaded across the grain, the cell walls bend, and then the properties depend on a power (3/2 or 2) of the density. That, plus the considerable anisotropy of the cell wall material (which is a directional composite of cellulose fibres in a hemicellulose/lignin matrix), explain the enormous difference between the modulus, strength and toughness along the grain and across it.

The properties of wood are generally inferior to those of metals. But the properties per unit weight are a different matter. Table 26.4 shows that the specific properties of wood are better than mild steel, and as good as many aluminium alloys (that is why, for years, aircraft were made of wood). And, of course, it is much cheaper.

Examples

26.1   Explain how the complex structure of wood results in large differences between the along-grain and across-grain values of Young's modulus, tensile strength and fracture toughness.

26.2   What functions do the polymers cellulose, lignin and hemicellulose play in the construction of the cells in wood?

26.3   Discuss, giving specific examples, how the anisotropic properties of wood are exploited in the design applications of this material.

26.4   Discuss, giving specific examples, how the anisotropic properties of wood are exploited in processes for manufacturing items from wood.

26.5   Chip-board and fibre-board are manufactured by glueing small chips or fibres of wood together. Why is the Young's modulus of these materials in the plane of the board (a) isotropic, (b) approximately mid-way between the along-grain and across-grain moduli of the source wood?

26.6   Plywood is made from alternate layers of wood (each some 2 mm thick) with the grain direction in successive layers arranged to produce a 0–90 laminate. Why is the Young's modulus in the 0 and 90 directions approximately 50% of the along-grain modulus of the source wood? How would you expect the splitting resistance of plywood to differ from that of the source wood?

26.7   The diagram shows the elevation of a wooden balustrade, together with a cross-section drawn on A–A looking towards the left-hand end of the balsutrade. The top rail of the balustrade is frequently subjected to a large horizontal force (in the direction shown by the arrow) from people learning against it. Why is the design of the vertical end post unsafe? Remember that wooden posts, planks, etc. always have the grain running along their length.

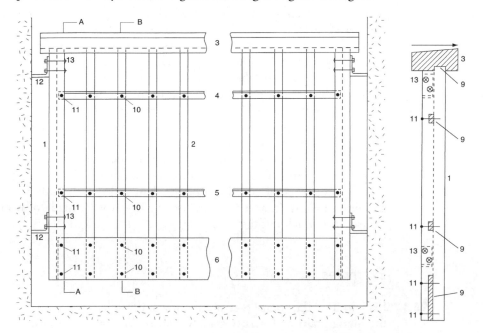

D. Designing with metals, ceramics, polymers and composites

Chapter 27
Design with materials

Introduction

Design is an iterative process. You start with the definition of a function (a pen, a hairdryer, a fuel pin for a nuclear reactor) and draw on your knowledge (the contents of this book, for instance) and experience (your successes and failures) in formulating a tentative design. You then refine this by a systematic process that we shall look at later.

Materials selection is an integral part of design. And because the principles of mechanics, dynamics and so forth are all well established and not changing much, whereas new materials are appearing all the time, innovation in design is frequently made possible by the use of new materials. Designers have at their disposal the range of materials that we have discussed in this book: metals, ceramics, polymers, and combinations of them to form composites. Each class of material has its own strengths and limitations, which the designer must be fully aware of. Table 27.1 summarises these.

At and near room temperature, metals have well-defined, almost constant, moduli and yield strengths (in contrast to polymers, which do not). And most metallic alloys have a ductility of 20% or better. Certain high-strength alloys (spring steel, for instance) and components made by powder methods, have less – as little as 2%. But even this is enough to ensure that an unnotched component yields before it fractures, and that fracture, when it occurs, is of a tough, ductile, type. But – partly because of their ductility – metals are prey to cyclic fatigue and, of all the classes of materials, they are the least resistant to corrosion and oxidation.

Historically, design with ceramics has been empirical. The great gothic cathedrals, still the most impressive of all ceramic designs, have an aura of stable permanence. But many collapsed during construction; the designs we know evolved from these failures. Most ceramic design is like that. Only recently, and because of more demanding structural applications, have design methods evolved.

In designing with ductile materials, a *safety-factor* approach is used. Metals can be used under static loads within a small margin of their ultimate strength with confidence that they will not fail prematurely. Ceramics cannot. As we saw earlier, brittle materials always have a wide scatter in strength, and the strength itself depends on the time of loading and the volume of material

Table 27.1   Design-limiting properties of materials

Material	Good	Poor
Metals High E, K_c Low σ_y	Stiff ($E \approx 100$ GPa) Ductile ($\varepsilon_f \approx 20\%$) — formable Tough ($K_c > 50$ MPa m$^{1/2}$) High MP ($T_m \approx 1000°C$) T-shock ($\Delta T > 500°C$)	Yield (pure, $\sigma_y \approx 1$ MPa) → alloy Hardness ($H \approx 3\sigma_y$) → alloy Fatigue strength ($\sigma_e = \frac{1}{2}\sigma_y$) Corrosion resistance → coatings

Ceramics
 	
High E, σ_y
 	
Low K_c
	Stiff ($E \approx 200$ GPa)
	
	Very high yield, hardness
	    ($\sigma_y > 3$ GPa)
	High MP ($T_m \approx 2000°C$)
	Corrosion resistant
	Moderate density
		Very low toughness
		    ($K_c \approx 2$ MPa m$^{1/2}$)
		T-shock ($\Delta T \approx 200°C$)
		
		Formability → powder methods

Polymers
Adequate σ_y, K_c
Low E
	Ductile and formable
	Corrosion resistant
	Low density
		Low stiffness ($E \approx 2$ GPa)
		Yield ($\sigma_y = 2 - 100$ MPa)
		Low glass temp
		    ($T_G \approx 100°C$) → creep
		Toughness often low
		    (1 MPa m$^{1/2}$)

Composites High E, σ_y, K_c but cost	Stiff ($E > 50$ GPa) Strong ($\sigma_y \approx 200$ MPa) Tough ($K_c > 20$ MPa m$^{1/2}$) Fatigue resistant Corrosion resistant Low density	Formability Cost Creep (polymer matrices)

under stress. The use of a single, constant, safety factor is no longer adequate, and the statistical approach of Chapter 18 must be used instead.

We have seen that the "strength" of a ceramic means, almost always, the fracture or crushing strength. Then (unlike metals) the compressive strength is 10 to 20 times larger than the tensile strength. And because ceramics have no ductility, they have a low tolerance for stress concentrations (such as holes and flaws) or for high contact stresses (at clamping or loading points, for instance). If the pin of a pin-jointed frame, made of metal, fits poorly, then the metal deforms locally, and the pin beds down, redistributing the load. But if the pin and frame are made of a brittle material, the local contact stresses nucleate cracks which then propagate, causing sudden collapse. Obviously, the process of design with ceramics differs in detail from that of design with metals.

That for polymers is different again. When polymers first became available to the engineer, it was common to find them misused. A "cheap plastic" product was one which, more than likely, would break the first time you picked it up. Almost always this happened because the designer used a polymer to replace a metal component, without redesign to allow for the totally different properties of the polymer. Briefly, there are three:

(a) Polymers have much lower moduli than metals – roughly 100 times lower. So elastic deflections may be large.
(b) The deflection of polymers depends on the time of loading: they creep at room temperature. A polymer component under load may, with time, acquire a permanent set.
(c) The strengths of polymers change rapidly with temperature near room temperature. A polymer which is tough and flexible at 20°C may be brittle at the temperature of a household refrigerator, 4°C.

With all these problems, why use polymers at all? Well, complicated parts performing several functions can be moulded in a single operation. Polymer components can be designed to snap together, making assembly fast and cheap. And by accurately sizing the mould, and using pre-coloured polymer, no finishing operations are necessary. So great economies of manufacture are possible: polymer parts really can be cheap. But are they inferior? Not necessarily. Polymer densities are low (all are near $1\,\mathrm{Mg\ m^{-3}}$); they are corrosion-resistant; they have abnormally low coefficients of friction; and the low modulus and high strength allows very large elastic deformations. Because of these special properties, polymer parts may be distinctly superior.

Composites overcome many of the remaining deficiencies. They are stiff, strong and tough. Their problem lies in their cost: composite components are usually expensive, and they are difficult and expensive to form and join. So, despite their attractive properties, the designer will use them only when the added performance offsets the added expense.

New materials are appearing all the time. New polymers with greater stiffness and toughness appear every year; composites are becoming cheaper as the volume of their production increases. Ceramics with enough toughness to be used in conventional design are becoming available, and even in the metals field, which is a slowly developing one, better quality control, and better understanding of alloying, leads to materials with reliably better properties. All of these offer new opportunities to the designer who can frequently redesign an established product, making use of the properties of new materials, to reduce its cost or its size and improve its performance and appearance.

Design methodology

Books on design often strike the reader as vague and qualitative; there is an implication that the ability to design is like the ability to write music: a gift given to few. And it is true that there is an element of creative thinking (as opposed to logical reasoning or analysis) in good design. But a design methodology can be formulated, and when followed, it will lead to a practical solution to the design problem.

Figure 27.1 summarises the methodology for designing a component which must carry load. At the start there are two parallel streams: materials selection and component design. A tentative material is chosen and data for it are assembled from data sheets like the ones given in this book or from data books (referred to at the end of this chapter). At the same time, a tentative component design is drawn up, able to fill the function (which must be carefully defined at the start); and an approximate stress analysis is carried out to assess the stresses, moments, and stress concentrations to which it will be subjected.

The two streams merge in an assessment of the material performance in the tentative design. If the material can bear the loads, moments, concentrated stresses (etc.) without deflecting too much, collapsing or failing in some other way, then the design can proceed. If the material cannot perform adequately, the first iteration takes place: either a new material is chosen, or the component design is changed (or both) to overcome the failing.

The next step is a detailed specification of the design and of the material. This may require a detailed stress analysis, analysis of the dynamics of the system, its response to temperature and environment, and a detailed consideration of the appearance and feel (the aesthetics of the product). And it will require better material data: at this point it may be necessary to get detailed material properties from possible suppliers, or to conduct tests yourself.

The design is viable only if it can be produced economically. The choice of production and fabrication method is largely determined by the choice of material. But the production route will also be influenced by the size of the production run, and how the component will be finished and joined to other components; each class of material has its own special problems here; they were discussed in Chapters 14, 19, 24 and 25. The choice of material and production route will, ultimately, determine the price of the product, so a second major iteration may be required if the costing shows the price to be too high. Then a new choice of material or component design, allowing an alternative production path, may have to be considered.

At this stage a prototype product is produced, and its performance in the market is assessed. If this is satisfactory, full-scale production is established. But the designer's role does not end at this point. Continuous analysis of the performance of a component usually reveals weaknesses or ways in which it could be improved or made more cheaply. And there is always scope for further innovation: for a radically new design, or for a radical change in the material which the component is made from. Successful designs evolve continuously, and only in this way does the product retain a competitive position in the market place.

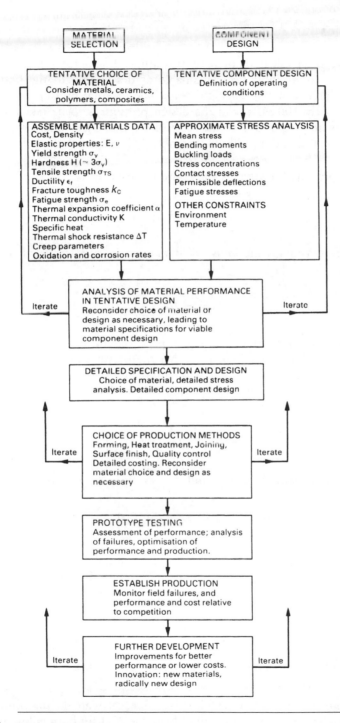

Figure 27.1   Design methodology.

Examples

27.1 You have been asked to prepare an outline design for the pressure hull of a deep-sea submersible vehicle capable of descending to the bottom of the Mariana Trench in the Pacific Ocean. The external pressure at this depth is approximately 100 MPa, and the design pressure is to be taken as 200 MPa. The pressure hull is to have the form of a thin-walled sphere with a specified radius r of 1 m and a uniform thickness t. The sphere can fail in one of two ways:

external-pressure buckling at a pressure p_b given by

$$p_b = 0.3E \left(\frac{t}{r}\right)^2,$$

where E is Young's modulus; yield or compressive failure at a pressure p_f given by

$$p_f = 2\sigma_f \left(\frac{t}{r}\right),$$

where σ_f is the yield stress or the compressive failure stress as appropriate.

The basic design requirement is that the pressure hull shall have the minimum possible mass compatible with surviving the design pressure.

By eliminating t from the equations, show that the minimum mass of the hull is given by the expressions

$$m_b = 22.9 r^3 p_b^{0.5} \left(\frac{\rho}{E^{0.5}}\right),$$

for external-pressure buckling, and

$$m_f = 2\pi r^3 p_f \left(\frac{\rho}{\sigma_f}\right),$$

for yield or brittle compressive failure. Hence obtain a merit index to meet the design requirement for each of the two failure mechanisms. [You may assume that the surface area of the sphere is $4\pi r^2$.]

Answers

$E^{0.5}/\rho$ for external-pressure buckling; σ_f/ρ for yield or brittle compressive failure.

27.2 For each material listed in the following table, calculate the minimum mass and wall thickness of the pressure hull of Example 27.1 for both failure mechanisms at the design pressure.

Material	E (GPa)	σ_f (MPa)	Density, ρ (Kg m^{-3})
Alumina	390	5000	3900
Glass	70	2000	2600
Alloy steel	210	2000	7800
Titanium alloy	120	1200	4700
Aluminium alloy	70	500	2700

Hence determine the limiting failure mechanism for each material. [Hint: this is the failure mechanism which gives the larger of the two values of t.]

What is the optimum material for the pressure hull? What are the mass, wall thickness and limiting failure mechanism of the optimum pressure hull?

Answers

Material	m_b (tonne)	t_b (mm)	m_f (tonne)	t_f (mm)	Limiting failure mechanism
Alumina	2.02	41	0.98	20	Buckling
Glass	3.18	97	1.63	50	Buckling
Alloy steel	5.51	56	4.90	50	Buckling
Titanium alloy	4.39	74	4.92	83	Yielding
Aluminium alloy	3.30	97	6.79	200	Yielding

The optimum material is alumina, with a mass of 2.02 tonne, a wall thickness of 41 mm and a limiting failure mechanism of external-pressure buckling.

27.3 Briefly describe the processing route which you would specify for making the pressure hull of Example 27.2 from each of the materials listed in the table. Comment on any particular problems which might be encountered. [You may assume that the detailed design will call for a number of apertures in the wall of the pressure hull.]

Chapter 28
Case studies in design

1. Designing with metals: conveyor drums for an iron ore terminal

Introduction

The conveyor belt is one of the most efficient devices available for moving goods over short distances. Billions of tons of minerals, foodstuffs and consumer goods are handled in this way every year. Figure 28.1 shows the essentials of a typical conveyor system. The following data are typical of the largest conveyors, which are used for handling coal, iron ore and other heavy minerals.

Capacity:	5000 tonne h^{-1}
Belt speed:	4 m s^{-1}
Belt tension:	5 tonne
Motor rating:	250 kW
Belt section:	1.5 m wide $\times$ 11 mm thick
Distance between centres of tail drum and drive drum:	200 m

It is important that conveyor systems of this size are designed to operate continuously for long periods with minimum "down-time" for routine maintenance: the unscheduled breakdown of a single unit in an integrated plant could lead to a total loss of production. Large conveyors include a number of critical components which are designed and built essentially as "one-offs" for a particular installation: it is doubly important to check these at the design stage because a failure here could lead to a damagingly long down-time while a harassed technical manager phones the length of the country looking for fabrication shops with manoeuvrable capacity, and steel merchants with the right sections in stock.

Tail drum design

The tail drum (Fig. 28.1) is a good example of a critical component. Figure 28.2 shows the general arrangement of the drum in its working environment and Fig. 28.3 shows a detailed design proposal. We begin our design check by looking at the stresses in the shaft. The maximum stress comes at the surface

326

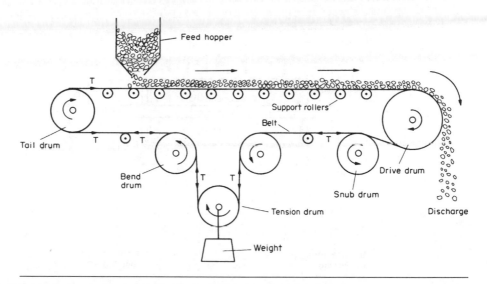

Figure 28.1   Schematic of a typical conveyor system. Because the belt tends to sag between the support rollers it must be kept under a constant tension *T*. This is done by hanging a large weight on the tension drum. The drive is supplied by coupling a large electric motor to the shaft of the drive drum *via* a suitable gearbox and overload clutch.

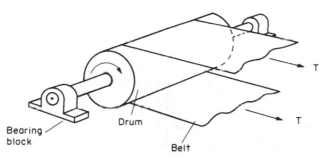

Figure 28.2   Close-up of the tail drum. The belt tension applies a uniformly distributed sideways loading to the drum.

of the shaft next to the shaft-plate weld (Fig. 28.4). We can calculate the maximum stress from the standard formula

$$\sigma_{max} = \frac{Mc}{I} \qquad (28.1)$$

where the bending moment *M* is given by

$$M = Fx \qquad (28.2)$$

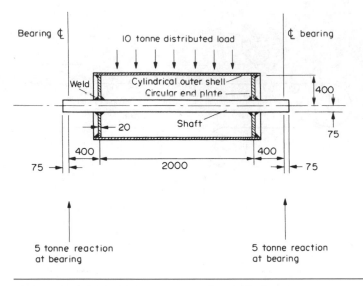

Figure 28.3   Cross-section through the tail drum. All dimensions are in mm. We have assumed a belt tension of 5 tonnes, giving a total loading of 10 tonnes.

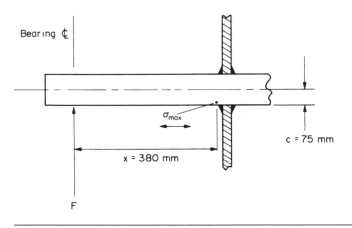

Figure 28.4   Shaft-plate detail.

and the second moment of area of the shaft is given by

$$I = \frac{\pi c^4}{4}. \tag{28.3}$$

Using values of $F = 5000 \times 9.81\,\text{N}$, $x = 380\,\text{mm}$, and $c = 75\,\text{mm}$, we get a value for σ_{max} of 56 MPa.

This stress is only a quarter of the yield stress of a typical structural steel, and the shaft therefore has an ample factor of safety against failure by *plastic overload*.

The second failure mode to consider is *fatigue*. The drum will revolve about once every second, and each part of the shaft surface will go alternately into tension and compression. The maximum fatigue stress range (of $2 \times 56 = 112$ MPa) is, however, only a quarter of the fatigue limit for structural steel (Fig. 28.5); and the shaft should therefore last indefinitely. But what about the welds? There are in fact a number of reasons for expecting them to have fatigue properties that are poorer than those of the parent steel (see Table 28.1).

Figure 28.6 shows the fatigue properties of structural steel welds. The fatigue limit stress range of 120 MPa for the best class of weld is a good deal less than the limiting range of 440 MPa for the parent steel (Fig. 28.5). And the worst class of weld has a limiting range of only 32 MPa!

The shaft-plate weld can be identified as a class E/F weld with a limiting stress range of 69 to 55 MPa. This is a good deal less than the stress range of 112 MPa experienced by the shaft. We thus have the curious situation where a weld which is merely an attachment to the shaft has weakened it so much that it will only last for about 2×10^6 cycles – or 1 month of operation. The obvious way of solving this problem is to remove the attachment weld from the surface of the shaft. Figure 28.7 shows one way of doing this.

We have seen that welds can be very weak in a fatigue situation, and we would be wise to check that the modified weld in Fig. 28.7 is adequate for the

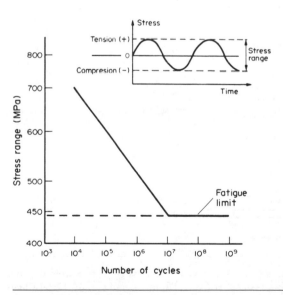

Figure 28.5   Fatigue data for a typical structural steel in dry air. Note that, if the fatigue stress range is less than 440 MPa (the *fatigue limit*) the component should last indefinitely. The data relate to a fatigue stress cycle with a *zero* mean stress, which is what we have in the case of our tail drum.

Table 28.1   Weld characteristics giving adverse fatigue properties

Characteristic	Comments
Change in section at weld bead.	Gives stress concentration. In the case of butt welds this can be removed by grinding back the weld until flush with the parent plates. Grinding marks must be parallel to loading direction otherwise they can initiate fatigue cracks.
Poor surface finish of weld bead.	Helps initiate fatigue cracks. Improve finish by grinding.
Contain tensile residual stresses which are usually as large as the yield stress.	Weld liable to fatigue *even when applied stress cycle is wholly compressive*. Reduce residual stresses by stress relieving, hammering or shot peening.
Often contain defects (hydrogen cracks, slag inclusions, stop–start marks).	Help initiate fatigue cracks. Critical welds must be tested non-destructively and defects must be gouged out.
Large differences in microstructure between parent metal, heat-affected zone and weld bead.	Sharp changes in mechanical properties give local stress concentrations.

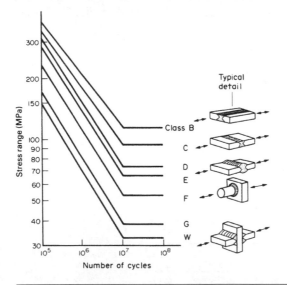

Figure 28.6   Fatigue data for welded joints in clean air. The class given to a weld depends critically on the weld detail and the loading direction. Classes B and C must be free from cracks and must be ground flush with the surface to remove stress concentrations. These conditions are rarely met in practice, and most welds used in general construction have comparatively poor fatigue properties.

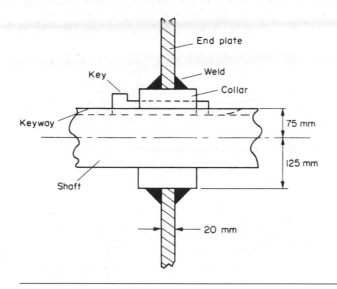

Figure 28.7   Modification to remove attachment weld from surface of shaft. The collar can be pressed on to the shaft and secured with a feather key, but we must remember that the keyway will weaken the shaft.

job. Although it is difficult to spot immediately, the weld is in fact subjected to large bending stresses (Fig. 28.8). These can be calculated as follows.

We begin by equating moments in Fig. 28.8 to give

$$F \times 0.4\,\text{m} = M_A + M_B. \qquad (28.4)$$

Here M_A is the couple which must be applied at the mid span of the shaft to make $(\mathrm{d}y/\mathrm{d}x)_{x=0} = 0$; and M_B is the couple that is needed to cause the out-of-plane deflection of the end discs. Using the standard beam-bending formula of

$$\frac{\mathrm{d}^2 y}{\mathrm{d}x^2} = \frac{M}{EI} \qquad (28.5)$$

we can write

$$\left.\frac{\mathrm{d}y}{\mathrm{d}x}\right)_B - \left.\frac{\mathrm{d}y}{\mathrm{d}x}\right)_A = \int_0^{1\text{m}} \frac{M}{EI}\mathrm{d}x. \qquad (28.6)$$

Setting $\left.\dfrac{\mathrm{d}y}{\mathrm{d}x}\right)_B = \theta_B$, $\left.\dfrac{\mathrm{d}y}{\mathrm{d}x}\right)_A = 0$, and $M\ (0 \leqslant x \leqslant 1\,\text{m} = M_A$ then gives

$$\theta_B = \frac{(F \times 0.4\,\text{m} - M_B)1\,\text{m}}{EI}. \qquad (28.7)$$

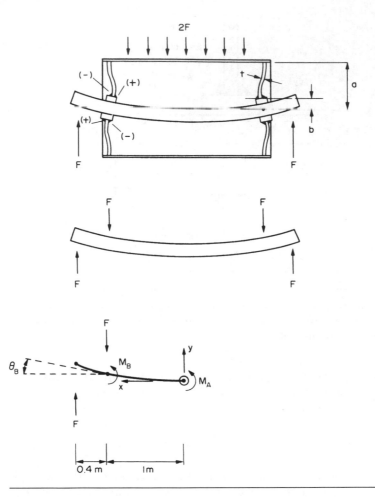

Figure 28.8   Exaggerated drawing of the deflections that occur in the loaded drum. The shaft deflects under four-point loading. This in turn causes the end plates to deflect out of plane, creating tensile (+) and compressive (−) stresses in the weld.

We can now turn our attention to the end plate. The standard formulae are

$$\sigma_{max} = \frac{\beta M_B}{a t^2} \tag{28.8}$$

and

$$\theta_B = \frac{\alpha M_B}{E t^3} \tag{28.9}$$

where σ_{max} is the maximum stress at the inner circumference of the plate, a is the outer radius of the plate and t is the plate thickness. β and α are dimensionless constants whose value depends on the ratio of the inner to the outer radius, b/a. Equations (28.7), (28.8) and (28.9) can be combined to give

$$\sigma_{max} = \frac{\beta(F \times 0.4\,\text{m} \times 1\,\text{m})t}{a(\alpha I + t^3 \times 1\,\text{m})}. \tag{28.10}$$

Values for α and β can be found from standard tables: for our value of b/a of 0.31 they are 0.46 and 2.8 respectively. Equation (28.10) then tells us that

$$\sigma_{max} = 142\,\text{MPa} \tag{28.11}$$

for our current values of $F = 5000 \times 9.81\,\text{N}$, $t = 20\,\text{mm}$, $a = 400\,\text{mm}$ and $I = (\pi/4)(75\,\text{mm})^4$.

This stress gives a fatigue range of 284 MPa which is about five times the limiting range for an E/F weld. This is obviously a case of going back to the drawing board! In fact, as Table 28.2 shows, the solution to this design problem is far from straightforward.

Table 28.2   Design modifications to overcome over-stressing of collar-plate weld

Modification	Comments
Make end plate thicker (increase t in eqn. 28.10).	Plate would have to be 70 mm thick: heavy; difficult to weld to collar; might overstress plate-shell weld; collar would become irrelevant.
Make end plate thinner (decrease t in eqn. 28.10).	Plate would have to be 2 mm thick: too thin to carry normal loading without risk of buckling.
Increase shaft diameter and decrease end plate thickness.	Increases I and decreases t in eqn. 28.10. Would need 250 mm diameter shaft + 15 mm thick end plate: very heavy and expensive. Would still need to check plate-shell weld.
Remove weld altogether. Make plate a running fit on shaft and drive it using splines.	Allows shaft to flex without bending plate. Continual flexing will wear the contacting surfaces and may lead to fretting fatigue.
Make shaft stiffer by supporting it with intermediate plates welded to both shaft and shell.	Alignment and assembly problems and higher cost.
Decrease shaft overhang – move bearings next to end plates.	Requires major modifications to support framing or over-long shells.
Use fixed (non-rotating) shaft and mount bearings in end plates.	Probably the best solution, but no good for drive drums.

Conclusions

Structural steelwork – from small items like our conveyor drums to large structures like ships and bridges – is usually fabricated by welding. It is relatively easy to design structures so that the *parent material* has an adequate resistance to elastic deformation, plastic collapse, fast fracture or fatigue. But the *welds* – with their stress concentrations, residual stresses, microstructural variations and hidden defects – are the Achilles' heel of any design. It has long been known that welds can be potent initiators for fast fracture. But it is less commonly known that most welds have abysmal fatigue strengths. Designers beware!

2. DESIGNING WITH CERAMICS: ICE FORCES ON OFFSHORE STRUCTURES

Introduction

To recover oil from the continental shelf of arctic Canada and Alaska, drilling and production platforms must be built some miles offshore, in roughly 40 m of water. This is not a great depth, and would present no new problems were it not that the sea freezes in winter to a depth of around 2 m. Wind blowing across the surface of the ice sheet causes it to move at speeds up to $1 \, \mathrm{m \, s^{-1}}$, pressing it against the structure. Ice is a ceramic. Like all ceramics it is weak in tension, but strong in compression. So the structure must be designed to withstand large ice forces.

Two possible structures are shown in Fig. 28.9. The first is a monopod: a slender pillar with a broad foot, presenting a small section (perhaps 10 m wide) at the water surface. The second (and favoured) design is a gravel island, with a width of 100 to 200 m. In both cases it is essential to compute the maximum force the ice can exert on the structure, and to design the structure to withstand it. We are concerned here with the first problem: the ice force.

Material properties of ice

Winter temperatures in the arctic range between $-50°C$ and $-4°C$. Expressed as a fraction of the melting point T_m of sea ice, these correspond to the range 0.82 to 0.99 T_m.

Above 0.5 T_m ceramics creep in exactly the same way that metals do. The strain-rate increases as a power of the stress. At steady state (see Chapter 17, eqn. 6) this rate is

$$\dot{\varepsilon}_{ss} = A\sigma^n \exp(-Q/RT). \qquad (28.12)$$

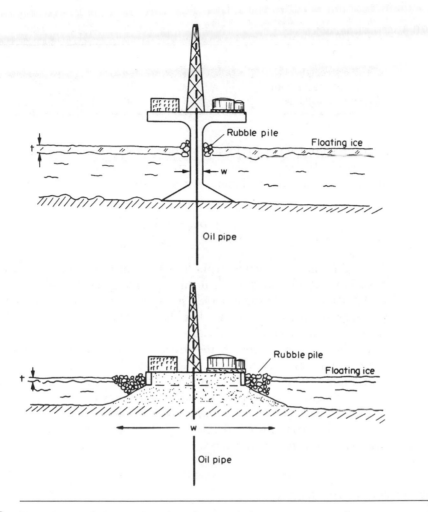

Figure 28.9 Two alternative designs for oil production platforms in ice-covered sea: a monopod and an artificial island.

When the ice moves slowly it creeps around the structure. The greater the velocity, the greater is the strain-rate of the ice as it deforms around the structure, and (from eqn. 28.12) the greater is the stress, and thus the load on the structure. But ice, like all ceramics, is a brittle solid. If deformed too fast, grain-size cracks nucleate throughout the body of ice, and these cracks propagate and link to give a crushing fracture (Chapter 17, Fig. 17.3). In practice, a *rubble pile* of broken ice blocks (Fig. 28.9) develops around test structures and natural and artificial islands in ice-covered seas, showing that the ice velocity is usually large enough to cause crushing. Then it is the crushing strength (not the creep strength) of the ice which determines the force on the structure.

Table 28.3   Properties of ice

Modulus	E (GPa)	9.1
Tensile strength	σ_{TS} (MPa)	1
Compressive strength	σ_c (MPa)	6
Fracture toughness	K_c (MPa m$^{1/2}$)	0.12

Material properties for ice are listed in Table 28.3. The fracture toughness is low (0.12 MPa m$^{1/2}$). The tensile strength of ice is simply the stress required to propagate a crack of dimensions equal to the grain size d, with

$$\sigma_{TS} = \frac{K_c}{\sqrt{\pi d/2}}. \tag{28.13}$$

Natural ice has large grains: typically 10 mm or more (it is like a casting which has solidified very slowly). Then this equation gives a tensile fracture strength of 1 MPa – precisely the observed strength. So when ice is loaded in tension, it creeps when the stresses are less than 1 MPa; the strength is limited to a maximum of 1 MPa by fast fracture.

In compression, of course, the strength is greater. Most ceramics are about fifteen times stronger in compression than in tension, for the reasons given in Chapter 17. For ice the factor is smaller, typically six, probably because the coefficient of friction across the crack faces (which rub together when the ceramic is loaded in compression) is exceptionally low. At stresses below 6 MPa, ice loaded in compression deforms by creep; at 6 MPa it crushes, and this is the maximum stress it can carry.

Forces on the structures

It is now easy to calculate the force on the narrow structure. If the pillar has a width $w = 10$ m where it passes through the ice sheet (thickness $t = 2$ m), it presents a section of roughly 20 m^2 on which ice presses. The maximum stress the ice can take is 6 MPa, so the maximum force it can exert on the structure is

$$F = tw\sigma_c = 120 \text{ MN (12,000 tonne).} \tag{28.14}$$

This is a large force – enough to demolish a structure which was merely designed to withstand large waves. Lighthouses have been lost in northern Sweden and arctic Canada in just this way.

The ice force on the large structure with a width w of 100–200 m would just seem, at first sight, to be that for the narrow structure, scaled up by the width. But gravel islands exist which (it is now known) could not take such a large force without shearing off, yet they have survived. The ice force on a large structure, evidently, is less than the simple crushing–strength argument predicts.

The force on the larger structure is less than expected because of *non-simultaneous failure*. As the ice pushes against the structure one heavily loaded corner of the ice breaks off (forming rubble). This allows others to come into contact with the structure, take up the load, and in turn break off (Fig. 28.10). The process involves a sequence of fractures and fragment movements, occurring at different points along the contact zone. You can reproduce the same effect by pressing a piece of rye crispbread, laid flat on a table, against the edge of an inverted coffee cup. First one fragment breaks away, then another, then a third. The contact remains jagged and irregular. The load on the structure (or the coffee cup) fluctuates violently, depending on how many contacts there are at any given moment. But it is never as large as two_c (the section times the crushing strength) because the brittle ice (or crispbread) never touches the structure (or the cup) simultaneously along its entire length.

This is another feature of brittleness. If the ice were ductile, point contacts would bed down by plastic flow until the ice sheet pressed uniformly along the entire contact zone. But because it is brittle some bits break off, leaving a gap, before other bits have yet come into contact. The force on the structure now

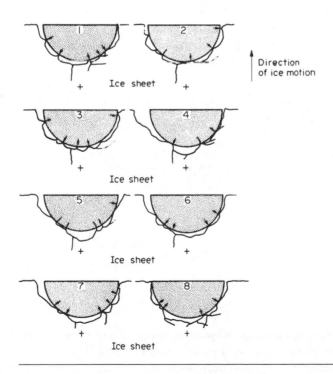

Figure 28.10   Successive contact profiles of a brittle sheet moving against a rigid structure. At any instant, contact is only made at discrete places.

depends on *the probability of contact*, that measures the number of true contacts per unit length of the contact zone. In Fig. 28.10 there are, on average, five contacts; though of course the number (and thus the total force) fluctuates as the brittle sheet advances. The peak force depends on the brittle process by which chunks break away from the ice sheet, since this determines the jagged shape and the number of contacts (it is probably a local bending-fracture). The point is that it is the *local*, not the *average*, strength which is important. It is a "weakest link" problem again: there are five contact points; the failure of the weakest one lets the entire ice-sheet advance. The statistics of "weakest link" problems was developed in Chapter 18. This problem is the same, except that the nominal area of contact

$$A = tw \tag{28.15}$$

replaces the volume V. Then, by analogy with eqn. (18.9), the probability of survival is given by

$$P_s(A) = \exp\left\{-\frac{A}{A_0}\left(\frac{\sigma}{\sigma_0}\right)^m\right\} \tag{28.16}$$

where σ and σ_0 are nominal section stresses. Data for the crushing of thin ice sheets by indenters of varying contact section A fit this equation well, with $m = 2.5$.

We can now examine the effect of structure size on the ice force. If the ice is to have the same survival probability at the edge of each structure then

$$\exp\left\{-\frac{tw_1}{A_0}\left(\frac{\sigma_1}{\sigma_0}\right)^m\right\} = \exp\left\{-\frac{tw_2}{A_0}\left(\frac{\sigma_2}{\sigma_0}\right)^m\right\} \tag{28.17}$$

or

$$\left(\frac{\sigma_1}{\sigma_2}\right) = \left(\frac{w_2}{w_1}\right)^{1/m}. \tag{28.18}$$

Thus

$$\sigma_1 = 6\,\text{MPa}\left(\frac{10\,\text{m}}{200\,\text{m}}\right)^{1/2.5}$$

$$= 1.8\,\text{MPa.} \tag{28.19}$$

These lower strengths give lower ice forces, and this knowledge is of immense importance in designing economic structures. Building large structures in the arctic, under difficult conditions and large distances from the source of raw materials, is very expensive. A reduction by a factor of 3 in the design load the structures must withstand could be worth $100,000,000.

3. Designing with polymers: a plastic wheel

Introduction

Small rotating parts – toy wheels, gears, pulleys – have been made of plastic for some years. Recently the use of polymers in a far more demanding application – wheels for bicycles, motorcycles and even automobiles – is receiving serious consideration. The design of heavily loaded plastic wheels requires careful thought, which must include properly the special properties of polymers. But it also presents new possibilities, including that of a self-sprung wheel. This case study illustrates many general points about designing with polymers.

Mould design

Polymers are not as stiff as metals, so sections have to be thicker. The first rule of mould design is to aim for *a uniform section throughout the component*. During moulding, hot polymer is injected or pressed into the mould. Solidification proceeds from the outside in.

Abrupt changes of section cause poor flow and differential shrinkage, giving sink marks (Fig. 28.11 – you can find them on the surface of many small polymer parts), distortion, and internal stress which can lead to cracks or voids. The way out is to design in the way illustrated in Fig. 28.12. Ribs, which are often needed to stiffen polymer parts, should have a thickness of no more than two-thirds of the wall thickness, and a height no more than three times the wall thickness. Corners are profiled to give a uniform section round the corner.

Polymers have a low fracture toughness. This is not as great a problem as it sounds: their moduli are also low, so (to avoid excessive deflections) the designer automatically keeps the stresses much lower than in a metal part. But at sharp corners or sudden changes of section there are stress concentrations, and here the (local) stress can be ten or even a hundred times greater than

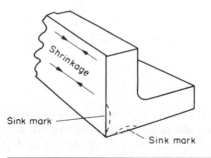

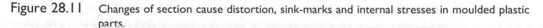

Figure 28.11   Changes of section cause distortion, sink-marks and internal stresses in moulded plastic parts.

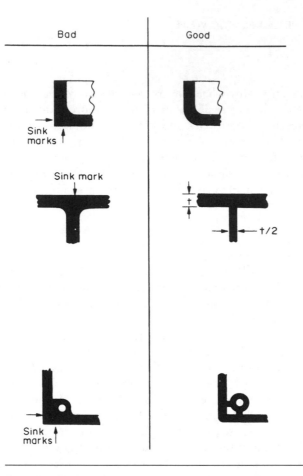

Figure 28.12   Example of good and bad design with polymers.

the nominal (average) level. So the second rule of mould design is to *avoid stress concentrations*. Figure 28.13 shows an example: a cantilever beam with a change in section which has local radius R. The general rule for estimating stress concentrations is that the local stress is bigger than the nominal stress by the factor

$$S \approx 1 + \left(\frac{t}{R}\right)^{1/2} \tag{28.20}$$

where t is the lesser of the section loss (t_2) and the section thickness (t_1). As the fillet becomes sharper, the local stress rises above that for a uniform beam, and these local stresses cause fracture. The rule-of-thumb is that radii should never be less than 0.5 mm.

The allowable dimensional variation (the *tolerance*) of a polymer part can be larger than one made of metal – and specifying moulds with needlessly

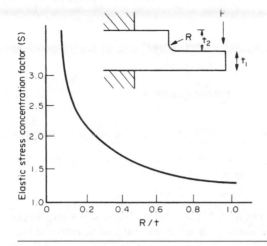

Figure 28.13 A sharp change of section induces stress concentrations. The local stress can be many times greater than the nominal stress.

high tolerance raises costs greatly. This latitude is possible because of the low modulus: the resilience of the components allows elastic deflections to accommodate misfitting parts. And the thermal expansion of polymers is almost ten times greater than metals; there is no point in specifying dimensions to a tolerance which exceeds the thermal strains.

Web and spoke design

From a moulder's viewpoint, then, the ideal wheel has an almost constant wall thickness, and no sharp corners. In place of spokes, the area between the hub and the rim can be made a solid web to give even flow of polymer to the rim and to avoid weld lines (places where polymer flowing down one channel meets polymer which has flowed down another). The wheel shown in Fig. 28.14 typifies this type of design. It replaces a die-cast part, saving both cost and weight. While the web is solid, axial stability is provided by the corrugated surface. The web is a better solution than ribs attached to a flat disc because the changes in section where the ribs meet the disc give shrinkage problems.

But in a large wheel (like that for a bicycle) spokes may be unavoidable: a solid web would be too heavy and would give too much lateral wind pressure. A design with few spokes leaves relatively large sectors of the rim unsupported. As the wheel rotates, the contact load where it touches the ground will distort the rim between spokes, but not beneath a spoke, giving a bumpy ride, and possible fatigue failure of the rim. Doubling the number of spokes reduces the rim deflection (since it is just a bending beam) by a factor of 8; and, of course,

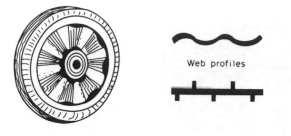

Figure 28.14   A design for a small plastic wheel. The upper, corrugated, web profile is better than the lower, ribbed, one because the section is constant.

the spokes must be ribbed in a way which gives a constant thickness, equal to the rim thickness; and the spokes must be contoured into the rim and the hub to avoid stress concentrations.

Rim design

Polymers, loaded for a long time, suffer creep. If a pneumatic tyre is used, the rim will be under constant pressure and the effect of creep on rim geometry must be taken into account. The outward force per unit length, F, exerted on the rim (Fig. 28.15) is roughly the product of the pressure p in the tyre and the radius of the tyre cross-section, R. This force creates bending stresses in the rim section which can be minimised by keeping the rim height H as small as possible: the left-hand design of Fig. 28.15 is poor; the right-hand design is better. Radial ribbing, added as shown, further stiffens the rim without substantially affecting the cross-section wall thickness.

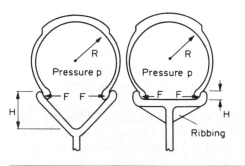

Figure 28.15   Rim design. The left-hand design is poor because the large bending moments will distort the rim by creep. The right-hand design is better: the bending moments are less and the ribs stiffen the rim.

Innovative design

Polymers have some obvious advantages for wheels. The wheel can be moulded in one operation, replacing a metal wheel which must be assembled from parts. It requires no further finishing, plating or painting. And its naturally low coefficient of friction means that, when loads are low, the axle may run on the polymer itself.

But a polymer should never be regarded simply as an inexpensive substitute for a metal. Its properties differ in fundamental ways – most notably its modulus is far lower. Metal wheels are designed as rigid structures: it is assumed that their elastic deflection under load is negligible. And – thus far – we have approached the design of a polymer wheel by assuming that it, too, should be rigid.

A good designer will think more broadly than this, seeking to exploit the special properties of materials. Is there some way to use the low modulus and large elastic deflections of the polymer? Could (for instance) the deflection of the spoke under radial load be made to compensate for the deflection of the rim, giving a smooth, sprung, ride, with fewer spokes? One possible resolution of such an assignment, for a bicycle wheel, is shown in Fig. 28.16. Each arc-shaped spoke is designed to deflect by the maximum amount acceptable to the wheel manufacturer (say 2.5 mm) when it is vertical, carrying a direct load. Of course, when two of the spokes straddle the vertical and share the load, the deflection would be only one-half this amount (1.3 mm). The cross-section of the rim could then be sized to deflect as a beam between spokes, providing the extra 1.3 mm of deflection necessary for a smooth ride. If the stresses are kept well below yield, permanent deflection and fatigue should not be a problem.

Stiffness, low creep and maximum toughness, are requirements for this application. Table 21.5 shows that nylon and polypropylene are tougher than most other polymers. The modulus of polypropylene is low, roughly one-third that of nylon; and its glass temperature is low also, so that it will creep more than nylon. PMMA has a modulus and glass temperature comparable with

Figure 28.16   Tentative design for a self-sprung wheel. The spokes and the rim both deflect. The design ensures that the sum of the deflections is constant.

nylon, but is much less tough. The best choice is nylon, a thermoplastic, which can conveniently be moulded or hot-pressed. And that is the choice of most manufacturers of wheels of this sort. To improve on this it would be necessary to use a composite (Chapter 25) and then the entire design must be rethought, taking account of the strengths and weaknesses of this material.

4. DESIGNING WITH COMPOSITES: MATERIALS FOR VIOLIN BODIES

Introduction

The violin (Fig. 28.17) is a member of the family of musical instruments which we call "string" instruments. Table 28.4 shows just how many different types of string instruments there are of European origin alone – not to mention the fascinating range that we can find in African, Asian or Oriental cultures.

String instruments all work on the same basic principle. A thin string, of gut or metal, is stretched tightly between two rigid supports. If the string is plucked, or hit, or bowed, it will go into sideways vibrations of precisely defined frequency which can be used as musical notes. But a string vibrating on its own can hardly be heard – the sideways-moving string cuts through the

Figure 28.17   A young child practises her violin.

Table 28.4   Some European string instruments

Violin family	Viol family	Guitar family	Harp family	Keyboard family
Violin	Treble	Guitar	Modern harp	Piano
Viola	Tenor	Lute	Folk harps	Harpsichord
Cello	Bass	Zither		Clavichord
Double bass		Balalaika		

air as a cheesewire cuts through cheese and the pressure wave that reaches our eardrums has scarcely any amplitude at all. For this reason all string instruments have a *soundboard*. This is forced into vibration by the strings and radiates strong pressure waves which can be heard easily.

But soundboards are much more than just radiating surfaces. They have their own natural frequencies of vibration and will respond much better to notes that fall within the resonance peaks than notes which fall outside. The soundboard acts rather like a selective amplifier, taking in the signal from the string and radiating a highly modified output; and, as such, it has a profound effect on the tone quality of the instrument.

Soundboards are traditionally made from wood. The leading violin makers can work wonders with this material. By taking a thin plate of wood and hollowing it out here and there they can obtain a reasonably even response over a frequency range from 200 to 5000 Hz. But the process of adjusting a soundboard is so delicate that a trained listener can tell that 0.1 mm has been removed from an area of the soundboard measuring only 20 mm × 20 mm.

Such skills are rare, and most students today can only afford rather indifferent mass-produced instruments. But the music trade is big business and there is a powerful incentive for improving the quality of the mass-market product.

One obvious way of making better violins would be to dismantle a fine instrument, make accurate thickness measurements over the whole of the soundboard, and mass-produce sound-boards to this pattern using computer-controlled machine tools. But there is a problem with this approach: because wood is a natural material, soundboard blanks will differ from one another to begin with, and this variability will be carried through to the finished product. But we might be able to make a good violin every time if we could replace wood by a synthetic material having reproducible properties. This case study, then, looks at how we might design an artificial material that will reproduce the acoustically important properties of wood as closely as possible.

Soundboard vibrations

In order to design a replacement for wood we need to look at the vibrational behaviour of soundboards in more detail. As Fig. 28.17 shows, the soundboard of a violin has quite an ornate shape and it is extremely difficult to analyse its

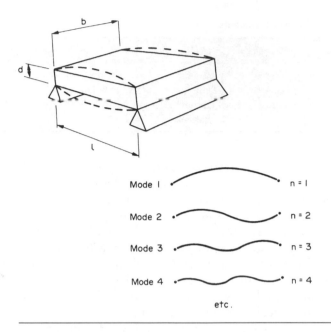

Mode 1   n = 1

Mode 2   n = 2

Mode 3   n = 3

Mode 4   n = 4

etc.

Figure 28.18   Idealized vibration modes of a soundboard. The natural frequencies of the modes are proportional to n^2.

behaviour mathematically. But an adequate approximation for our purposes is to regard the soundboard as a rectangular panel simply supported along two opposing edges and vibrating from side to side as shown in Fig. 28.18.

The natural frequencies are then given by

$$f_n = n^2 \left(\frac{\pi}{2l^2} \right) \left(\frac{EI}{\rho bd} \right)^{1/2} \tag{28.21}$$

where E is Young's modulus, $I\,(= bd^3/12)$ is the second moment of area of the section, and ρ is the density of the soundboard material. We get the lowest natural frequency when the panel vibrates in the simplest possible way (see Fig. 28.18 with $n = 1$). For more complex vibrations, with mode numbers of 2, 3, 4 and so on, f scales as the square of the mode number.

Making the soundboard out of wood introduces a complication. As we can see in Chapter 26, wood has a much bigger modulus along the grain than across the grain. A wooden soundboard therefore has both along-grain and across-grain vibrations (Fig. 28.19).

The frequencies of these vibrations are then

$$f_\parallel = n^2 \left(\frac{\pi}{2l_\parallel^2} \right) \left(\frac{E_{w\parallel} I_w}{\rho_w b_\parallel d_w} \right)^{1/2} \tag{28.22}$$

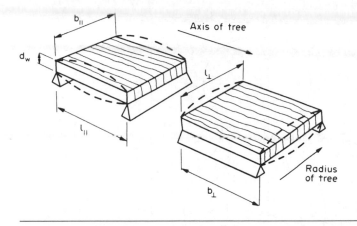

Figure 28.19   A wooden soundboard has both along-grain and across-grain vibrations. Although not shown here, both types of vibration have a full set of modes, with $n = 1, 2, 3 \ldots$

and

$$f_\perp = n^2 \left(\frac{\pi}{2l_\perp^2} \right) \left(\frac{E_{w\perp} I_w}{\rho_w b_\perp d_w} \right)^{1/2} \tag{28.23}$$

where $E_{w\parallel}$ is the axial modulus and $E_{w\perp}$ is the radial modulus.

In order to estimate the soundboard frequencies, we set $d_w = 3\,\text{mm}$, $l_\parallel = 356\,\text{mm}$, and $l_\perp = 93$ or $123\,\text{mm}$ (Fig. 28.20). Violin soundboards are usually made from spruce, which typically has $E_{w\parallel} = 11.6\,\text{GPa}$, $E_{w\perp} = 0.71\,\text{GPa}$ and $\rho_w = 0.39\,\text{Mg m}^{-3}$. Equations (28.22) and (28.23) then give us the following data.

$$f_\parallel(\text{s}^{-1}) \approx 59 \quad 236 \quad 531 \quad 944 \quad 1475 \quad 2124 \quad 2891 \quad \text{etc.}$$
$$f_\perp(\text{S}^{-1}) \approx \begin{cases} 217 & 868 & 1953 & 3472 & \text{etc.} \\ 124 & 496 & 1116 & 1984 & 3100 \quad \text{etc.} \end{cases}$$

For all the crudity of our calculations, these results show clearly that wooden violin soundboards have an impressive number of natural frequencies.

Replacement materials

Spruce soundboards have a Young's modulus anisotropy of about $(11.6\,\text{GPa}/0.71\,\text{GPa}) = 16$. A replacement material must therefore have a similar anisotropy. This requirement immediately narrows the choice down to composites (isotropic materials like metals or polymers will probably sound awful).

Because it is fairly cheap, we should begin by looking at GFRP. The moduli of glass fibres and resin matrices are $72\,\text{GPa}$ and $3\,\text{GPa}$ respectively, giving us

$$E_{c\parallel} = 72 V_f + 3(1 - V_f) \tag{28.24}$$

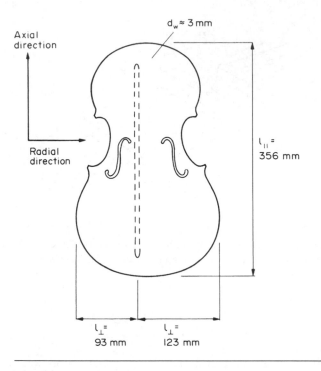

Axial direction

$d_w \approx 3\,mm$

Radial direction

$l_{||} = 356\,mm$

$l_\perp = 93\,mm$

$l_\perp = 123\,mm$

Figure 28.20   Dimensions of a violin soundboard. The dashed outline marks the position of the *bass bar* – a wooden "girder" stuck underneath the soundboard in an off-centre position. This effectively divides the soundboard into two different across-grain regions.

and

$$E_{c\perp} = \left(\frac{V_f}{72} + \frac{1-V_f}{3} \right)^{-1}. \tag{28.25}$$

Now, as Fig. 28.21 shows, $E_{c\parallel}/E_{c\perp}$ can never be greater than 6.6 for GFRP; and this is far too small to give a useful soundboard.

We must therefore hope that CFRP can give us the required anisotropy. The modulus of type-1 carbon fibres is 390 GPa along the fibre axis (although it is only 12 GPa at right angles to this). So

$$E_{c\parallel} = 390V_f + 3(1-V_f) \tag{28.26}$$

and

$$E_{c\perp} = \left(\frac{V_f}{12} + \frac{1-V_f}{3} \right)^{-1}. \tag{28.27}$$

This gives a maximum ratio of 43, which is more than enough. In fact, to bring the anisotropy down to the target figure of 16 we need to reduce the volume fraction of fibres to only 0.13!

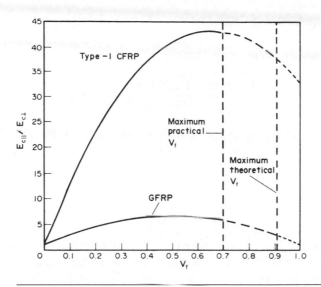

Figure 28.21 Modulus anisotropies $E_{c\|}/E_{c\perp}$ for aligned GFRP and CFRP composites.

Having matched $f_\|/f_\perp$ in this way we must now go on to match the frequencies themselves. We can see from eqn. (28.22) that this requires

$$n^2 \left(\frac{\pi}{2l_\|^3}\right) \left(\frac{E_{w\|}b_\| d_w^2}{\rho_w b_\| d_w 12}\right)^{1/2} = n^2 \left(\frac{\pi}{2l_\|^2}\right) \left(\frac{E_{c\|}b_\| d_c^3}{\rho_c b_\| d_c 12}\right)^{1/2} \tag{28.28}$$

which tells us that the composite plate must have a thickness of

$$d_c = d_w \left(\frac{E_{w\|}\rho_c}{E_{c\|}\rho_w}\right)^{1/2}. \tag{28.29}$$

Now we already know three of the terms in this equation: d_w is 3 mm, $E_{w\|}$ is 11.6 GPa and ρ_w is 0.39 Mg m^{-3}. The last two unknowns, $E_{c\|}$ and ρ_c, can be calculated very easily. When $V_f = 0.13$ eqn. (28.24) tells us that $E_{c\|} = 53$ GPa. ρ_c can be found from the simple rule of mixtures for density, with

$$\rho_c = (0.13 \times 1.9 + 0.87 \times 1.15) \, \text{Mg m}^{-3}$$
$$= 1.25 \, \text{Mg m}^{-3}. \tag{28.30}$$

Equation (28.29) then tells us that $d_c = 2.52$ mm.

This value for the thickness of the substitute CFRP soundboard sets us a difficult problem. For the *mass* of the CFRP plate is greater than that of the spruce plate by the factor

$$\frac{2.52 \, \text{mm} \times 1.25 \, \text{Mg m}^{-3}}{3 \, \text{mm} \times 0.39 \, \text{Mg m}^{-3}} = 2.69. \tag{28.31}$$

This excessive mass is quite unacceptable – the CFRP plate will need far too much energy to get it vibrating. But how can we reduce the mass without shifting the vibration frequency?

Sectional composites

The solution that has been adopted by makers of composite soundboards is to fabricate a sandwich structure where a layer of high-quality cardboard is glued between two identical layers of CFRP (Fig. 28.22). The philosophy of this design modification is to replace some CFRP by a much lighter material in those regions that contribute least to the overall stiffness of the section.

The density of the cardboard layer is around $0.2 \, \text{Mg m}^{-3}$. To match the mass of the sandwich to that of the spruce soundboard we must then have

$$0.39 d_w = 1.25(d_2 - d_1) + 0.2 d_1 \tag{28.32}$$

or

$$d_w = 3.21 d_2 - 2.69 d_1. \tag{28.33}$$

In order to formulate the criterion for frequency matching we can make the very reasonable assumption that the CFRP dominates the stiffness of the sandwich section. We also simplify eqn. (28.22) to

$$f_\parallel = n^2 \left(\frac{\pi}{2 l_\parallel^{3/2}} \right) \left(\frac{E_{w\parallel} I_w}{m_w} \right)^{1/2} \tag{28.34}$$

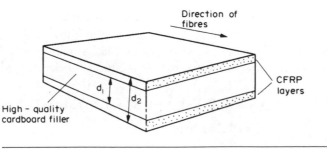

Figure 28.22   Sandwich-type sectional composites give a much-improved stiffness-to-mass ratio.

where $m_w = \rho_w b_\| l_\| d_w$ is the total mass of the wooden soundboard. Then, because we want $m_w = m_{\text{sandwich}}$, frequency matching requires

$$E_{w\|} I_w = E_{c\|} I_c \tag{28.35}$$

or

$$E_{w\|} \frac{b_\| d_w^3}{12} = E_{c\|} b_\| \frac{(d_2^3 - d_1^3)}{12}. \tag{28.36}$$

We know that $E_{c\|}/E_{w\|} = 53/11.6 = 4.6$ so that eqn. (28.36) reduces to

$$d_w^3 = 4.6(d_2^3 - d_1^3). \tag{28.37}$$

Finally, combining eqns (28.33) and (28.37) gives us

$$(3.21 d_2 - 2.69 d_1)^3 = 4.6(d_2^3 - d_1^3). \tag{28.38}$$

This result can be solved numerically to give $d_1 = 0.63 d_2$; and, using eqn. (28.33), we can then show that $d_2 = 0.66 d_w$.

Conclusions

This design study has shown that it is possible to design a sectional composite that will reproduce both the vibrational frequencies and the mass of a traditional wooden soundboard. For a soundboard made out of spruce the equivalent composite is a sandwich of cardboard glued between two identical layers of aligned CFRP with a fibre volume fraction of 0.13. If the wooden soundboard is 3 mm thick the replacement composite must be 1.98 mm thick with a cardboard core of 1.25 mm.

Chapter 29
Engineering failures and disasters – the ultimate test of design

Introduction

As we have seen in our two books *Engineering Materials 1* and *Engineering Materials 2*, engineering materials can fail by a wide range of mechanisms (conveniently divided into elastic deformation, plastic deformation, ductile fracture, fast fracture, brittle fracture, fatigue, creep, environmental attack and wear). The critical failure process depends on the properties of the material (strongly determined by whether it is a metal, a ceramic, a polymer, or a composite) and also – most importantly – on the way in which the material is used in the final component or structure. Although all failures should be avoidable by applying existing knowledge properly, life in the engineering world (as in life in general) is rarely that simple.

Until the nineteenth century, all design was essentially empirical, based on a process of trial and error and intuition, and not backed up by any proper calculations of collapse loads. As might be expected, major structural collapses were not uncommon, especially in the case of the more adventurous buildings. The wonderful thirteenth-century cathedral at Ely, near Cambridge (England), is a good example. Shortly after completion, the main central tower (made of stone) collapsed under its huge self-weight, and had to be replaced by a much lighter timber structure (which survives to this day – although modern structural analysis has shown that this, too, would probably have collapsed if it had been built only slightly differently). Earlier (and generally more monolithic) structures, such as the Egyptian and Aztec temples, have (predictably) fared much better.

Now that engineers are able to design everything using rigorous numerical methods (at least in principle), one would expect engineering failures and collapses to be a thing of the past. The fact that this is far from the case is due to a number of factors. One is the constant temptation to push the "design envelope" as far as possible, driven by the demands of scale and cost (and encouraged by the natural tendency of engineers and architects constantly to challenge what has gone before). Then there is a danger that the technology can run ahead of the theoretical understanding. This was graphically illustrated with the new generation of lightweight long-span box-girder bridges in the 1960s, which led to two spectacular collapses – the river Cleddau crossing near Milford Haven in Wales (June 1970, five deaths) and the river Rhine crossing at Koblenz in Germany (November 1971, 12 deaths).

Another factor is failure on the part of the manufacturer or fabricator to comply with all the details of the engineering drawings and specifications. A classic example of poor fabrication led directly to the collapse of one of the spans of the Sungsoo Grand Bridge in Seoul, Korea (October 1994, 32 deaths). In this case, one of only four links from which the whole span was suspended failed due to fatigue from a grossly substandard weld. However, one can argue that in this case – as in many others – the designers should have taken more care to identify the critical load-bearing details, should have supervised the welding operations directly, and should have insisted on rigorous non-destructive flaw detection after the welds had been made.

One of the most common factors leading to failures and disasters is poor management. This may be on a global scale (for example, failure to transfer technology from one area of engineering to another) or on a local scale (for example, failure to pass on vital information, or incompetent design work which is not discovered). As an example of the first, bridge designers (and designers of other large structures such as cranes and excavators) have known since the 1960s that welds possess poor fatigue properties (and have designed accordingly), whereas many mechanical engineers remain ignorant of the weld fatigue codes (so that fatigue failures are still common in rotating machinery fabricated by welding). As an example of the second, one need look no further than the disaster to the space shuttle Challenger on 28 January 1986. This was caused by a faulty O-ring seal in one of the two solid rocket boosters, which started to leak flame on the launch pad. Shortly after launch, the flames spread to all three rockets, which then exploded. The manufacturer of the boosters (Moreton Thiokol, Utah) knew that the O-rings would suffer serious loss of sealing capability at low temperature, yet the launch went ahead on a freezing cold day in spite of the reservations of Moreton's engineering team – because they could not actually *prove* that the seals would fail, their directors (subject to huge commercial and political pressures) lacked the courage to play safe.

It should now be clear why learning from past mistakes is an essential part of safe and efficient design. However, enough of generalities – the best way to appreciate what can be learned from engineering failures is to examine case studies of real accidents and disasters. We give four detailed examples below – a classic bridge collapse, a classic air crash, a recent railway disaster, and a fatal bungee jumping accident (which will also have immediate relevance for some of our readers).

Case study 1: the Tay Bridge railway disaster – 28 December 1879

North of Edinburgh (Scotland) lie two wide tidal estuaries, the Firth of Forth and the Firth of Tay. These presented a major barrier to communications up the east coast of Scotland, and an Act of Parliament was passed in 1870 authorising the construction of a single-track railway bridge over the Tay. Construction was begun in May 1871, and the bridge was opened to traffic in

May 1878. The designer was the freelance civil engineering consultant Thomas Bouch.

A photograph of the bridge is shown in Fig. 29.1. In order to allow shipping to pass up the Tay a clearance of 90 ft was required between the girders of the bridge and the high water mark in the middle of the firth. On the south shore of the firth, at Wormit, the land rose steeply to a height of about 200 ft and this provided an ideal jumping-off point for the bridge. After leaving the shore on a short curved alignment the track climbed gradually at 1 in 490 until it reached pier 29. It then ran level to pier 36. After passing pier 37 the track fell rapidly, at 1 in 74, until it reached the north shore at Dundee. At pier 53 the track entered a large, sweeping curve which took it in alongside the shore and down to a height of about 40 ft above high water. The overall length of the bridge was 10,300 feet, which at the time made it the longest iron bridge in the world.

The spans of the bridge were supported on a total of 85 piers. The first 14 were made from brick and looked fairly substantial. The rest, fabricated from iron, looked rather spindly in comparison. Over most of the bridge the track ran on top of the girders. But between piers 28 and 41 the construction was different. This was the place where the navigation channel lay and where the bridge had to have the full headroom of 90 ft. To get the extra height the piers were extended to bring their tops up to the level of the track. Thirteen spans of lattice-work box section were then placed end-to-end on top of the piers and the track was carried on the floor of the box. The 13 spans (11 of 245 ft and two of 227 ft) were aptly called the high girders.

Figure 29.2 shows the construction of the high girders and their supporting piers. Each pier consisted of six vertical columns of cast iron tied together with bracing bars made from wrought iron. A single column was built up from seven flanged cast-iron pipes bolted together through the flanges. The bracing bars were located in lugs, which were cast integrally with the flanges. The horizontal bars were made from channel-section iron. The diagonal tie-bars were made from iron flats with a cross-section measuring $4.5'' \times 0.5''$. Each tie was held by a single bolt at one end and a tapered cotter at the other. In theory this arrangement meant that each tie could be pulled up tight after assembly to make it take its share of the load. An original working drawing of the bracing arrangements is shown in Fig. 29.3.

Construction began in 1871 and the first service train ran on 31 May 1878. The morning of Sunday 28 December 1879 was quiet. When Captain Wright took his ferry boat, the *Dundee*, across the firth at 1.15 pm he noted that the weather was good and the water was calm. The 4.15 pm crossing was just as uneventful but the captain noted that the wind had freshened. By 5.15 pm a gale was moving in from the west and the river, in the words of the captain, "was getting up very fast". The local shuttle train left Newport at 5.50 pm and arrived in Dundee station shortly after six. The passengers had a worrying crossing. Their carriages were buffeted by the growing force of the storm and lines of sparks flew from the flanges of the wheels under the sideways force of the wind.

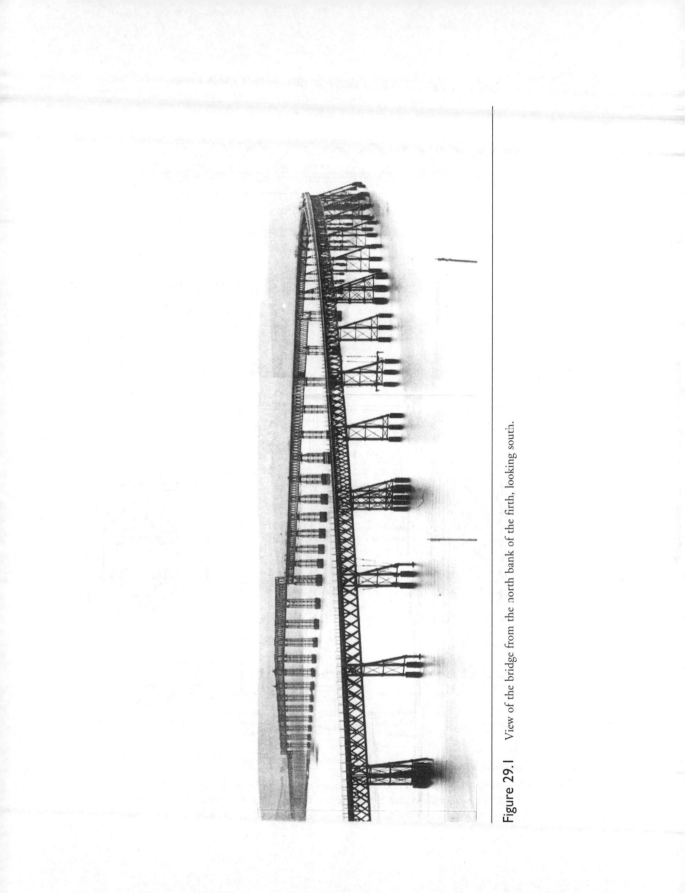

Figure 29.1 View of the bridge from the north bank of the firth, looking south.

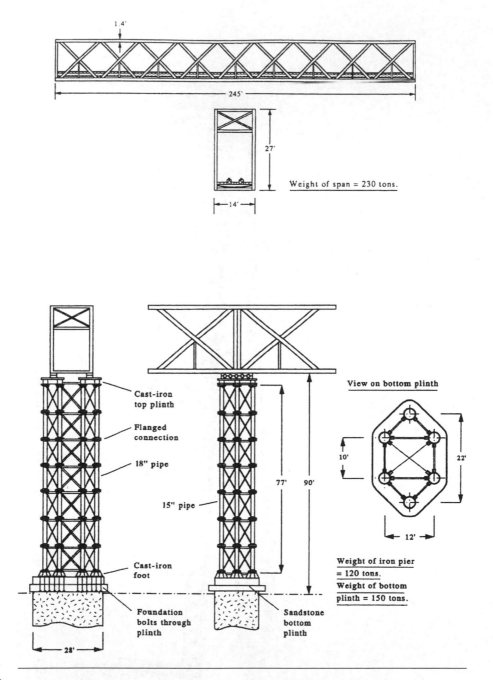

Figure 29.2   (Top) Simplified elevation and cross-section of a span of the high girders. (Bottom) Simplified elevations and plan of the piers in the high girders length of the bridge. Dimensions in feet and inches.

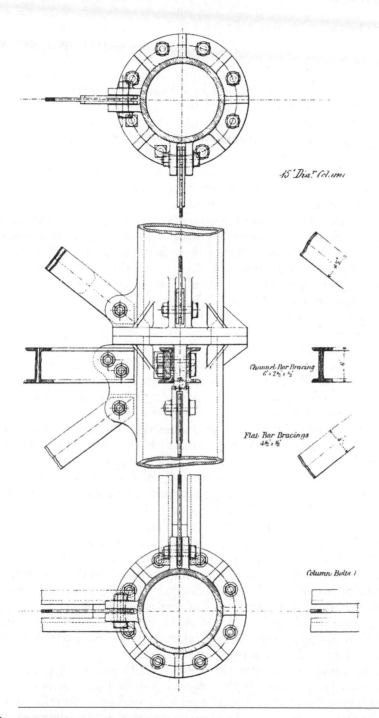

Figure 29.3   Details of the flanges and lugs of the cast-iron columns. Dimensions in inches.

The mail train from Edinburgh had left Burntisland at 5.20 pm and by the time the local had arrived in Dundee the mail had reached Thornton Junction, 27 miles south of the bridge. The last station was St Fort, which lay in a small depression 2 miles short of the bridge. The station staff collected the tickets of the passengers who were going on to Dundee. In addition to three men on the footplate of the locomotive there were 72 passengers. By 7.13 pm the train had reached the signal cabin at Wormit. The driver slowed the locomotive down to walking pace so his fireman could take the staff for the single line from the signalman. Then he opened up the regulator and took the train out onto the bridge and into the teeth of the westerly gale. The signalman returned to the shelter of his cabin and sent the "train entering section" signal to his opposite number in the signal box at Dundee.

The train receded into the darkness and the light of the three red tail lamps grew dimmer. Sparks flew from the wheels and merged into a continuous sheet that was dragged to the lee of the bridge by the wind. Eye witnesses saw a bright glow of light from the direction of the train just after it must have entered the high girders and then all went dark. The train was timed to pass the Dundee box at 7.19 pm. When it failed to arrive the signalman tried to telegraph the Wormit box but to no avail. The obvious conclusion was that the telegraph wires had been severed where they passed over the bridge. James Roberts, the locomotive foreman at the Dundee engine sheds, walked out along the bridge to investigate. Although at times he was forced to crawl on all fours by the force of the gale he eventually made his way to the end of the low-level girders. Further progress was impossible: the whole length of the high girders had disappeared into the river taking the train with it.

Photographs taken after the disaster (Fig. 29.4) show that the iron piers which had supported the high girders had fractured. Pier 1 had broken off at the flanges 22 ft above the base of the columns. The upper part of the pier had fallen into the river. In the lower part of the pier, which was still anchored to its masonry base, four of the ties were left hanging free. Pier 3 had broken away at the flanges 11 ft above the base. The remaining piers (numbers 2 and 4 to 12) had broken off where the columns were bolted to the base and had fallen into the river. Ties were also hanging loose from piers 28 and 41. Divers found that the high girders were lying on their sides on the river bed to the east of the bridge. Most of the train was lying inside the fifth span counting from the south end of the high girders. When the locomotive was raised from the river bed it was found that the regulator was fully open, the reversing lever was in the third notch from mid-gear and the brakes were fully off. This is the normal position for the controls when an engine is up to speed and the driver could have had no warning of the disaster.

As shown in Fig. 29.5, the lateral wind loading on high girders plus train generates tensile forces in the diagonal tie-bars. Estimates using modern wind-loading codes produce a maximum wind loading of 60 tons on a single pier. The pier is a statically indeterminate structure, which is most readily analysed by constructing a plastic mechanism in which both the flanged joints and

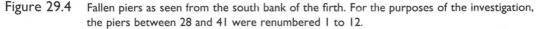

Figure 29.4   Fallen piers as seen from the south bank of the firth. For the purposes of the investigation, the piers between 28 and 41 were renumbered 1 to 12.

the diagonal ties are assumed to deform plastically. The wind load required to mobilise the pier plastically is estimated to be 100 tons. However, this is considerably more than the estimated wind force of 60 tons.

Experiments performed for the original court of inquiry found that the cast-iron lugs failed at about 24 tons. By comparison, a diagonal tie would have yielded at about 29 tons. Accordingly, the pier could not shake down plastically, and hence could not develop its full strength. Unequal load sharing between the diagonal ties was further exacerbated by the fact that the tapered cotters tended to work loose in service. The failure of the piers was caused by the fracture of the cast-iron lugs (see Fig. 29.6) at a wind force only 60% of that required to reach the plastic limit state.

The fracture area of a lug (see Fig. 29.6) is about 20 sq. in. Tests on specimens taken from the cast-iron columns gave a tensile strength of 9 tons/sq. in. This suggests that the lug should have been able to carry a tensile load of 180 tons. The discrepancy with the actual failure load of 24 tons can be explained by invoking a stress concentration factor of 7.5 at the perimeter of the bolt hole. Stress concentration factors of this magnitude have recently been calculated for bolted connections of similar geometry.

In conclusion, although the failure was primarily caused by a contemporary lack of knowledge of the effects of wind loading, from a materials standpoint the cast-iron lugs were totally inadequate. Because of the brittle fracture behaviour of cast iron, the lugs were unable to deform locally so as to relieve the large

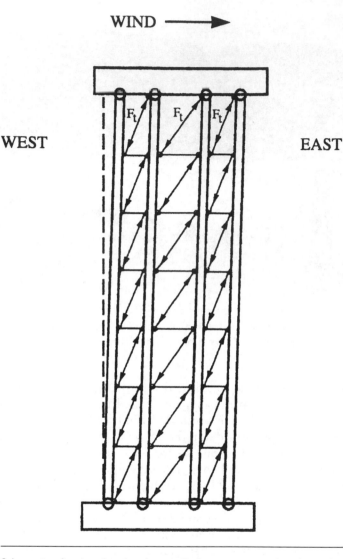

Figure 29.5   Schematic of a pier showing the tensile forces generated in the tie-bars by the lateral wind loading on the high girders plus train.

stress concentration at the perimeter of the bolt hole. Had a ductile material such as wrought iron been used instead, the lugs would have had a breaking load well in excess of the yield load of the diagonal ties, allowing the limit state to be achieved. However, we should remember that stress concentrations, brittle fracture and the plastic design of structures were poorly understood 130 years ago.

Faced with the loss of its major rail link, the railway company began work almost immediately on a replacement bridge across the Tay (not surprisingly,

Figure 29.6   Broken lug on cast-iron column.

a different civil engineer – W. H. Barlow – was commissioned to do the new design). As a predictable over-reaction to the slenderness of the old bridge, the piers of the new were fabricated from iron plates in the form of a massive tubular arch (and therefore had a very large factor of safety with respect to lateral wind forces). The new bridge runs parallel to the course of the old one, and the stumps of the original piers can still be seen from a passing train. Many of the girders from the old bridge were incorporated into the structure of its successor and they survive to the present day. When the Firth of Forth was eventually bridged in 1890, it marked a new dimension in bridge construction. The main crossing is 5330 ft long and has a headroom above high water of 157 ft. It consists of three huge double cantilevers fabricated from steel with a maximum height above high water of 361 ft. The Forth Bridge is as massive as the old Tay Bridge was slender – again the designer (this time Sir Benjamin Baker) was taking no chances.

However, over-design on this scale could not be sustained indefinitely, and bridge designers gradually pared back their margins of safety. There is elegance and economy in having the lightest structure compatible with function. But history has a habit of repeating itself. In 1940, a new suspension bridge with a central span of 2800 ft was built over the Tacoma narrows in the USA. It was soon noticed that the bridge deck was prone to oscillate in certain winds – the vertical oscillation could be as much as 5 ft, but the bridge was never closed to traffic. Four months after the opening, the deck went out of control in a 40 mph gale, and literally shook itself to pieces (fortunately this took some time, so people in the road vehicles which had become stranded on the bridge were able to crawl to safety in time). The bridge had been a victim of "flutter", caused by the inadequate torsional stiffness of the bridge deck. The designers might have avoided disaster if they had realised that the Old Chain Pier in Brighton, England, had collapsed for the very same reason way back in 1836!

Case study 2: the Comet air disasters – 10 January and 8 April 1954

Figure 29.7 is a drawing of the Comet 1 aircraft. These were designed and built in England by de Havilland, and were put into service with the British Overseas Airways Corporation (BOAC) in May 1952. The aircraft was powered by four de Havilland Ghost turbo-jet engines, and had a cruising altitude of 35,000 ft. It was intended for long-distance high-speed flights on passenger and freight routes with a maximum all-up weight of 49 tons. By comparison, existing civil aircraft used turbo-prop engines and had an altitude ceiling of about 17,000 ft, so the Comet was the most advanced civil aircraft of its time.

Unfortunately, a disturbing crash occurred on 10 January 1954. Comet G-ALYP left Rome airport at 0931 bound for London. The crew kept in contact with control after take-off and at 0950 reported that they were over Orbotello. The flight plan indicated that the plane should have climbed to 26,000 ft at this stage in the ascent. A message from the Comet to another BOAC plane was broken off at 0951 in mid-sentence. At the same time, eyewitnesses on Elba saw debris fall into the sea to the south of the island.

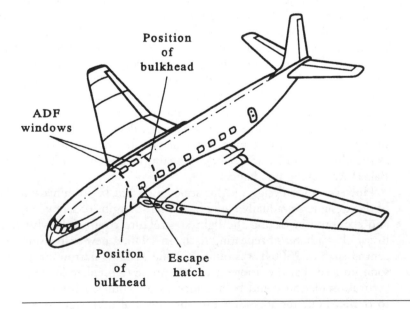

Figure 29.7   The Comet I aircraft.

None of the 29 passengers and six crew survived. The weather conditions were generally good, so there was no evidence that bad weather had caused the crash.

On 8 April 1954, Comet G-ALYY left Rome at 1832 bound for Cairo. At 1857 the crew reported that they were alongside Naples, and were approaching the cruising altitude of 35,000 ft. At 1905 the crew radioed Cairo to give their estimated arrival time. This was the last transmission from the aircraft, and neither Rome nor Cairo was able to make contact again. Some light wreckage was recovered from the sea the next day, and it was presumed that the 14 passengers and seven crew had perished. Again, the weather conditions were reasonable, so there was no indication that bad weather had caused the crash. There was now mounting evidence that something was wrong with the structural integrity of the aircraft, so BOAC grounded the whole fleet.

The first priority was to recover as much of the wreckage of the two aircraft as possible. Unfortunately, G-ALYY had gone down in water 1000 m deep, and there was no realistic prospect of recovering any wreckage from such depths with the undersea technology of the time. G-ALYP had sunk in water 180 m deep, and recovery was feasible if difficult. The search was carried out by two naval vessels which carried a lifting grab and an underwater inspection chamber fitted with a CCTV system (the first ever use of TV in an underwater salvage operation). By August 1954 the recovery team had raised 70% of the structure, 80% of the engines and 50% of the equipment.

The wreckage of the fuselage was reconstructed and the failure was traced to a crack that had started near the corner of a window in the cabin roof. The window was one of a pair which were positioned one behind the other just aft of the leading wing spar. The windows housed the antennae for the automatic direction finding (ADF) system. As Fig. 29.8 shows, the crack had started near the rear starboard corner of the rear ADF window, and had then run backwards along the top of the cabin parallel to the axis of the fuselage. Circumferential cracks had developed from the main longitudinal crack and as a result whole areas of the skin had peeled away from the structure. A second longitudinal crack then formed at the forward port corner of the rear window, and ran into the forward ADF window. Finally, two more cracks nucleated at the forward corners of the forward window and developed into a pair of circumferential fractures. As one might expect, the circumferential cracks all followed the line of a transverse frame member. Figure 29.9 is a close-up of the rear starboard corner of the rear window showing where the first crack probably originated. Fatigue markings were found on the fracture surface at this location: they had probably started at the edge of a countersunk hole, which had been drilled through the skin to take a fastener.

Passengers cannot be carried at high altitude unless the cabin is pressurised. The Comet cabin was pressurised in service to 0.57 bar gauge, 50% more than the pressure in other civil aircraft at the time. During flight, the fuselage functions as a pressure vessel, and is subjected both to an axial and a circumferential tensile stress as a result. With each flight the fuselage experiences a single cycle of pressure loading. Comets G-ALYP and G-ALYY had

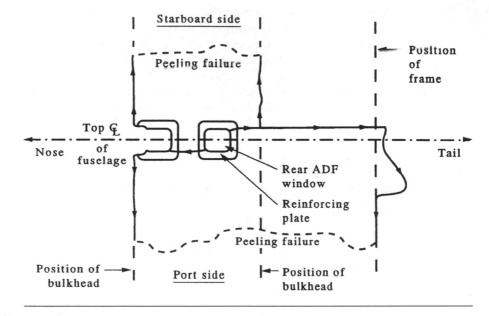

Figure 29.8   The failure on Comet G-ALYP.

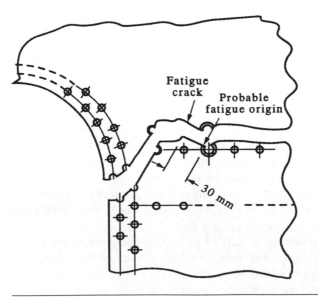

Figure 29.9   Close-up showing the origin of the failure on Comet G-ALYP.

flown 1290 and 900 flights respectively. The inescapable conclusion was that this comparatively small number of cycles had generated a fatigue crack, which in time had become long enough to cause fast fracture.

The skin of the fuselage was made from aluminium alloy DTD 546 sheet (now obsolete, but related to the 2000 series), with yield and tensile strengths of approximately 350 and 450 MN m^{-2}. Approximate hand calculations by de Havilland (at normal cabin pressure) gave an average stress of about 195 MN m^{-2} near the corner of a typical window. The rules of the International Civil Aviation Organisation (ICAO) contained two major requirements for pressurised cabins: there was to be no deformation of the structure when the cabin was pressurised to 33% above normal; and the maximum stress in the structure at a pressure 100% above normal was to be less than the tensile strength. De Havilland decided to go one better – they increased the "design" pressure from 100% to 150% above normal, and decided to pressure test a section of the fuselage to 100% above normal. At this test pressure, the maximum stress near the windows should have gone up to 390 MN m^{-2} if the stress calculations were to be believed. This was still less than the tensile strength, and de Havilland were sure that the cabin would stand up to this increased test pressure without excessive deformation. In the event, the test section withstood two pressure tests to 100% above normal without any problems, showing that there was an ample margin of safety on static pressure loading.

The ICAO rules did not consider the possibility of fatigue caused by repeated cycles of pressure loading. But there was a growing awareness among aircraft designers in the UK that this was a potential problem. As a result, de Havilland decided to determine the likely fatigue life of the fuselage. Between July and September 1953 (more than a year after the aircraft had entered service!), the test section was subjected to a continuous fatigue test, which involved cycling the pressure between zero and the normal operating value. After 18,000 cycles, the section failed from a fatigue crack, which had initiated at a small defect in the skin next to the corner of a window. The design life of the Comet was 10,000 flights, so de Havilland considered that they had demonstrated a reasonable margin of safety against fatigue from pressure cycling.

The air accident investigators decided to do a full-scale simulation test on a Comet aircraft, in order to confirm that the failure had indeed been caused by fatigue from repeated cycles of cabin pressure. The aircraft chosen was G-ALYU, which had already made 1230 flights. The aim was to cycle the cabin pressure between zero and the normal operating value until it failed by fast fracture. This could not be done using air, because the energy released when the cabin broke up would be equivalent to a 225 kg bomb. It was decided instead to pressurise the cabin using water. In order to balance the weight of water in the cabin the whole of the fuselage was put into a large tank and immersed in water at atmospheric pressure. Finally, the required pressure differential was generated by pumping a small volume of water into the fuselage.

The cabin failed after 1830 cycles in the tank, giving a total of $1230 + 1830 = 3060$ "flights". The crack started at an escape hatch, which was positioned in

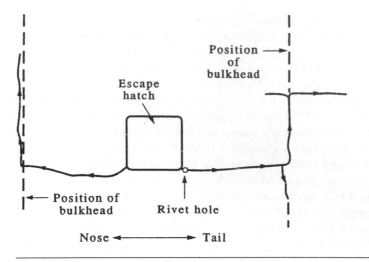

Figure 29.10   The failure on Comet G-ALYU produced by the simulation test.

the port side of the cabin just forward of the wing. As shown in Fig. 29.10, the fatigue crack initiated at the bottom corner of the hatch. Once the crack had reached the critical length it ran backwards along the axis of the fuselage. The crack was diverted in a circumferential direction at the site of a transverse bulkhead. A second crack then nucleated at the forward bottom corner of the hatch. This crack ran forward along the axis of the fuselage until it was diverted into a circumferential crack at another transverse bulkhead. The panel of skin detached by the cracks was pushed out of the fuselage by a few centimetres, relieving the pressure inside the cabin and arresting the failure process. If the cabin had been filled with air, the panel would have blown right out, and the whole fuselage would have exploded.

The fuselage was repaired and strain gauges were stuck on to the skin immediately next to the edge of a window. The cabin was taken up to normal operating pressure and the stresses were found from the strain gauge readings. The maximum stress appeared at the corner of the window, and had a value of $297\,\mathrm{MN\,m^{-2}}$. The background stress at the location of the critical bolt hole would obviously have been much less, since it was located some distance away from the edge of the window. However, the bolt hole would have acted as a stress raiser, and this could have increased the local stress by as much as a factor of 3 above the background stress.

The designers should have been aware that introducing bolt holes near the windows was bound to lead to high local stresses, and that under repeated cycles of cabin pressurisation these bolt holes were likely to be preferential sites for fatigue crack initiation. In addition, the investigators were able to point to two crucial fallacies in the de Havilland figure for the fatigue life. The first was purely statistical. Data for low-cycle fatigue were already known to show a large amount of scatter. The probability that a single Comet would fail by fatigue

...tigue fracture surface from the broken tyre.

The stress geometry is shown in Fig. 29.14. The most important load on wheel is the vertical load Q at the wheel-rail interface. This is produced by adweight of the carriage (78 kN per wheel) plus a multiplier of 1.25 to or average dynamic loads, giving a total design load of 98 kN. Looking 29.14, one can see that because the tyre is supported on a relatively er foundation, it will deflect above the loading point, and this will

could only be found if both the mean and the standard deviation of the test data were known. This information could not have been obtained from one solitary fatigue test. The second was that the test section had been pressurised to 100% above normal before the fatigue test, whereas the production aircraft themselves had only been proof tested to 33% above normal. When the test section was taken to 100% above normal, the bolt holes could have yielded in tension. In this case, when the structure was let down to atmosphere again, the elastic spring-back would have put the holes into compression. When, later on, the test section was taken up to normal operating pressure during the fatigue test, the metal around the holes would have experienced a reduced tensile stress and consequently an increased fatigue life.

The final irony is that the doublers and other stiffeners around the window openings were not meant to be attached to the skin with bolts and rivets at all, but instead with glued joints. However, the design was revised at a late stage because it was felt that it would be difficult and expensive to make glued joints in these geometrically complex locations (although glued joints were extensively used elsewhere in the airframe – yet another pioneering innovation by de Havilland, and one much used in modern aircraft).

In conclusion, although the enquiry established the technical reasons for the failures, in reality the most important factors were probably that de Havilland tried to introduce too many innovations at once, and as a medium-sized company overstretched their design department. It is no surprise that civil aircraft construction today is dominated by huge companies or consortia (most notably Boeing in the USA and Airbus in Europe) because of the sheer range and complexity of design, manufacturing and maintenance capability needed to support modern civil aircraft fleets. Even so, air crashes caused by the failure of structures or components continue to occur (although nowadays many are due to faulty maintenance procedures rather than design errors, and the numbers are much smaller in relation to the size of aircraft fleets than they used to be).

Case study 3: the Eschede railway disaster – 5 June 1998

On 5 June 1998, a high-speed train travelling from Munich to Hamburg (Germany) derailed near the village of Eschede (see Fig. 29.11). The speed of the train was approximately 250 km h^{-1} at the time. The train consisted of two power cars (one at each end) plus 12 carriages. The steel tyre of one of the wheels suffered a fatigue fracture, came away from the wheel, and jammed under the floor of the carriage. The broken tyre then became stuck in a set of points (or switch), and switched them from the running line to the junction line. The broken tyre came from the rear bogie of the leading carriage, so although the front power car was still on the running line, the rear of the leading carriage was diverted on to the junction line, and it and the remainder of the train were derailed as a result. By a very rare coincidence, the derailment occurred just before an over-bridge. The derailed part of the train (still travelling at high speed) demolished one of the supporting piers of the bridge, which promptly collapsed on to the train. This led to the almost total destruction of much of the train. One hundred people were killed in the crash, and more than 100 injured. Criminal proceedings for negligent homicide were commenced against the operators of the railway and the manufacturer of the wheel. The subsequent technical investigations involved a total of 13 experts from five different countries.

Of the 112 wheels in the train, 36 were of the monobloc type (consisting of a one-piece steel wheel with an integral tyre). However, the remaining 76 were rubber-sprung wheels (see Fig. 29.12) in which rubber blocks were fitted between the tyre and the wheel centre to help damp out the vibrations generated by wheel-rail contact. This type of wheel had been introduced into the high-speed trains in Germany in 1991, although similar wheels had been used successfully in commuter trains and trams for some time previously.

The fracture surface of the broken tyre is shown in Fig. 29.13. It has class fatigue beach marks, which show that the fatigue crack initiated at the bo the tyre near the position of maximum thickness. The area proportion fast fracture is small, indicating a relatively low maximum stress.

Figure 29.11 Photograph of the wreckage of the high-speed train after the crash.

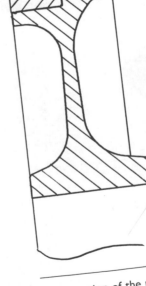

Figure 29.12 Construction of the rubber-sp

Figure 29.13

the
the
allow
at Fig
soft rub

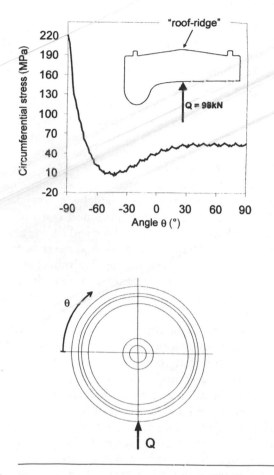

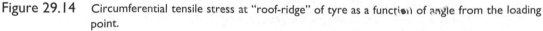

Figure 29.14   Circumferential tensile stress at "roof-ridge" of tyre as a function of angle from the loading point.

generate a tensile stress at the bore of the tyre. The effect of this deflection is felt all around the circumference of the tyre. The stress is a maximum above the loading point, as would be expected, and has a value of 220 MN m^{-2}. However, moving around the circumference away from the loading point, the stress falls off rapidly, reaching a minimum of 6 MN m^{-2} after an angular rotation of 45 degrees. Moving around to the top of the wheel, the stress increases again (but to a maximum of only 55 MN m^{-2}). The bore of the tyre is thus subjected to one complete fatigue cycle with each revolution of the wheel, with stress amplitude 107 MN m^{-2} and mean stress 113 MN m^{-2}. However, there are two equal stress troughs per cycle, and two unequal stress peaks per cycle.

It should be noted that the rubber blocks did not fill all the space between the tyre and the wheel centre. Each wheel was fitted with a total of 34 rubber blocks, which were equally spaced around the circumference of the wheel centre, with circumferential gaps between them. Had there not been adequate circumferential gaps between the blocks, they would not have been able to expand circumferentially when compressed by the radial load Q (they were not able to expand axially because they were a close fit to the width of the channel in the wheel centre). Because Poisson's ratio for rubber is 0.5, it must be allowed to expand sideways in at least one principal direction if it is to be "springy". If sideways expansion were prevented in both principal directions, the block would behave as a very stiff structure indeed, and be useless as a damper (see Example 3.1 in *Engineering Materials 1*).

The stress plot in Fig. 29.14 was generated using an immensely complex finite-element model. A full three-dimensional model was generated, the elastic behaviour of the rubber was modelled with a non-linear constitutive equation, and the frictional contact between the rubber block and the steel was modelled with a non-linear frictional law. Such an analysis had never been done at the design stage, which is not surprising given the rapid advances in finite-element modelling and computing power in recent years. Interestingly, the model was so good that the gaps between adjacent rubber blocks gave rise to a periodic "wobble" in the stress plot.

Turning to the fatigue strength of the tyre, tests on the steel gave a tensile strength of $828\,\mathrm{MN\,m^{-2}}$, so as a first approximation one would expect a fatigue strength for infinite life (fatigue limit) of approximately 45% of this figure (i.e. $370\,\mathrm{MN\,m^{-2}}$ stress amplitude) in laboratory tests. However, when designing the actual wheel, the fatigue limit was reduced to only $200\,\mathrm{MN\,m^{-2}}$ stress amplitude. This was to allow for the effect of tensile mean stress (Goodman diagram) and also to input a numerical safety factor (low probability of failure). According to the published data, therefore, the tyre should not have failed (because the stress amplitude in service should only have been about half the fatigue strength of the tyre with safety factor included). The experts could not agree on the cause for the failure, and as a result, the criminal proceedings were dropped.

The tyres of railway wheels wear as a result of contact with the running rails, and as a result they need to be re-profiled regularly. To do this, the wheel-set is removed from the bogie, and the outer circumference of the tyre is turned in a lathe to give the correct profile again. Over time, repeated re-profiling reduces the outer diameter of the tyre, and the failed tyre (diameter 860 mm) had nearly reached the diameter for scrapping. New tyres had an outer diameter of 920 mm. After the accident, the lower limit on diameter was increased to 880 mm. Obviously, the larger the diameter, the thicker the tyre, and the lower the fatigue stress for a given applied loading.

The disaster raises a number of interesting issues. The first is that a fatigue failure *did* occur, and therefore there *must* be an explanation. The implication is that the investigators must have missed something. This is not to be in any way critical of them. The standard of proof required in a criminal case is to prove beyond all reasonable doubt that the defendants were grossly negligent, which is a much more limited brief than the technical investigation required to ensure the future safety of a fleet of trains. Some technical issues that come to mind are: (a) what was the actual service load spectrum (as distinct from that assumed in the design); (b) was there any fretting between the rubber blocks and the tyre; (c) did the rubber blocks "creep" around the circumference of the wheel with service time, resulting in inadequate support? Finally, when an innovative design of wheel brings with it such serious technical difficulties, is it any wonder that the railway industry is so conservative?

Case study 4: a fatal bungee-jumping accident

The sport of bungee jumping has become very popular worldwide over a short space of time, and as a result the equipment has evolved in an empirical way, using rubber cords originally developed for military applications, together with attachments intended for use by climbers. Many sporting organisations and government agencies have established codes of practice for bungee jumping, but these are essentially empirical, and are not based on a quantitative materials engineering analysis of the forces generated in the load train in relation to the strength of the components. It is therefore not surprising that accidents occur from time to time, and a recent accident resulted in the death of the jumper.

Figure 29.15 shows the general set-up for the bungee jump. The jumper (height 1.83 m and weight 132 kg) jumped from a crane-mounted cage, which had its floor approximately 53 m above ground level. The bungee rope consisted of three nominally identical cords used in parallel, and taped together with insulating tape at regular intervals. The inboard end of the rope was secured to a pair of snap hooks mounted on the vertical centreline of the cage, and positioned approximately 1.35 m above the cage floor. The rope passed vertically down through a large circular hole in the floor of the cage. Before the jump commenced, the rope would have turned back up again so that the outboard end would have entered the cage through the access gateway and lain on the cage floor.

The jumper was attached to the outboard end of the rope by means of a pair of cuffs pulled tight around the lower legs. Each cuff was attached to the end of the rope by a webbing strap. Measurements of the cuffs indicated that, when a jumper was hanging upside-down from the outboard end of the rope, the soles of his feet would have been approximately 0.36 m below the end of the rope. As a safety measure, the jumper was independently attached to the end of the rope by a length of webbing tape (which was knotted to his body harness). The nominal strength of the tape was 12,500 N, but in practice the breaking strength would have been much less because of the weakening effect of the knot.

In the accident, the jumper jumped off the edge of the cage through the open gateway, taking the outboard end of the bungee rope with him in the normal way. During the descent, he moved from an upright to an inverted

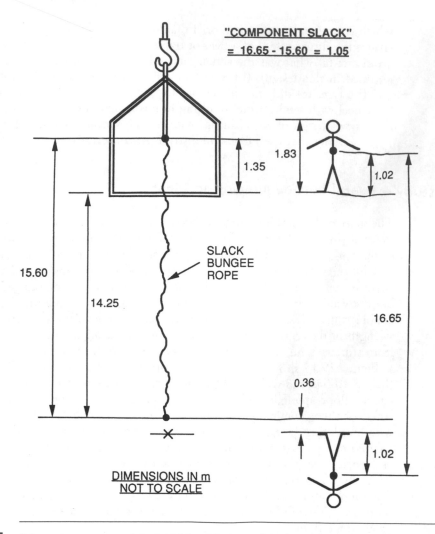

Figure 29.15   Schematic geometry of the initial freefall phase of the bungee-jumping accident.

position, and began to apply force to the rope. However, with the rope vertical and under tension, his legs pulled out of the cuffs. Tension was then applied to the safety webbing, which immediately snapped at the knot. The jumper then descended by freefall to the ground below.

The bungee rope was laid out flat on the floor and pulled straight but free of tension. The unstretched length of the assembly from the webbing loop at the inboard end to the karabiner at the outboard end measured 15.60 m. The bungee rope consisted of three apparently identical cords set side-by-side and taped together with insulating tape at regular intervals. Each cord had a braided sheath (19 mm OD) containing a large number of fine, parallel rubber filaments.

The tapes were removed, and the three cords separated. One of the cords was tested to an extension of 16.0 m (100% strain). At this extension, the measured force was 135 kg (1324 N). The force/extension curve for the complete assembly of three cords was obtained by multiplying the forces measured in the single cord by a factor of three. As shown in Figure 29.16, the curve has a highly non-linear shape, which is primarily determined by the complex extensional behaviour of the braided sheath. Note, that at the extension of 16.0 m, the force rises almost vertically with extension. The breaking strength of the rope was not determined on safety grounds, but is likely to be much higher than the maximum force of 3972 N obtained from the test. Nevertheless, the rope had become a very stiff structural element at an extension of 16.0 m, and a small additional extension would probably have caused failure of the braid. The extension of 16.0 m can therefore be considered as being the "limit of extension" in practical terms. The area under the force/extension curve represents the stored strain energy in the complete rope assembly. The fact that the unloading curve falls well below the loading curve means that only a proportion of the stored strain energy is released on unloading. This hysteresis energy loss explains why in a normal bungee jump, the jumper rebounds to a height significantly less than the height of the cage.

A tensile test to fracture was carried out on a webbing assembly identical to that which failed in the accident. The graph of the force/extension curve is shown in Fig. 29.17; the sling broke at only 5000 N near a knot.

The mechanics of the jump were analysed using an energy-based approach, as shown in Fig. 29.18. In the first stage of the descent, the jumper is in freefall. He progressively loses potential energy, which goes into progressively increasing his kinetic energy (and hence his speed). However, once the rope becomes taut and then stretches, it absorbs strain energy. In this second stage of the descent, the loss in the potential energy of the jumper is converted into strain energy as well as kinetic energy. This has the effect of slowing the jumper down. However, the jumper will only be arrested safely if all his potential energy can be absorbed as strain energy in the rope.

The calculations show that the maximum strain energy which the bungee rope can absorb before it reaches the limit of extension (the area under the loading curve in Fig. 29.16) is 28,660 J. However, the potential energy released by the jumper as he falls to the limit of extension is 42,344 J. This leaves a surplus of 13,684 J as kinetic energy, so he continues to travel downwards (at a speed of 14.4 m/s) rather than being arrested by the rope.

The shape of the force/extension curve in Fig. 29.16 indicates that the force in the rope at the moment of release could have been anywhere between 3972 N (405 kg) and the breaking strength of the rope. In this context, it is hardly surprising that the cuffs were pulled off the jumper's legs. Once the legs have pulled out of the cuffs, the remaining kinetic energy of the falling jumper (13,684 J) must be absorbed by the safety webbing if arrest is to occur. The energy absorbed by the safety webbing was calculated from the area under the force/extension curve (see Fig. 29.17), and was found to be 940 J – only

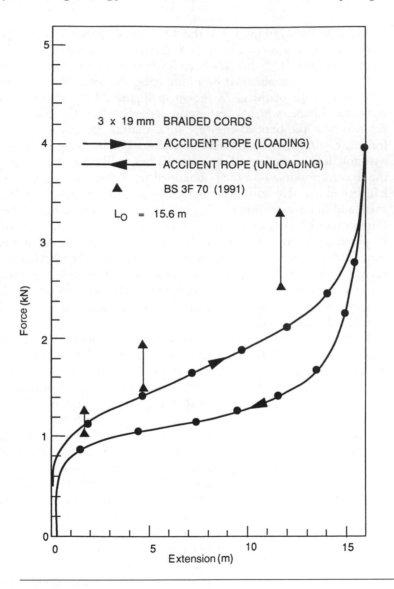

Figure 29.16   Force/extension curve for the bungee rope, showing both loading and unloading curves. The force/extension characteristics for a new rope consisting of three 19 mm cords, as specified by BS 3F 70, are shown as triangular data points.

7% of the 13 684 J required for arrest. In other words, the webbing was totally inadequate in terms of static strength and, most crucially, energy absorbing capacity. A dedicated energy-absorbing device should have been used instead, capable of absorbing the kinetic energy of the falling jumper without applying excessive force to his body.

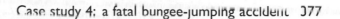

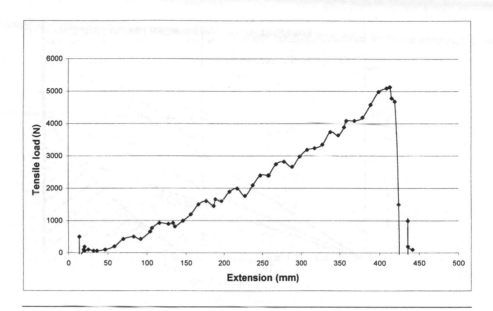

Figure 29.17   Force/extension curve for the webbing.

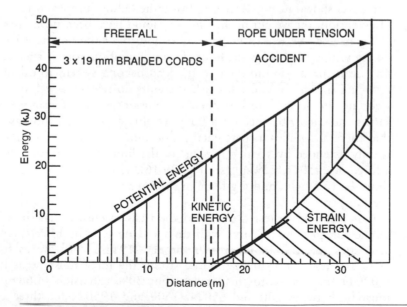

Figure 29.18   Energy diagram for the bungee-jumping process (accident case).

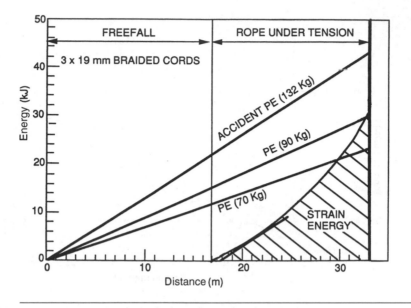

Figure 29.19   Energy diagram for the bungee-jumping process (accident case, plus variants for reduced body weights of 90 and 70 kg).

In spite of the defects of the cuffs and the safety webbing, the root cause of the accident was the use of a bungee rope with an unsuitable force/extension curve. As shown in Fig. 29.19, to bring the falling jumper to rest at the limit of extension, the weight of the jumper should have been no more than 90 kg.

To have a *safe* arrest, the rope must absorb all the potential energy of the falling jumper well *before* the limit of extension of the rope is reached. This will also limit the maximum force in the load-bearing system to calculable levels, reducing the risk of cuff release. If the cuffs do release, a well-designed safety webbing should not break in these circumstances. As an example, Fig. 29.19 shows the energy balance for a body weight of 70 kg. In this case, it is clear that there is significant spare energy-absorbing capacity in the rope. The rope is only extended by 13 m, 3 m short of the limit of extension. This produces a force of 2250 N (229 kg) (see Fig. 29.16). However, even this force is 3.27 times body weight, and a g-force of 3.27 may well cause physical damage to the jumper.

Cords should be tested on a periodic basis to ensure that there is no drift in the force/extension curve with time and usage. The force/extension characteristics specified in BS 3F 70 for new 19 mm cord are as follows. For 10% extension (1.56 m in the case study) the force must be at least 340 N (1020 N for three cords in parallel). For 30% extension (4.68 m) the force must be between 500 and 650 N (1500 and 1950 N for three cords). For 70% extension (11.70 m) the force must be between 850 and 1100 N (2550 and 3300 N for three cords). The minimum total extension must be 105%

(16.38 m). These data points are plotted in Fig. 29.16. Comparison of the specified and actual data shows that the bungee rope has suffered significant degradation of energy-absorbing capacity as the result of repeated use.

In conclusion, the technical cause of the accident was the use of a bungee rope with an unsuitable force/extension curve, only able to absorb 68% of the potential energy of the falling jumper at the limit of extension. This was partly due to degradation of the cords, presumably as the result of repeated use. As a result, the rope did not arrest the falling jumper, who was consequently subjected to a large force, sufficient to pull the cuffs off his legs. Subsequent to cuff release, the safety webbing took the full force of the rope. Because of its low strength and totally inadequate energy-absorbing capacity, the webbing broke and the jumper fell to the ground with a speed of impact equivalent to a freefall from a height of 31 m. However, had the relevant authorities imposed a mandatory code derived from a quantitative mechanics-based analysis (such as that done here), the accident could not have occurred.

On a final note, braided ropes of the type encountered in this case study are generally used when it is important to impose a well-defined limit on the distance fallen (essentially so that novice jumpers falling from low platforms do not bang their heads on the ground). Although the rubber filaments themselves are capable of very large extensions (typically 400%), the much stiffer braided sheath imposes a limit of extension on the assembly of only 105% (or a little more). On the other hand, unbraided ropes are generally used when much greater percentage elongations are required, as in the more extreme sporting situations. Although the analysis described here can in principle be applied to unbraided ropes, they have a very much "softer" force/extension curve than braided ropes. Then, the critical parameter is not the maximum force, but the maximum extension. One of the authors (DRHJ) has witnessed jumpers descending head first from the deck of the Victoria Falls railway bridge (on the border between Zambia and Zimbabwe) to within a few feet of the Zambesi river, some 360 feet below. In such situations, very minor changes in the force/extension characteristics of the rope can have tragic results (not because the jumper breaks away from the rope, but because the jumper smashes their head on a rock in the river bed).

Appendix 1
Teaching yourself phase diagrams

To the student

If you work through this course, doing the problems, you will get a working knowledge of what a phase diagram (or equilibrium diagram) means, and how to use it. Don't rush it: learn the definitions and meditate a little over the diagrams themselves, checking yourself as you go. Some parts (the definitions, for instance) are pretty concentrated stuff. Others (some of the problems, perhaps) may strike you as trivial. That is inevitable in a "teach yourself" course which has to accommodate students with differing backgrounds. Do them anyway. The whole thing should take you about 4 hours. The material given here broadly parallels the introductory part of the excellent text by Hansen and Beiner (referenced below); if you can read German, and want to learn more, work through this.

Phase diagrams are important. Whenever materials engineers have to report on the properties of a metallic alloy, or a ceramic, the first thing they do is reach for the phase diagram. It tells them what, at equilibrium, the structure of the alloy or ceramic is. The real structure may not be the equilibrium one, but equilibrium structure gives a base line from which other (non-equilibrium) structures can be inferred.

Where do you find these summaries-of-structure? All engineering libraries contain:

Source books

M. Hansen and K. Anderko, *Constitution of Binary Alloys*, McGraw-Hill, 1958; and supplements, by R. P. Elliott, 1965, and F. A. Shunk, 1969.

J. Hansen and F. Beiner, *Heterogeneous Equilibrium*, De Gruyter, 1975.

W. Hume-Rothery, J. W. Christian and W. B. Pearson, *Metallurgical Equilibrium Diagrams*, Institute of Physics, 1952.

E. M. Levin, C. R. Robbins and H. F. McMurdie, *Phase Diagrams for Ceramicists*, American Ceramic Society, 1964.

Smithells' Metals Reference Book, 7th edition, Butterworth-Heinemann, 1992.

ASM Metals Handbook, 10th edition, ASM International, 1990, vol. 8.

TEACHING YOURSELF PHASE DIAGRAMS, PART I COMPONENTS, PHASES AND STRUCTURES

Definitions are enclosed in boxes and signalled by "*DEF.*" These you have to learn. The rest follows in a logical way.

Alloys

> *DEF.*   A *metallic alloy* is a mixture of a metal with other metals or non-metals. Ceramics, too, can be mixed to form alloys.

Copper (Cu) and zinc (Zn), when mixed, form the alloy *brass*. Magnesia (MgO) and alumina (Al_2O_3) when mixed in equal proportions form *spinel*. Iron (Fe) and carbon (C) mix to give *carbon steel*.

Components

Alloys are usually made by melting together and mixing the components.

> *DEF.*   The *components* are the chemical elements which make up the alloy.

In *brass* the components are Cu and Zn. In *carbon steel* the components are Fe and C. In *spinel*, they are Mg, Al and O.

> *DEF.*   A *binary alloy* contains two components. A *ternary alloy* contains three; a *quaternary*, four, etc.

Symbols

Components are given capital letters: A, B, C or the element symbols Cu, Zn, C.

Concentration

An alloy is described by stating the components and their concentrations.

DEF.   The weight % of component A:

$$W_A = \frac{\text{weight of component A}}{\Sigma \text{ weights of all components}} \times 100$$

The atom (or mol) % of component A:

$$X_A = \frac{\text{number of atoms (or mols) of component A}}{\Sigma \text{ number of atoms (or mols) of all components}} \times 100$$

(Weight in g)/(atomic or molecular wt in g/mol) = number of mols.

(Number of mols) × (atomic or molecular wt in g/mol) = weight in g.

Questions*

1.1   (a) Calculate the concentration in wt% of copper in a brass containing 40 wt% zinc.
              Concentration of copper, in wt%: W_{Cu} = ---------------------------
       (b) 1 kg of an α-brass contains 0.7 kg of Cu and 0.3 kg of Zn.
              The concentration of copper in α-brass, in wt%: W_{Cu} = --------------
              The concentration of zinc in α-brass, in wt%: W_{Zn} = ----------------
       (c) The atomic weight of copper is 63.5 and of zinc 65.4.
              The concentration of copper in the α-brass, in at%: X_{Cu} = -----------
              The concentration of zinc in the α-brass, in at%: X_{Zn} = -------------

1.2   A special brazing alloy contains 63 wt% of gold (Au) and 37 wt% of nickel (Ni). The atomic weight of Au (197.0) is more than three times that of Ni (58.7). At a glance, which of the two compositions, in at%, is likely to be the right one?

       (a) $X_{Au} = 0.34$, $X_{Ni} = 0.66$.
       (b) $X_{Au} = 0.66$, $X_{Ni} = 0.34$.

1.3   Your favourite vodka is 100° proof (49 wt% of alcohol). The molecular weight of water is 18; that of ethyl alcohol – C_2H_5OH – is 46. What is the mol% of alcohol in the vodka?
       Mol% of alcohol: $X_{C_2H_5OH}$ = ---

---

* Answers are given at the end of each section. But don't look at them until you have done your best to answer *all* the questions in a given group.

1.4 An alloy consists of X_A at% of A with an atomic weight a_A, and X_B at% of B with an atomic weight of a_B. Derive an equation for the concentration of A in wt%. By symmetry, write down the equation for the concentration of B in wt%.

Structure

Alloys are usually made by melting the components and mixing them together while liquid, though you can make them by depositing the components from the vapour, or by diffusing solids into each other. No matter how you make it, a binary alloy can take one of four forms:

(a) a single solid solution;
(b) two separated, essentially pure, components;
(c) two separated solid solutions;
(d) a chemical compound, together with a solid solution.

How can you tell which form you have got? By examining the *microstructure*. To do this, the alloy is cut to expose a flat surface which is then polished, first with successively finer grades of emery paper, and then with diamond pastes (on rotating felt discs) until it reflects like a brass doorknob. Finally, the polished surface is etched, usually in a weak acid or alkali, to reveal the microstructure – the pattern of grains and phases; brass doorknobs often show this pattern, etched by the salts from sweaty hands. Grain boundaries show up because the etch attacks them preferentially. The etch also attacks the crystals, leaving densely packed crystallographic planes exposed; light is reflected from these planes, so some grains appear light and others dark, depending on whether the light is reflected in the direction in which you are looking. Phases can be distinguished, too, because the phase boundaries etch, and because many etches are designed to attack one phase more than another, giving a contrast difference between phases.

The Al–11 wt% Si casting alloy is typical of (b): the Si separates out as fine needles ($\approx 1\,\mu$m diameter) of essentially *pure* Si in a matrix of *pure* Al. The Cd–60 wt% Zn alloy typifies (c): it consists of a *zinc-rich phase* of Zn with 0.1 wt% Cd dissolved in it plus a *cadmium-rich phase* of Cd with 0.8 wt% Zn dissolved in it. Finally, slow-cooled Al–4 wt% Cu is typical of (d) (p. 384).

Questions

1.5 List the compositions of the alloy and the phases mentioned above.

	wt% Cd	wt% Zn
Cadmium-zinc alloy		
Zinc-rich phase		
Cadmium-rich phase		

Phases

> *DEF.*   All parts of an alloy with the same physical and chemical properties and the same composition are parts of a *single phase*.

The Al–Si, Cd–Zn and Al–Cu alloys are all made up of two phases.

Questions

1.6   You heat pure copper. At 1083°C it starts to melt. While it is melting, solid and liquid copper co-exist. Using the definition above, are one or two phases present? ---
Why? ---

1.7   Three components A, B and C of an alloy dissolve completely when liquid but have no mutual solubility when solid. They do not form any chemical compounds. How many phases, and of what compositions, do you think would appear in the solid state?
Phases ---
Compositions ---

The constitution of an alloy

> *DEF.*   The *constitution* of an alloy is described by:
>   (a) The phases present.
>   (b) The weight fraction of each phase.
>   (c) The composition of each phase.

The properties of an alloy (yield strength, toughness, oxidation resistance, etc.) depend critically on its constitution and on two further features of its structure: the *scale* (nm or μm or mm) and *shape* (round, or rod-like, or plate-like) of the phases, not described by the constitution. The constitution, and the scale and shape of the phases, depend on the thermal treatment that the material has had.

EXAMPLE

The alloy aluminium–4 wt% copper forms the basis of the 2000 series (Duralumin, or Dural for short). It melts at about 650°C. At 500°C, solid Al dissolves as much as 4 wt% of Cu completely. At 20°C its equilibrium solubility is only 0.1 wt% Cu. If the material is slowly cooled from 500°C to 20°C, 4 wt% − 0.1 wt% = 3.9 wt% copper separates out from the aluminium as large lumps of a new phase: not pure copper, but of the compound $CuAl_2$. If, instead, the material is quenched (cooled very rapidly, often by dropping it into cold water) from 500°C to 20°C, there is not time for the dissolved copper atoms to move together, by diffusion, to form $CuAl_2$, and the alloy remains a solid solution.

At room temperature, diffusion is so slow that the alloy just stays like this, frozen as a single phase. But if you heat it up just a little – to 160°C, for example – and hold it there ("ageing"), the copper starts to diffuse together to form an enormous number of very tiny (nm) plate-like particles, of composition roughly $CuAl_2$. On recooling to room temperature, this new structure is again frozen in.

The yield strength and toughness of Dural differ enormously in these three conditions (slow-cooled, quenched, and quenched and aged); the last gives the highest yield and lowest toughness because the tiny particles obstruct dislocations very effectively.

It is important to be able to describe the constitution and structure of an alloy quickly and accurately. So do the following, even if they seem obvious.

Questions

1.8   In the example above:

   (a) How many phases are present at 500°C? ----------------------------
   (b) How many phases after slow cooling to 20°C? ----------------------
   (c) How many phases after quenching to 20°C? -------------------------
   (d) How many phases after quenching and ageing? ----------------------

1.9   An alloy of 120 g of lead (Pb) and 80 g of tin (Sn) is melted and cast. At 100°C, two phases are found. There is 126.3 g of the lead-rich phase and 73.7 g of the tin-rich phase. It is known that the lead-rich phase contains $W_{Pb} = 95\%$ of lead. The constitution of the alloy at room temperature is described by:

   (a) Number of phases --
   (b) Weight% of lead-rich phase --------------------------------------
       Weight% of tin-rich phase ---------------------------------------
   (c) Composition of lead-rich phase, in wt%: $W_{Pb} =$ -------------------
       $W_{Sn} =$ ---------------------
   (d) Composition of tin-rich phase, in wt%: $W_{Pb} =$ -------------------
       $W_{Sn} =$ ---------------------

Equilibrium constitution

The Al- 4 wt% Cu alloy of the example can exist at 20°C in three different states. Only one – the slowly cooled one – is its equilibrium state, though given enough time the others would ultimately reach the same state. At a given temperature, then, there is an *equilibrium constitution* for an alloy, to which it tends.

> *DEF.*   A sample has its *equilibrium constitution* when, at a given, constant temperature T and pressure p, there is no further tendency for its constitution to change with time. This constitution is the stable one.

Alloys can exist in non-equilibrium states – the Al–Cu example was an illustration. But it is always useful to know the equilibrium constitution. It gives a sort of base-line for the constitution of the real alloy, and the likely non-equilibrium constitutions can often be deduced from it.

State variables

Ten different samples with the same composition, held at the same T and p, have the same equilibrium constitution. Ten samples each of different composition, or each held at different T or p values, have ten different equilibrium constitutions.

> *DEF.*   The independent *constitution variables* or *state variables* are T, p and composition.

EXAMPLE FOR THE AL-CU ALLOY (DESCRIBED ON PAGE p. 384):

Values of the state variables   *Equilibrium constitution*

(a) $T = 500°C$
$p = 1$ atm.
$W_{Al} = 96\%$
$W_{Cu} = 4\%$ $\rightarrow$ Single-phase solid solution of copper in aluminium

(b) $T = 20°C$
$p = 1$ atm.
$W_{Al} = 96\%$
$W_{Cu} = 4\%$ $\rightarrow$ Two-phases: Al containing 0.1 wt% Cu, and $CuAl_2$

They are *equilibrium* constitutions because they are the ones reached by very slow cooling; slow cooling gives time for equilibrium to be reached.

Certain thermodynamic relations exist between the state variables. In general for a binary alloy we choose p, T and X_B (the at% of component B) as the *independent variables* – though presently we shall drop p. The volume V and the composition X_A ($= 1 - X_B$) are then determined: they are the *dependent variables*. Of course, the weight percentages W_A and W_B can be used instead.

Equilibrium constitution (or phase) diagrams

The equilibrium constitution of an alloy can be determined experimentally by metallography and thermal analysis (described later). If the pressure is held constant at 1 atm., then the independent variables which control the constitution of a binary alloy are T and X_B or W_B.

> *DEF.* An equilibrium-constitution diagram or *equilibrium diagram* for short (or, shorter still, *phase diagram*), is a diagram with T and X_B (or W_B) as axes. It shows the results of experiments which measure the equilibrium constitution at each T and X_B (or W_B).

Figure A1.1 shows a phase diagram for the lead–tin system (the range of alloys obtained by mixing lead and tin, which includes soft solders). The horizontal axis is composition X_{Pb} (at%) below and W_{Pb} (wt%) above. The vertical axis is temperature in °C. The diagram is divided into *fields*: regions in which the number of phases is constant. In the unshaded fields the equilibrium constitution is single phase: liquid (above), or tin containing a little dissolved lead (left), or lead containing a little dissolved tin (right). In the shaded fields the equilibrium constitution has two phases: liquid plus solid Sn, or liquid plus solid Pb, or solid Pb mixed with solid Sn (each containing a little of the other in solution).

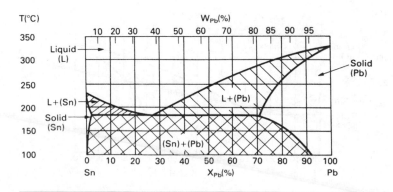

Figure A1.1

> *DEF.*   The diagram shows the equilibrium constitution for all the binary alloys that can be made of lead and tin, in all possible proportions, or, in short, for the lead–tin *system.*
> A *binary system* is a system with two components.
> A *ternary system* is a system with three components.

The constitution point

The state variables define a point on the diagram: the "constitution point". If this point is given, then the equilibrium number of phases can be read off. So, too, can their composition and the quantity of each phase – but that comes later. So the diagram tells you the entire constitution of any given alloy, at equilibrium. Refer back to the definition of *constitution* (p. 384) and check that this is so.

Questions

1.10   Figure A1.2 shows the Pb–Sn diagram again, but without shading.

    (a) What is the composition and temperature (the state variables) of point 1?
        Composition ----------------- at% Pb and ----------------- at% Sn
        Temperature -------------------°C
    (b) Mark the constitution point for a Pb–70 at% Sn alloy at 250°C onto Fig. A1.2.

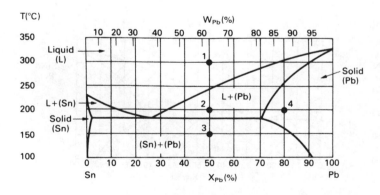

Figure A1.2

What does the alloy consist of at 250°C? ----------------------------
How many phases? ---

(c) Mark the point for a Pb–30 at% Sn at 250°C.
What does it consist of? ---
How many phases? --

(d) Describe what happens as the alloy corresponding initially to the constitution point 1 is cooled to room temperature.
At which temperatures do changes in the number or type of phases occur?

What phases are present at point 2? -----------------------------------
What phases are present at point 3? -----------------------------------

(e) Describe similarly what happens when the alloy corresponding to the constitution point 4 is cooled to room temperature.
Initial composition and temperature ------------------------------------
Initial number of phases ---
Identify initial phase(s) --
Temperature at which change of phase occurs ----------------------------
Number of phases below this temperature --------------------------------
Identify phases --

Part 1: final questions

1.11 Is a mixture of a metal and a non-metal called an alloy?
Yes   No

1.12 Pernod is a transparent yellow fluid consisting of water, alcohol and Evil Esters. The Evil Esters dissolve in strong water–alcohol solutions but precipitate out as tiny whitish droplets if the solution is diluted with more water. It is observed that Pernod turns cloudy at 60 wt% water at 0°C, at 70 wt% water at 20°C, and at 85 wt% water at 40°C. Using axes of T and concentration of water in wt%, sketch an approximate phase diagram (Fig. A1.3) for the Pernod–water system, indicating the single-phase and two-phase fields.

1.13 A micrograph reveals 10 black-etching needles and 8 globular regions that etch grey, set in a white-etching matrix.

(a) How many phases would you judge there to be? ---------------------
(b) Does the constitution of the alloy depend on the shape of the phases?

(c) Can the constitution of the alloy depend on its thermal history? -------
(d) What do you call the entire range of alloys which can be made of lead and tin? --

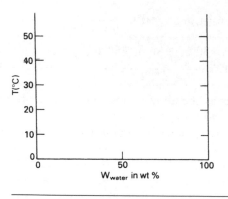

Figure A1.3

Answers to questions: part I

1.1   (a) $W_{Cu} = 60\%$.
      (b) $W_{Cu} = 70\%$, $W_{Zn} = 30\%$.
      (c) $X_{Cu} = 71\%$, $X_{Zn} = 29\%$.

1.2   (a) is the correct composition.

1.3   Your vodka contains 27 mol% of alcohol.

1.4 $$W_A = \frac{a_A X_A}{a_A X_A + a_B X_B}.$$

      $$W_B = \frac{a_B X_B}{a_A X_A + a_B X_B}.$$

1.5

	wt% Cd	wt% Zn
Cadmium–zinc alloy	40	60
Zinc-rich phase	0.1	99.9
Cadmium-rich phase	99.2	0.8

1.6   Two phases: liquid and solid. Although they have the same chemical composition, they differ in *physical* properties.

1.7   Three phases: pure A, pure B and pure C.

1.8   (a) 1.
      (b) 2.
      (c) 1.
      (d) 2.

1.9   (a) 2.
     (b) 63%, 37%.
     (c) 95%, 5%.
     (d) 0%, 100%.

1.10   (a) 50%, 50%, 300°C.
        (See Fig. A1.4.)

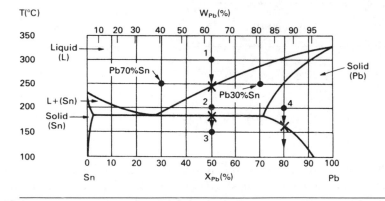

Figure A1.4

     (b) Liquid; 1.
     (c) Liquid plus lead-rich solid; 2.
     (d) 240°C, 183°C.
        At point 2: liquid plus solid (Pb).
        At point 3: two solids, (Sn) and (Pb).
     (e) $X_{Pb} = 80\%$, $T = 200°C$.
        1 phase.
        Lead-rich solid.
        155°C.
        Two phases: lead-rich solid (Pb) and tin-rich solid (Sn).

1.11   *Yes* (see definition, on p. 381).

1.12   (See Fig. A1.5.)

1.13   (a) 3.
     (b) Not at all.
     (c) Yes (see example of Dural on p. 384).
     (d) The lead–tin alloy system.

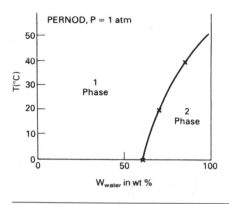

Figure A1.5

TEACHING YOURSELF PHASE DIAGRAMS PART 2 ONE AND TWO COMPONENT SYSTEMS

Phase diagrams are mostly determined by *thermal analysis*. We now discuss one-component systems to show how it works. The more complicated diagrams for binary, ternary or quaternary alloys are determined by the same method.

Reminder

One-component systems	independent variables p and T
Binary $(A+B)$ systems	independent variables p, T and X_B
Ternary $(A+B+C)$ systems	independent variables p, T, X_B and X_C
Quaternary $(A+B+C+D)$ systems	independent variables p, T, X_B, X_C and X_D

One-component systems

The equilibrium constitution of a one-component system is fixed by the variables p and T and so the equilibrium phases can be shown on a diagram with p and T axes. The one shown in Fig. A1.6 has only one solid phase. Some, like ice, or iron, have several.

Single-phase regions are *areas*.
Two phases co-exist along *lines*.
Three phases co-exist at a *point*: the triple point.

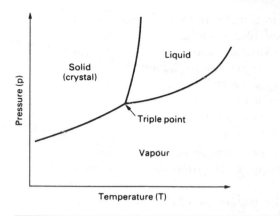

Figure A1.6

The behaviour at constant p is given by a horizontal cut through the diagram. The solid *melts* at T_m and *vaporises* at T_v. The phase diagram at constant pressure is a line (shown on the right) along which the span of stability of each phase is marked, as shown in Fig. A1.7.

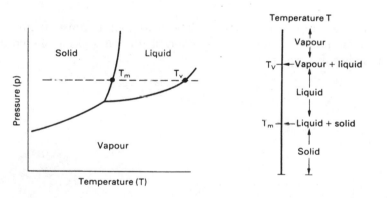

Figure A1.7

Questions

2.1 List the phases shown on the $T-p$ diagram (Fig. A1.7) ------------------------
----

2.2 If the pressure is increased, does the melting point of the material of the diagram increase --------------------, decrease --------------------- or stay constant ------------------------?

2.3   At 1 atmosphere, iron melts at 1536°C and boils at 2860°C. When it solidifies (a phase change), it does so in the b.c.c. crystal structure and is called δ-iron. On cooling further it undergoes two further phase changes. The first is at 1391°C when it changes to the f.c.c. crystal structure, and is then called γ-iron. The second is at 914°C when it changes *back* to the b.c.c. crystal structure, and is called α-iron.

(a) Construct the one-dimensional phase diagram at constant pressure (1 atmosphere) for iron (Fig. A1.8).

(b) Mark on it the regions in which each phase is stable. Label them with the names of the phases.

(c) Indicate with arrows the points or regions where two phases coexist in equilibrium.

Figure A1.8

Cooling curves

If a one-component system is allowed to cool at constant pressure, and the temperature is recorded as a function of time, it looks as shown in Fig. A1.9. It shows five regions:

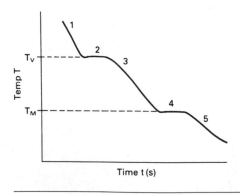

Figure A1.9

1. Vapour;
2. Vapour-to-liquid phase change;
3. Liquid;
4. Liquid-to-solid phase change;
5. Solid.

The system is a single phase in regions 1, 3 and 5. Phase changes occur at the temperatures corresponding to regions 2 and 4. When a phase transformation takes place, the *latent heat* of the transformation is released (on cooling) or absorbed (on heating). Because of this the temperature stays almost constant during the transformation, giving shelves 2 and 4; cooling continues only when the transformation is complete.

Phase transformations in the solid state (like those in iron), too, have latent heats. They may be small, but with sensitive equipment for measuring cooling curves or heating curves, they are easily detected.

The shelves of the cooling curve are called *arrest points*. The two shown in the picture are at the boiling point and the melting point of the material, at the given pressure.

Differential thermal analysis, DTA

Even in complicated, multi-component alloys, phase changes can be determined by cooling (or heating) a sample, recording temperature as a function of time, and observing the arrest points. Greater accuracy is possible with *differential thermal analysis*. A sample with the same thermal mass as the test sample, but showing no phase transformations, is cooled (or heated) side-by-side with the test sample, and the *difference* ΔT between the cooling (or heating) curves is plotted. Sometimes the difference in power needed to heat the two samples at the *same rate* is measured instead: there is a sudden difference in power at the phase transformation. Both are just sophisticated ways of getting the information shown in the cooling curve.

Questions

2.4 Construct the general shape of a cooling curve for iron, starting at 3000°C, and ending at 0°C, using the data given in Example 2.3 (see Fig. A1.10).

2.5 Which phases do you expect at each of the following constitution points, for the lead–tin system?

(a) $X_{Pb} = 40\%$, $T = 175°C$ ---
(b) $W_{Pb} = 15\%$, $T = 200°C$ ---
(c) $W_{Sn} = 10\%$, $T = 200°C$ ---
(d) $W_{Pb} = 35\%$, $T = 200°C$ ---

2.6 You cool a sample of $X_{Pb} = 80\%$ from 325°C to 125°C sufficiently slowly that equilibrium is maintained. Mark the two points on Fig. A1.11 and join them to show the *cooling path*. In sequence, what phases appear as the alloy is cooled?

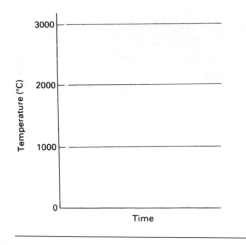

Figure A1.10

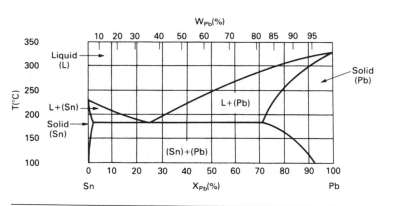

Figure A1.11

(a) ---
(b) ---
(c) ---

2.7   Figure A1.12 shows the phase diagram for ice. (The pressures are so large that steam appears only at the extreme upper left.) There are eight different solid phases of ice, each with a different crystal structure.

Current ideas of the evolution of the large satellite of Jupiter, Ganymede, assume it to be largely made of ice. The pressure caused by gravitational forces rises about linearly from the surface to the centre, reaching a peak of around 2 GPa. Radioactive decay causes the centre to have a temperature of about 30°C, but at the surface the temperature is below −100°C. Assuming a linear temperature gradient from the surface to the centre, which phases of ice would be found in Ganymede?

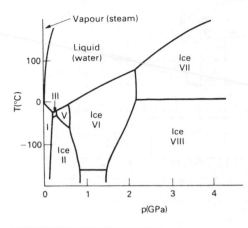

Figure A1.12

Binary systems

What defines the *constitution* of an alloy? If you can't remember, refer back to the definition on p. 384 and revise. The phase diagram gives all three pieces of information. The first you know already. This section explains how to get the other two.

At first sight there is a problem in drawing phase diagrams for binary systems: there are three state variables (p, T and X_B).

The pressures on the ice phase diagram (above) are enormous, and when they are this large they do affect phase equilibrium. But, almost always, we are interested only in pressures near atmospheric (maybe up to 100 atmospheres), and such small pressures have almost no effect on phase diagrams for solids and liquids. So we can drop p as a variable and plot the phase diagram in two dimensions, with T and X_B as axes. For ternary (or higher) systems there is no way out – even after dropping p there are three (or more) variables. Then we can only show *sections* through the phase diagram of constant T or of constant X_C, for example.

Composition of the phases

The phase diagram for a binary alloy (Fig. A1.13) shows *single-phase fields* (e.g. liquid) and *two-phase fields* (e.g. liquid plus A). The fields are separated by *phase boundaries*. When a phase boundary is crossed, a phase change starts, or finishes, or both.

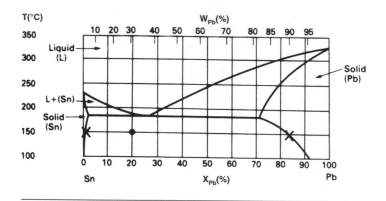

Figure A1.13

> DEF.   When the constitution point lies in a single-phase
> region, the alloy consists of a single, homogeneous,
> phase. Its composition must (obviously) be that of
> the alloy. The *phase composition* and the *alloy
> composition* coincide in single-phase fields.

When the constitution point lies in a two-phase region the alloy breaks up into two phases which do not have the same composition as the alloy (though a properly weighted *mean* of the two compositions must equal that of the alloy).

What are these compositions? Well, if the alloy is at a temperature T, the two phases (obviously) are at this temperature. Consider a Sn–20 at% Pb alloy, at 150°C. Figure A1.13 shows that, at this temperature, tin dissolves 1 at% of lead and lead dissolves 17 at% of tin. The compositions are shown as Xs on the 150°C isotherm (horizontal line): they are the *equilibrium phases* at 150°C. The line joining them is called a *tie line*.

> DEF.   When the constitution point for an alloy lies in the
> two-phase field the alloy breaks up into a mixture of
> two phases. The composition of each phase is
> obtained by constructing the *tie line* (the isotherm
> spanning the two-phase region, terminating at the
> nearest phase boundary on either side). The
> *composition of each phase* is defined by the ends of the
> tie line.

Any alloy which lies on the tie line breaks up into the same two phases at its ends. The *proportions* of each phase, of course, depend on the alloy composition.

Questions

2.8 (a) The constitution point for a Sn–60 at% Pb alloy at 250°C lies in a two-phase field. Construct a tie line on Fig. A1.14 and read off the two phases and their compositions.

The phases are --

The composition of phase 1 is --

The composition of phase 2 is --

(b) The alloy is slowly cooled to 200°C. At 200°C:

The phases are --

The composition of phase 1 is --

The composition of phase 2 is --

(c) The alloy is cooled further to 150°C. At this temperature:

The phases are --

The composition of phase 1 is --

The composition of phase 2 is --

(d) Indicate with arrows on the figure the lines along which:

1. The composition of phase 1 moves
2. The composition of phase 2 moves

The overall composition of the alloy stays the same, of course. How can the compositions of the phases change?

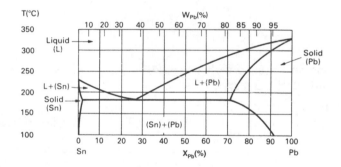

Figure A1.14

Proportions of phases in two-phase alloys

You can get the *relative amounts* of each phase in a two-phase alloy from the phase diagram.

> *DEF.* The weight fraction of phase α is designated W_α.
> That of phase β is W_β. In a binary alloy $W_\alpha + W_\beta = 1$.

Prove this result to yourself. Notice that, for the first time in these notes, we have not given a parallel result for atomic (or mol) fraction. You cannot have an atom fraction of a *phase* because a phase (as distinct from a pure element, or a chemical compound of a *specific* composition) does not have an atomic weight. Its composition can vary within the limits given by the phase diagram.

To find the relative amounts of each phase, start off by constructing a tie line through the constitution point and read off the compositions of the phases (Fig. A1.15).

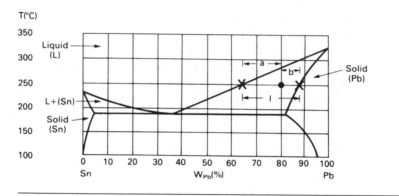

Figure A1.15

EXAMPLE

At 250°C the Sn–Pb alloy with $W_{Pb} = 80\%$ consists of two phases:
Liquid with $W_{Pb}^{LIQ} = 65\%$;
Solid with $W_{Pb}^{SOL} = 87\%$.
The *weight fractions* of each phase, W_{LIQ} and W_{SOL}, are fixed by the requirement that matter is conserved. Then:

> The weight fraction of solid in the alloy is
> $$W_{SOL} = \frac{a}{l}$$
> and the weight fraction of liquid is
> $$W_{LIQ} = \frac{b}{l}$$
> $\left.\begin{array}{c} \\ \\ \\ \\ \\ \end{array}\right\}$ the "lever" rule

and $W_{SOL} + W_{LIQ} = 1$ as they obviously must.

The easiest way to understand this result is to notice that, if the constitution point coincides with the left-hand X (the left-hand end of the tie line), the alloy is all liquid; and when it coincides with the right-hand X it is all solid; *provided the phase diagram has a linear wt% scale*, the weight fraction of each phase is proportional to distance measured along the tie line. Figure A1.16

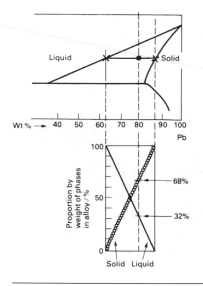

Figure A1.16

shows how this relates to our example of an 80 wt% Pb alloy. The weight percentages of other alloys along the tie line can be found in exactly the same way: the results can be calculated using the lever rule, or, more approximately, can be read straight off the linear graphs in the figure.

When the wt% scale is *not* linear (often it is not) you get the weight fractions by writing

$$l = W_{Pb}^{SOL} - W_{Pb}^{LIQ}$$
$$a = W_{Pb} - W_{Pb}^{LIQ}$$
$$b = W_{Pb}^{SOL} - W_{Pb}$$

where W_{Pb}^{LIQ} is the percentage of lead in the liquid, and W_{Pb}^{SOL} is that in the solid (as before). These can be read off the (non-linear) wt% scale, and the results used in the lever rule.

Questions

2.9 A lead–tin alloy with composition $W_{Pb} = 80\%$ is held at a temperature T.

(a) At $T = 280°C$ which is the dominant phase?
(b) At $T = 200°C$ which is the dominant phase?

Indicate by arrows on Fig. A1.17 the changes in the compositions of the liquid phase and the solid phase as the alloy is cooled from 280°C to 200°C.

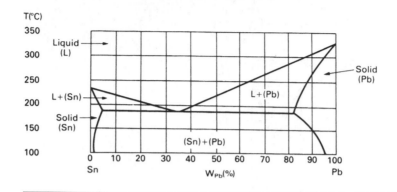

Figure A1.17

2.10　The alloy is cooled to 150°C.

(a) How many phases are present? ------------------------------------
(b) List the approximate composition of the phase(s) -------------------
(c) Which is the dominant phase? ------------------------------------
(d) What (roughly) are the proportions by weight of each phase? ----------
----

Gibbs' phase rule

> *DEF.*　The number of phases P which coexist in equilibrium is given by the phase rule
>
> $$F = C - P + 2$$
>
> where C is the number of *components* (p. 339) and F is the number of free *independent state variables* (p. 344) or "degrees of freedom" of the system. If the pressure p is held constant (as it usually is for solid systems) then the rule becomes
> $F = C - P + 1$.

A one-component system $(C = 1)$ has two independent state variables $(T$ and $p)$. At the triple point three phases (solid, liquid, vapour) coexist at equilibrium, so $P = 3$. From the phase rule $F = 0$, so that at the triple point, T and p are fixed – neither is free but both are uniquely determined. If T is free but p depends on T (a sloping *line* on the phase diagram) then $F = 1$

and $P = 2$ that is, two phases, solid and liquid, say, co-exist at equilibrium. If both p and T are free (an *area* on the phase diagram) $F = 2$ and $P = 1$; only one phase exists at equilibrium (see Fig. A1.18).

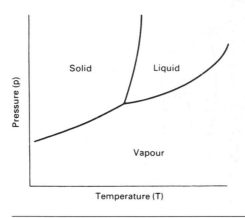

Figure A1.18

Questions

2.11   For a binary A–B alloy:

(a) The number of components
     C = --
(b) The independent state variables are

(c) If pressure is held fixed at atmospheric pressure, there are at most ------
     --- degrees of freedom.
(d) If $F = 0$, how many phases can coexist in equilibrium at constant p? ---
(e) If $F = 1$, how many phases coexist at constant p? -------------------
(f) If $F = 2$, how many phases coexist at constant p? -------------------
(g) In a single-phase field, how many degrees of freedom are there at constant
     p? --
(h) In a two-phase field, how many degrees of freedom are there at constant p?

Part 2: final question

2.12   Figure A1.19 shows the phase diagram for the copper–zinc system. It is more complicated than you have seen so far, but all the same rules apply. The Greek letters (conventionally) identify the single-phase fields.

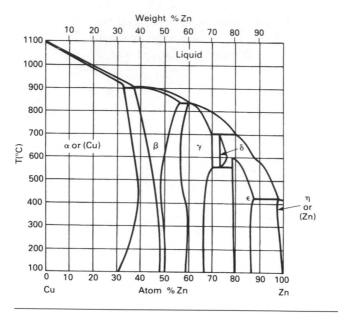

Figure A1.19

(a) Shade in the two-phase fields. Note that single-phase fields are always separated by two-phase fields except at points.

(b) The two common commercial brasses are:

$$70/30 \; brass: \; W_{Cu} = 70\%;$$

$$60/40 \; brass: \; W_{Cu} = 60\%.$$

Mark their constitution points onto the diagram at 200°C (not much happens between 200°C and room temperature).
What distinguishes the two alloys? ----------------------------------

(c) What, roughly, is the melting point of 70/30 brass? ------------------

(d) What are the phase(s) in 60/40 brass at 200°C? ---------------------
What are the phase(s) in 60/40 brass at 800°C? ---------------------
What has happened to the other phase? ----------------------------

Answers to questions: part 2

2.1 Crystalline solid, liquid and vapour.

2.2 It increases.

2.3 (See Fig. A1.20.)

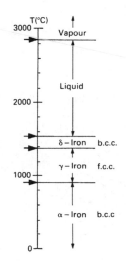

Figure A1.20

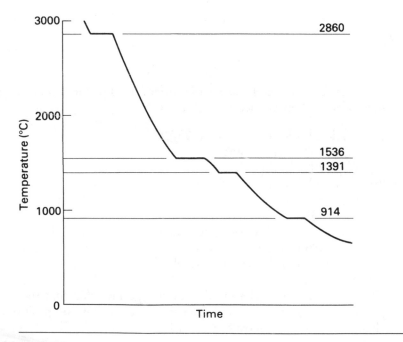

Figure A1.21

2.4   (See Fig. A1.21.)

2.5   (a) Tin-rich solid plus lead-rich solid.
       (b) Liquid plus tin-rich solid.

(c) Lead-rich solid only.
(d) Liquid only.

2.6   (a) Liquid plus lead-rich solid between 1 and 2.
    (b) Lead-rich solid between 2 and 3.
    (c) Lead-rich solid plus tin-rich solid below 3 (see Fig. A1.22).

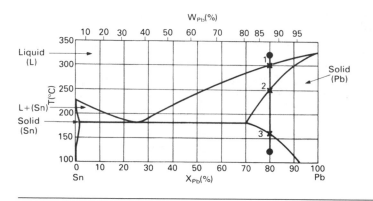

Figure A1.22

2.7   Ice VI at the core, ice II nearer the surface, ice I at the surface, possibly a thin shell of ice V between ice II and ice VI.

2.8   (a) Liquid plus lead-rich solid at 250°C
       $X_{Pb} = 53\%$ in the liquid
       $X_{Pb} = 79\%$ in the solid.
    (b) Liquid plus lead-rich solid at 200°C
       $X_{Pb} = 33\%$ in the liquid
       $X_{Pb} = 73\%$ in the solid.
    (c) Tin-rich solid plus lead-rich solid at 150°C
       $X_{Pb} = 1\%$ in the tin-rich solid
       $X_{Pb} = 83\%$ in the lead-rich solid.
    (d) (See Fig. A1.23.)
The compositions of the phases can change provided that their relative proportions change so as to lead to the same overall alloy composition. In practice changes in phase composition occur by diffusion.

2.9   (a) Liquid.
    (b) Solid.
       See Fig. A1.24.

2.10   (a) 2.
    (b) $W_{Pb} = 2\%$, $W_{Pb} = 90\%$.

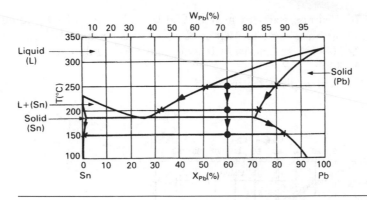

Figure A1.23

(c) The lead-rich solid.
(d) Tin-rich solid 11% of total weight.
Lead-rich solid 89% of total weight (see Fig. A1.24).

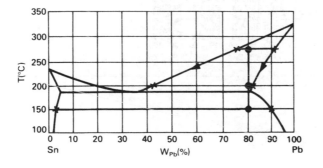

Figure A1.24

2.11 (a) $C = 2$.
(b) Intensive variables p, T, X_B or X_A.
(c) 2.
(d) 3.
(e) 2.
(f) 1.
(g) 2.
(h) Only 1. The compositions of the phases are given by the ends of the tie-lines so that T and X_B (or X_A) are dependent on one another.

2.12 (a) See Fig. A1.25.
(b) 70/30 brass is single-phase, but 60/40 brass is two-phase.

(c) 70/30 brass starts to melt at 920°C and is completely liquid at 950°C. 60/40 brass starts to melt at 895°C and is completely liquid at 898°C.
(d) At 200°C : α (copper-rich solid) and β (roughly CuZn).
At 800°C : β.
The α has dissolved in the β.

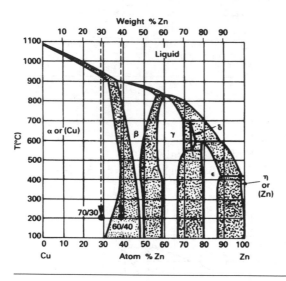

Figure A1.25

TEACHING YOURSELF PHASE DIAGRAMS, PART 3 EUTECTICS, EUTECTOIDS AND PERITECTICS

Eutectics and eutectoids are important. They are common in engineering alloys, and allow the production of special, strong, microstructures. Peritectics are less important. But you should know what they are and what they look like, to avoid confusing them with other features of phase diagrams.

Eutectics

The Pb–Sn system has a *eutectic*. Look at the Pb–Sn phase diagram (Fig. A1.26). Above 327°C, liquid lead and liquid tin are *completely miscible*, that is, the one dissolves in the other completely. On cooling, solid first starts to appear when the lines (or boundaries) which limit the bottom of the liquid field are reached.

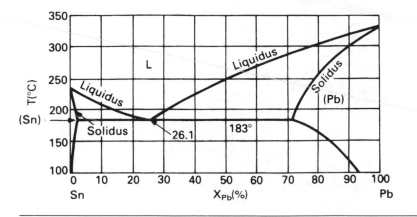

Figure A1.26

> *DEF.* The phase boundary which limits the bottom of the
> liquid field is called the *liquidus line*. The other
> boundary of the two-phase liquid–solid field is called
> the *solidus line*.

The liquidus lines start from the melting points of the pure components.
Almost always, alloying lowers the melting point, so the liquidus lines *descend*
from the melting points of the pure components, forming a shallow V.

> *DEF.* The bottom point of the V formed by two liquidus
> lines is the *eutectic point*.

In the lead–tin system it is the point $X_{Pb} = 26.1$ wt%, $T = 183°C$.

Most alloy systems are more complicated than the lead–tin system, and show
intermediate phases: compounds which form between components, like $CuAl_2$,
or Al_3Ni, or Fe_3C. Their melting points are, usually, lowered by alloying also,
so that eutectics can form between $CuAl_2$ and Al (for example), or between
Al_3Ni and Al. The eutectic point is always the apex of the more or less shallow
V formed by the liquidus lines.

Figure A1.27 shows the unusual silver–strontium phase diagram. It has four
intermetallic compounds. Note that it is just five simple phase diagrams, like
the Pb–Sn diagram, stuck together. The first is the $Ag–SrAg_5$ diagram, the
second is the $SrAg_5–Sr_3Ag_5$ diagram, and so on. Each has a eutectic. You can
always dissect complicated diagrams in this way.

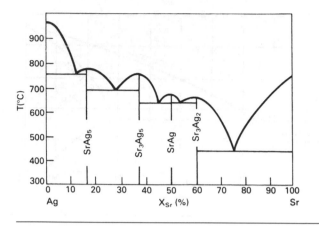

Figure A1.27

Questions

3.1   The three phase diagrams, or parts of diagrams, shown in Fig. A1.28, all have a eutectic point. Mark the point with an arrow and list the eutectic temperature and composition in wt% (the co-ordinates of the point).

Phase reactions

When an alloy is cooled the constitution point for the alloy drops vertically on the phase diagram. In a single-phase field the composition of the phase is, of course, that of the alloy. In a two-phase region the compositions of the two phases are related by the tie line through the constitution point (p. 383: check it if you've forgotten); the phase compositions are given by the two ends of the tie line. These do *not* (in general) fall vertically; instead they run along the phase boundaries. The compositions of the two phases then change with temperature.

> *DEF.*   When the compositions of the phases change with temperature, we say that a *phase reaction* takes place.

Cooling *without* a phase reaction occurs:

(a) in a single-phase field,
(b) when both phase boundaries on either side of the constitution point are vertical.

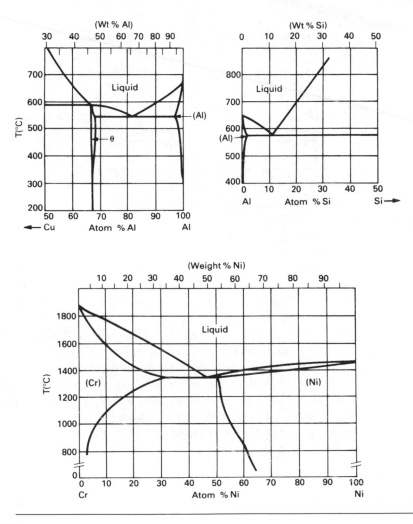

Figure A1.28

Cooling *with* a phase reaction occurs when the constitution point lies in a two-phase region, and at least one of the phase boundaries is not vertical.

Figure A1.29 shows the cooling of a lead–tin alloy with $X_{Pb} = 80\%$. On cooling from 350°C the following regimes appear.

1. *From 350°C to 305°C.* Single-phase liquid; no phase reaction.
2. *From 305°C to 255°C.* The liquidus line is reached at 305°C; the reaction liquid → solid (Pb-rich solid solution) starts. The solid contains less tin than the liquid (see first tie line), so the liquid becomes richer in tin and the composition of the liquid moves down the liquidus line as shown by the arrow. The composition of the solid in equilibrium with this liquid also changes, becoming richer in tin also, as shown by the arrow on the solidus

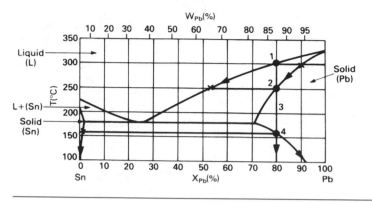

Figure A1.29

line: a *phase reaction* is taking place. The *proportion* of liquid changes from 100% (first tie line) to 0% (second tie line).

3. *From 255°C to 160°C*. Single-phase solid, with composition identical to that of the alloy. No phase reaction.

4. *From 160°C to room temperature*. The lead-rich phase becomes unstable when the phase boundary at 160°C is crossed. It breaks down into *two solid phases*, with compositions given by the ends of the tie line through point 4. On further cooling the composition of the two solid phases changes as shown by the arrows: each dissolves less of the other. A phase reaction takes place. The *proportion* of each phase is given by the lever rule. The *compositions* of each are read directly from the diagram (the ends of the tie lines).

The eutectic reaction

Consider now the cooling of an alloy with 50 at% lead. Starting from 300°C, the regions are shown in Fig. A1.30.

1. *From 300°C to 245°C*. Single-phase liquid; no phase reactions.
2. *From 245°C to 183°C*. The liquidus is reached at 245°C, and solid (a lead-rich solid solution) first appears. The composition of the liquid moves along the liquidus line, that of the solid along the solidus line. This regime ends when the temperature reaches 183°C. Note that the alloy composition in *weight* % (64) is roughly half way between that of the solid (81 wt%) and liquid (38 wt%); so the alloy is about half liquid, half solid, by weight.
3. *At 183°C*. The liquid composition has reached the *eutectic point* (the bottom of the V). This is the lowest temperature at which liquid is stable. At this temperature all the remaining liquid transforms to two solid phases:

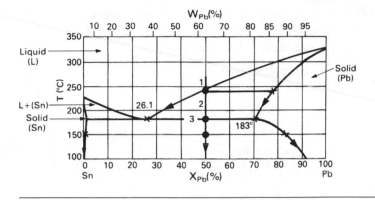

Figure A1.30

a tin-rich α phase, composition $X_{Pb} = 1.45\%$ and a lead-rich β phase, composition $X_{Pb} = 71\%$. This reaction:

$$Liquid \rightarrow \alpha + \beta$$

at constant temperature is called a *eutectic reaction*.

> *DEF.* A *eutectic reaction* is a three-phase reaction, by which, on cooling, a liquid transforms into two solid phases at the same time. It is a phase reaction, of course, but a special one. If the bottom of a liquid-phase field closes with a V, the bottom of the V is a eutectic point.

At the eutectic point the three phases are in equilibrium. The compositions of the two new phases are given by the ends of the line through the eutectic point.

4. *From 183°C to room temperature.* In this two-phase region the compositions and proportions of the two solid phases are given by constructing the tie line and applying the lever rule, as illustrated. The compositions of the two phases change, following the phase boundaries, as the temperature decreases, that is, a further phase reaction takes place.

Questions

3.2 Check, using the phase rule, that three phases can coexist only at a point (the eutectic point) in the lead–tin system at constant pressure. If you have trouble, revise the phase rule on p. 388.

3.3 Not all alloys in the lead–tin system show a eutectic: pure lead, for example, does not. Examine the Pb–Sn phase diagram and list the composition range for which a eutectic reaction is possible.

3.4 We defined a eutectic reaction (e.g. that of the lead–tin system) as a three-phase reaction by which, on cooling, a liquid transforms into two solids. In general:

$$\left.\begin{array}{c} L \rightarrow \alpha + \beta \\ \text{Liquid (Pb–Sn)} \rightarrow \text{(Pb)} + \text{(Sn)} \end{array}\right\} \text{on cooling}$$

or, for the lead–tin system

What happens on heating?

Eutectic structure

The aluminium *casting alloys* are mostly based on the Al–Si system (phase diagram Fig. A1.31). It is a classic eutectic system, with a eutectic point at about 11% Si and 577°C. Consider the cooling of an Al–6% Si casting alloy. The liquidus is reached at about 635°C, when solid (Al) starts to separate out

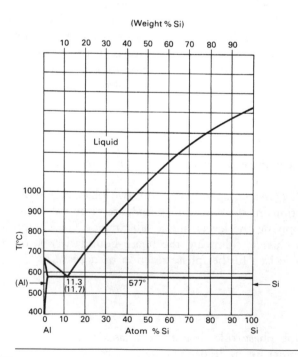

Figure A1.31

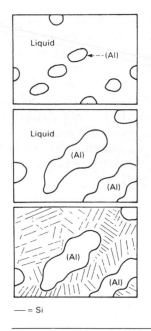

— = Si

Figure A1.32

(top of Fig. A1.32). As the temperature falls further the liquid composition moves along the liquidus line, and the amount of solid (Al) increases. When the eutectic temperature (577°C) is reached, about half the liquid has solidified (middle of Fig. A1.32). The solid that appears in this way is called *primary solid*, primary (Al) in this case.

At 577°C the eutectic reaction takes place: the liquid decomposes into solid (Al) mixed with solid Si, but on a finer scale than before (bottom of Fig. A1.32). This intimate mixture of *secondary* (Al) with *secondary* Si is the eutectic structure.

On further cooling to room temperature the composition of the (Al) changes – it dissolves less silicon at the lower temperature. So silicon must diffuse out of the (Al), and the amount of Si must increase a little. But the final structure still looks like the bottom of Fig. A1.32.

Dendrites

When a metal is cast, heat is conducted out of it through the walls of the mould. The mould walls are the coldest part of the system, so solidification starts there. In the Al–Si casting alloy, for example, primary (Al) crystals form on the mould wall and grow inwards. Their composition differs from that of the liquid: it is purer, and contains less silicon. This means that silicon is *rejected* at the surface of the growing crystals, and the liquid grows richer in silicon: that is why the liquid composition moves along the liquidus line.

The rejected silicon accumulates in a layer just ahead of the growing crystals, and *lowers* the melting point of the liquid there. That slows down the solidification, because more heat has to be removed to get the liquid in this layer to freeze. But suppose a protrusion or bump on the solid (Al) pokes through the layer (Fig. A1.33). It finds itself in liquid which is *not* enriched with silicon, and *can* solidify. So the bump, if it forms, is unstable and grows rapidly. Then the (Al) will grow, not as a sphere, but in a branched shape called a *dendrite*. Many alloys show *primary dendrites* (Fig. A1.34); and the eutectic, if it forms, fills in the gaps between the branches.

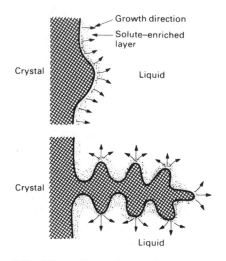

Figure A1.33

Segregation

If an 80 at% Pb alloy is cooled, the first solid appears at 305°C, and is primary (Pb) with a composition of about 90% Pb (see Fig. A1.35). From 305 to 255°C the amount of primary (Pb) increases, and its composition, which (at equilibrium) follows the solidus line, changes: it becomes richer in tin. This means that lead must diffuse *out* of the solid (Pb), and tin must diffuse *in*.

This diffusion takes time. If cooling is slow, time is available and equilibrium is maintained. But if cooling is rapid, there is insufficient time for diffusion, and, although the new primary (Pb), on the outside of the solid, has the proper composition, the inside (which solidified first) does not. The inside is purer than the outside; there is a *composition gradient* in each (Pb) grain, from the middle to the outside. This gradient is called *segregation*, and is found in almost all alloys (see Fig. A1.36).

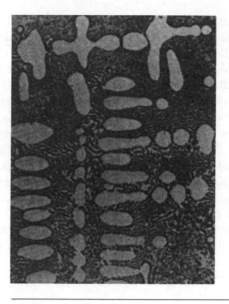

Figure A1.34 Dendrites of silver in a copper–silver eutectic matrix, ×330. (After G. A. Chadwick, *Metallography of Phase Transformations*, Butterworth, 1972.)

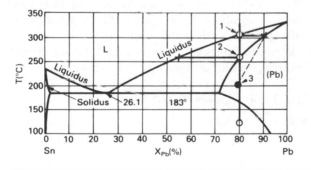

Figure A1.35

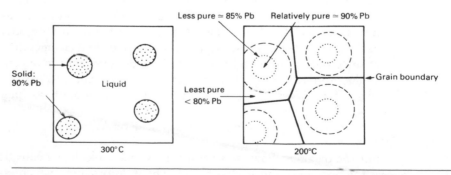

Figure A1.36

The phase diagram describes the equilibrium constitution of the alloy – the one given by very slow cooling. In the last example all the liquid should have solidified at the point marked 2 on Fig. A1.35, when all the solid has moved to the composition $X_{Pb} = 80\%$ and the temperature is 255°C. Rapid cooling prevents this; the solid has not had time to move to a composition $X_{Pb} = 80\%$. Instead, it has an average composition about half-way between that of the first solid to appear ($X_{Pb} = 90\%$) and the last ($X_{Pb} = 80\%$), that is, an average composition of about $X_{Pb} = 85\%$. This "rapid cooling" solidus lies to the right of the "equilibrium" solidus; it is shown as a broken line on Fig. A1.35. If this is so, the alloy is *not* all solid at 260°C. The rule for calculating the amounts of each phase still applies, using the "rapid cooling" solidus as one end of the tie line: it shows that the alloy is completely solid only when point 3 is reached. Because of this, the liquid composition overshoots the point marked X, and may even reach the eutectic point – so eutectic may appear in a rapidly cooled alloy even though the equilibrium phase diagram says it shouldn't.

Eutectoids

Figure A1.37 shows the iron–carbon phase diagram up to 6.7 wt% carbon (to the first intermetallic compound, Fe_3C). Of all the phase diagrams you, as an engineer, will encounter, this is the most important. So much so that you simply have to learn the names of the phases, and the approximate regimes of composition and temperature they occupy. The phases are:

Ferrite: α(b.c.c.) iron with up to 0.035 wt% C dissolved in solid solution.

Austenite: γ(f.c.c.) iron with up to 1.7 wt% C dissolved in solid solution.

δ-iron: δ(b.c.c.) with up to 0.08 wt% C dissolved in solid solution.

Cementite: Fe_3C, a compound, at the right of the diagram.

Ferrite (or α) is the low-temperature form of iron. On heating, it changes to austenite (or γ) at 914°C when it is pure, and this form remains stable until it reaches 1391°C when it changes to δ-iron (if you have forgotten this, check back to p. 351). The phase diagram shows that carbon changes the temperatures of these transitions, stabilising γ over a wider temperature interval.

The iron–carbon system has a eutectic: find it and mark it on the diagram (Fig. A1.37). At the eutectic point the phase reaction, on cooling, is

$$Liquid \rightarrow austenite + cementite.$$

But the diagram shows another feature which looks like a eutectic: it is the V at the bottom of the austenite field. The transformation which occurs there is very like the eutectic transformation, but this time it is a *solid*, austenite, which

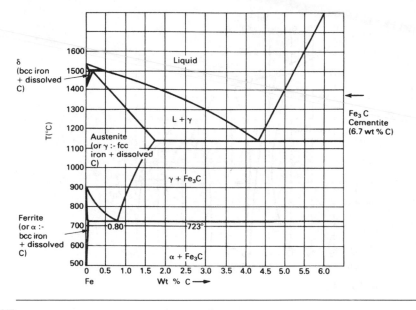

Figure A1.37

transforms on cooling to two other solids. The point at the base of the V is called a *eutectoid point*.

> *DEF.* A *eutectoid reaction* is a three-phase reaction by which, on cooling, a solid transforms into two other solid phases at the same time. If the bottom of a single-phase solid field closes (and provided the adjacent two-phase fields are solid also), it does so with a eutectoid point.

The compositions of the two new phases are given by the ends of the tie line through the eutectoid point.

Questions

3.5 The copper–zinc system (which includes brasses) has one eutectoid reaction. Mark the eutectoid point on the phase diagram (Fig. A1.38).

3.6 The copper–tin system (which includes *bronzes*) has four eutectoids (Fig. A1.39). One is obvious; the other three take a little hunting for. Remember that, if the bottom of the single-phase field for a solid closes, then it does

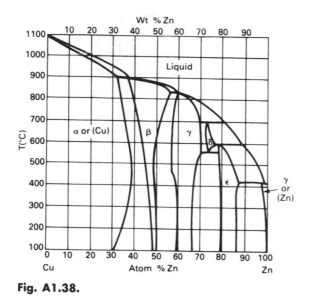

Fig. A1.38.

Figure A1.38

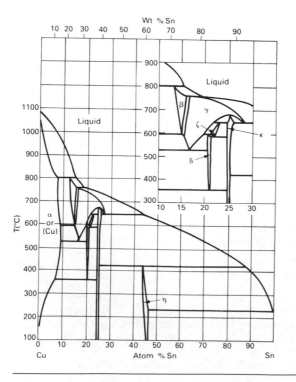

Figure A1.39

so with a eutectoid. Try to locate (and ring carefully) the four eutectoid points on the copper–tin phase diagram.

Eutectoid structures

Eutectoid structures are like eutectic structures, but much finer in scale. The original solid decomposes into two others, both with compositions which differ from the original, and in the form (usually) of fine, parallel plates. To allow this, atoms of B must diffuse away from the A-rich plates and A atoms must diffuse in the opposite direction, as shown in Fig. A1.40. Taking the eutectoid decomposition of iron as an example, carbon must diffuse to the carbon-rich Fe_3C plates, and away from the (carbon-poor) α-plates, just ahead of the interface. The colony of plates then grows to the right, consuming the austenite (γ). The eutectoid structure in iron has a special name: it is called *pearlite* (because it has a pearly look). The micrograph (Fig. A1.41) shows pearlite.

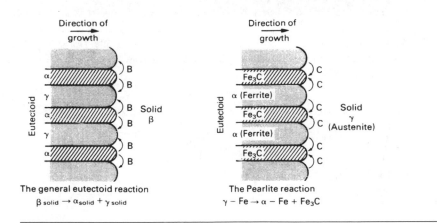

Figure A1.40

Peritectics

Eutectics and eutectoids are common features of engineering alloys. At their simplest, they look like a V resting on a horizontal line (see Fig. A1.42). The phase reactions, on cooling, are

$$\text{Liquid L} \rightarrow \alpha + \beta \text{ (eutectic)}$$

$$\text{Solid } \beta \rightarrow \alpha + \gamma \text{ (eutectoid)}.$$

Figure A1.41 Pearlite in a eutectoid-composition plain-carbon steel, ×500. (After K. J. Pascoe, *An Introduction to the Properties of Engineering Materials*, Van Nostrand Reinhold, London, 1978.)

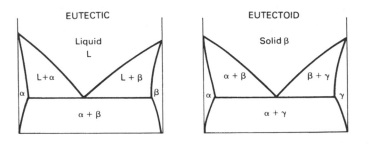

Figure A1.42

Many phase diagrams show another feature. It looks like an upside-down V (i.e. a ∧) touching a horizontal line. It is a *peritectic reaction*, and the tip of the ∧ is a *peritectic point* (see Fig. A1.43).

> *DEF.* A *peritectic reaction* is a three-phase reaction by which, on cooling, two phases (one of them liquid) react to give a single new solid phase.

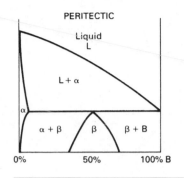

Figure A1.43

On Fig. A1.43, the peritectic reaction is

$$\text{Liquid} + \text{solid } \alpha \rightarrow \text{solid } \beta.$$

The composition of the β which forms (in this example) is 50 at% B.

Questions

3.7   The iron–carbon diagram (Fig. A1.37) has a peritectic point. Ring it on the diagram.

3.8   The copper–zinc system shown in Fig. A1.38 has no fewer than five peritectic reactions. Locate them and ring the peritectic points. (Remember that when a single-phase field closes above at a point, the point is a peritectic point.)

Peritectoids

> DEF.   A peritectoid is a three-phase reaction by which, on cooling, two *solid* phases react to give a single new solid phase.

On Fig. A1.44 the peritectoid reaction is

$$A + B \rightarrow \delta.$$

Answers to questions: part 3

3.1   (See Fig. A1.45.) 550°C, 67%; 580°C, 11%; 1350°C, 49%.

3.2   The reduced (constant pressure) phase rule is

$$F = C - P + 1.$$

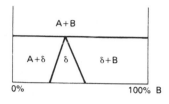

Figure A1.44

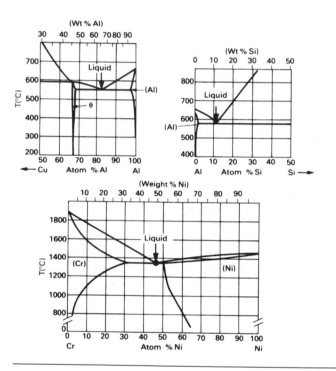

Figure A1.45

There are two components; the three phases (two solids and one liquid) coexist. So $F = 0$, that is, the three phases can coexist only at a point (the eutectic point).

3.3 From $X_{Pb} = 1.45\%$ to $X_{Pb} = 71\%$.

3.4 Remember that this is an *equilibrium* diagram. Any point on the diagram corresponds to a unique constitution. So, on heating, the reaction simply goes in reverse. The two solids "react" to give a single liquid. In general:

$$\left. \begin{array}{c} \alpha + \beta \rightarrow \text{Liquid} \\[6pt] \text{or, for the lead–tin system} \quad (Pb) + (Sn) \rightarrow \text{Liquid} (Pb - Sn) \end{array} \right\} \text{on heating}$$

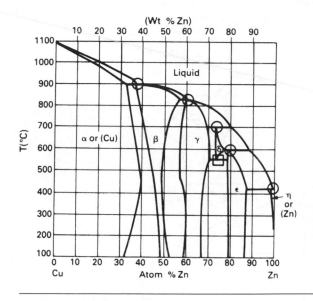

Figure A1.46   ○ = peritectic; □ eutectoid.

3.5   (Also 3.8) (See Fig. A1.46.)

3.6   Eutectoids ringed with solid circles (see Fig. A1.47).

3.7   (See Fig. A1.48.)

TEACHING YOURSELF PHASE DIAGRAMS, PART 4: FINAL PROBLEMS

This last section allows you to test your understanding of the use of phase diagrams.

Questions

4.1   When the temperature or pressure is decreased very rapidly, high temperature or high pressure phases can be "trapped", and are observed at atmospheric temperature and pressure (diamond, for instance, is a high-pressure form of carbon. It is only metastable at atmospheric pressure: the stable form is graphite.)

A roughly spherical meteorite of pure iron passes through the Earth's atmosphere, causing surface heating, and impacts in the Mill pool (a local pond) creating a (uniform) pressure wave, and a certain amount of consternation. The meteorite is recovered and sectioned. It shows signs of having melted externally, and of having had an outer shell of γ-iron, an inner shell of ε-iron and a core of α-iron. Use the p–T phase diagram for iron to deduce the approximate magnitude of the pressure wave. Express the result in atmospheres (see Fig. A1.49).

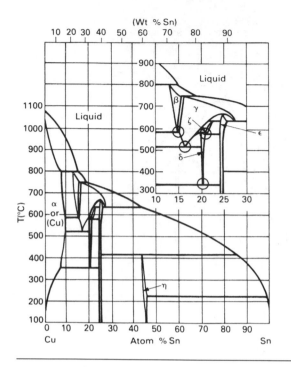

Figure AI.47

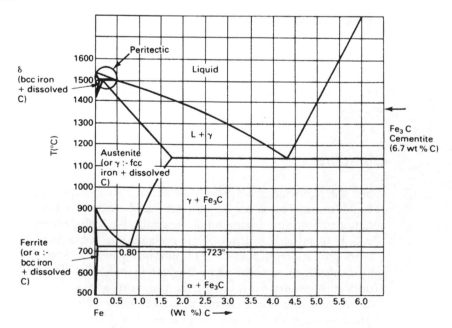

Figure AI.48

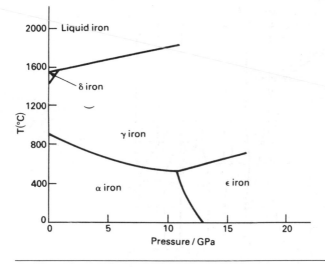

Figure A1.49

4.2   Your ancient granny dies and leaves you her most prized possession: an Urn of Pure Gold. One afternoon, while mixing paint-remover in the urn, you are disturbed to note that it has turned an evil green in colour. Whipping out your magnifying glass, you observe that the paint remover, in attacking the urn, has also etched it, clearly revealing the presence of two phases. This (of course) raises in your mind certain nagging doubts as to the Purity of the Gold. Your friend with an electron microprobe analyser performs a quick chemical analysis for you, with the distressing result:

Copper 60   at%;
Zinc   40   at%;
Gold < 0.001 at%;

Figure A1.50 shows the appropriate phase diagram. Assuming the Urn to be at equilibrium (though Granny might not have been):

(a) Mark the constitution point onto the diagram (assume that the constitution at room temperature is the same as at 200°C).
(b) Is it in a single- or two-phase region? -------------------------------
(c) What phase(s) are present? --
(d) List the approximate phase composition(s) ---------------------------
(e) Calculate approximately the proportions of each phase ---------------

Well done! You have determined the Constitution of Granny's Urn.

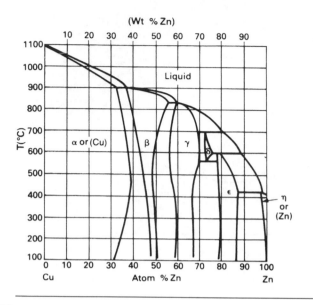

Figure AI.50

4.3   Describe, using the copper–nickel ("monel") system as an example, the process of zone-refining. (Figure A1.51 shows a system with *complete solid solubility*.) How many phases are present in an alloy of 60 wt% Ni and 40 wt% Cu at:

(a) 1400°C;
(b) 1300°C;
(c) 1000°C.

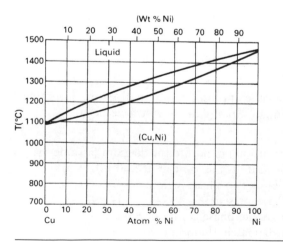

Figure AI.51

4.4 Figure A1.52 shows the Ti–Al phase diagram (important for the standard commercial alloy Ti–6% Al–4% V. It shows two *peritectic reactions*, at each of which liquid reacts with a solid phase to give an intermetallic compound. (a) Ring the peritectics and give the (approximate) chemical formula for the two compounds. (b) Shade all two-phase fields. (c) At what temperature does a Ti–6 wt% Al alloy start to melt? (d) Over what temperature range does it change from the α (c.p.h.) to the β (b.c.c.) structure?

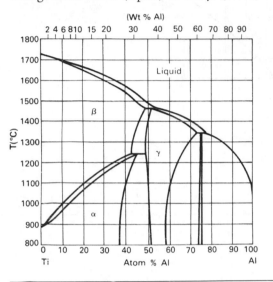

Figure A1.52

4.5 Figure A1.53 shows the aluminium–silicon system, basis of most aluminium casting alloys.

(a) What is the eutectic composition and temperature?
(b) How many phases are present in an alloy of eutectic composition at 1000°C, and at 400°C?
(c) Describe the solidification of an alloy of eutectic composition, and the resulting structure.
(d) Compare and contrast this with the formation of a eutectoid structure.

4.6 A hypothetical equilibrium diagram between two elements A and B shows the following features:

A has three solid allotropic forms with change temperatures of 800°C and 1150°C and melts at 1980°C. These form solid solutions α, β and γ containing B, α being the low-temperature one.

An intermediate compound A_2B_3 melts at 1230°C. It has a limited solid solubility for A, forming solid solution ϵ and no solid solubility for B.

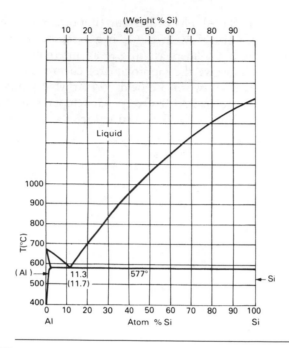

Figure A1.53

B melts at 800°C and has negligible solid solubility for A.

Eutectic reactions:
at 1000°C, liquid (55% B) → β(25% B) + ε(60% B)
at 650°C, liquid (90% B) → A_2B_3 + B.

Peritectic reaction at 1300°C:
γ(8% B) + liquid (35% B) → β(15% B).

Eutectoid reaction at 600°C:
β(12% B) → α(5% B) + ε(65% B).

Peritectoid reaction at 300°C:
α(3% B) + ε(69% B) → δ(40% B).

At 0°C the solubilities of B in A and A in A_2B_3 are negligible and the δ phase extends from 35% to 45% B.

All percentages given are by weight. The atomic weight of B is twice that of A.

Draw the equilibrium diagram assuming all phase boundaries are straight lines.

For an alloy containing 30% B describe the changes that occur as it is cooled from 1600°C to 0°C. Give the proportions of phases present immediately above and immediately below each temperature at which a reaction occurs.

Answers to questions: part 4

4.1   Between about 11.5 and 13.0 GPa or $1.14 \times 10^5 - 1.28 \times 10^5$ atm.

4.2   (a) See (Fig. A1.54).
      (b) Two-phase region.
      (c) α (copper-rich solid) and β (the compound CuZn).
      (d) $W_{Zn} \approx 33\%$, $W_{Zn} \approx 48\%$.
      (e) Very roughly, 50–50; more precisely:

$$\frac{\text{wt\% of } \alpha}{\text{wt\% of } \beta} = \frac{48-40}{40-33} = \frac{8}{7}.$$

(Wt % Zn)

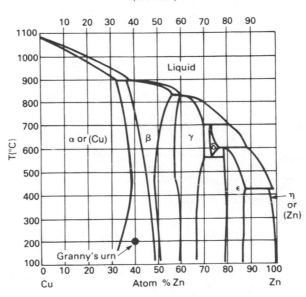

Figure A1.54

4.3   The solid which first appears on cooling is higher in nickel. Repeated directional remelting and solidification "zones" the copper up to the end of the bar, and leaves most of the bar increasingly pure in nickel.

      (a) 1.
      (b) 2.
      (c) 1.

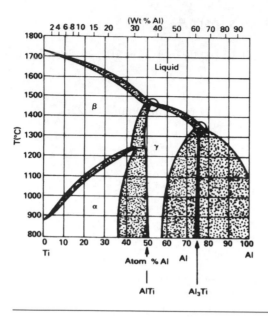

Figure A1.55

4.4   (a) AlTi, Al₃Ti.
         (b) See Fig. A1.55.
         (c) 1680°C.
         (d) 980°C to 1010°C.

4.5   (a) 11.7 wt% Si, 577°C.
         (b) One phase at 1000°C, two phases at 400°C.
         (c) See pp. 398, 400.
         (d) Eutectoid structure produced by the decomposition of a *solid* phase, not a
              liquid.

4.6 A_2B_3 contains $\frac{3\times2}{2\times1+3\times2} = 75\%$ B by weight. Hence equilibrium diagram
         is as given in Fig A1.56. On cooling 30% B mixture from 1600°C: at
         1397°C, solidification commences by separation of γ crystals. Just above
         1300°C $\frac{22}{27}$ (= 81.5%) liquid (35% B) + $\frac{5}{27}$ (= 18.5%) γ (8% B). At
         1300°C, all γ+ some liquid form β in peritectic reaction. Just below
         1300°C $\frac{15}{20}$ (= 75%) liquid (35% B) + $\frac{5}{20}$ (= 25%) β (15% B). 1300°C →
         1000°C, more β separates. Just above 1000°C $\frac{5}{30}$ (= 17%) liquid (55% B) +
         $\frac{25}{30}$ (= 83%) β (25% B). At 1000°C all liquid forms β and ε in eutectic reac-
         tion. Just below 1000°C $\frac{5}{35}$ (= 14.3%) ε (60% B) + $\frac{30}{35}$ (= 85.7%) β (25% B).
         1000°C → 600°C, β precipitates ε and ε precipitates β. Just above
         600°C $\frac{18}{53}$ (= 34%) ε (65% B) + $\frac{35}{53}$ (= 66%) β (12% B). At 600°C all β forms
         α and ε in eutectoid reaction. Just below 600°C $\frac{25}{60}$ (= 42%) ε (65% B) +
         $\frac{35}{60}$ (= 58%) α (5% B). 600°C → 300°C, α precipitates ε and ε precipitates α.

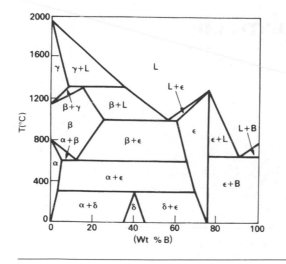

Figure A1.56

Just above 300°C $\frac{27}{66}$ ($= 41\%$) ε (69% B) $+ \frac{39}{66}$ ($= 59\%$) α (3% B). At 300°C all ε and some α form δ in peritectoid reaction. Just below 300°C $\frac{27}{37}$ ($=$ 73%) δ (40% B) $+ \frac{10}{37}$ ($= 27\%$) α (3% B). 300°C $\rightarrow$ 0°C, amount of α decreases and δ increases. At 0°C $\frac{30}{35}$ ($= 86\%$) δ (35% B) $+ \frac{5}{35}$ ($=$ 14%) α (0% B).

Appendix 2
Symbols and formulae

List of principal symbols

Symbol	Meaning (units)
Note:	Multiples or sub-multiples of basic units indicate the unit suffixes normally used in materials data.
a	lattice parameter (nm)
a	crack length (mm)
A	availability (J)
A_1	eutectoid temperature (°C)
A_3	first ferrite temperature (°C)
A_{cm}	first Fe_3C temperature (°C)
b	Burgers vector (nm)
c	height of c.p.h. unit cell (nm)
C	concentration (m^{-3})
CCR	critical cooling rate (°C s^{-1})
DP	degree of polymerisation (dimensionless)
E	Young's modulus of elasticity (GPa)
f	force (N)
F	force (N)
g	acceleration due to gravity on the Earth's surface ($m\,s^{-2}$)
G	shear modulus (GPa)
G	Gibbs function (J)
G_c	toughness ($kJ\,m^{-2}$)
H	hardness (GPa)
ΔH	latent heat of transformation (J)
I	second moment of area of structural section (mm^4)
k	ratio of C_{solid}/C_{liquid} on phase diagram (dimensionless)
k	Boltzmann's constant ($J\,K^{-1}$)
k	shear yield strength (MPa)
K_c	fracture toughness ($MPa\,m^{1/2}$)
L	liquid phase

(*Continued*)

Symbol	Meaning (units)
m	mass (kg)
m	Weibull modulus (dimensionless)
M	bending moment (N m)
M_F	martensite finish temperature (°C)
M_S	martensite start temperature (°C)
n	time exponent for slow crack-growth (dimensionless)
p	pressure (Pa)
P_f	failure probability (dimensionless)
P_S	survival probability (dimensionless)
q	activation energy per atom (J)
Q	activation energy per mole (kJ mol^{-1})
r^*	critical radius for nucleation (nm)
R	universal gas constant (J K^{-1} mol^{-1})
T	absolute temperature (K)
T_e	equilibrium temperature (K)
T_G	glass temperature (K)
T_m	melting temperature (K)
ΔT	thermal shock resistance (K)
v	velocity (m s^{-1})
V	volume (m^3)
V	volume fraction (dimensionless)
W_A	weight % (dimensionless)
W_f	free work (J)
X_A	mol % (dimensionless)
α	linear coefficient of thermal expansion (MK^{-1})
γ	energy of interface (J m^{-2}) or tension of interface (N m^{-1})
δ	elastic deflection (mm)
ε	true (logarithmic) strain (dimensionless)
ε_f	(nominal) strain after fracture; tensile ductility (dimensionless)
$\dot{\varepsilon}_{ss}$	steady-state tensile strain-rate in creep (s^{-1})
η	viscosity (P, poise)
ν	Poisson's ratio (dimensionless)
ρ	density (Mg m^{-3})
σ	true stress (MPa)
σ_c	(nominal) compressive strength (MPa)
σ_r	modulus of rupture (MPa)
σ_{TS}	(nominal) tensile strength (MPa)
σ_y	(nominal) yield strength (MPa)

Greek letters are used to label the phases on phase diagrams.

Summary of principal formulae and magnitudes

Chapter 3 and Teaching yourself phase diagrams: phase diagrams

Composition is given by

$$W_A = \frac{\text{weight of A}}{\text{weight of A} + \text{weight of B}} \times 100$$

in weight %, and by

$$X_A = \frac{\text{atoms (mols) of A}}{\text{atoms (mols) of A} + \text{atoms (mols) of B}} \times 100$$

in atom (mol) %.

$$W_A + W_B = 100\%; \quad X_A + X_B = 100\%.$$

Three-phase reactions

$$\text{Eutectic: } L \rightleftharpoons \alpha + \beta$$
$$\text{Eutectoid: } \beta \rightleftharpoons \alpha + \gamma$$
$$\text{Peritectic: } L + \alpha \rightleftharpoons \beta$$
$$\text{Peritectoid: } A + B \rightleftharpoons \delta$$

Chapter 4: Zone refining

$$C_s = C_0 \left\{ 1 - (1 - k) \exp\left(-\frac{kx}{l}\right) \right\}.$$

C_s = concentration of impurities in refined solid; C_0 = average impurity concentration; $k = C_{\text{solid}}/C_{\text{liquid}}$; x = distance from start of bar; l = zone length.

Chapter 5: Driving forces

Driving force for solidification

$$W_f = -\Delta G = -\frac{\Delta H}{T_m}(T_m - T).$$

ΔH = latent heat of solidification; T_m = absolute melting temperature; T = actual temperature (absolute).

Driving force for solid-state phase change

$$W_f = -\Delta G = -\frac{\Delta H}{T_e}(T_e - T).$$

ΔH = latent heat of transformation; T_e = equilibrium temperature (absolute).

Chapter 6: Kinetics of diffusive transformations

Speed of interface

$$v \propto e^{-q/kT} \Delta T.$$

q = activation energy per atom; k = Boltzmann's constant; T = absolute temperature; ΔT = difference between interface temperature and melting or equilibrium temperature.

Chapter 7: Nucleation

Nucleation of solids from liquids: critical radius for homogeneous *and* heterogeneous nucleation

$$r^* = \frac{2\gamma_{SL}T_m}{|\Delta H|(T_m - T)}.$$

γ_{SL} = solid–liquid interfacial energy; T_m = absolute melting temperature; ΔH = latent heat of solidification; T = actual temperature (absolute).

Chapter 8: Displacive transformations

Overall rate of *diffusive* transformation

$$\propto \text{no. of nuclei} \times \text{speed of interface.}$$

Chapter 10: The light alloys

Solid solution hardening

$$\sigma_y \propto \varepsilon_s^{3/2} C^{1/2}.$$

C = solute concentration; ε_s = mismatch parameter.

Work-hardening

$$\sigma_y \propto \varepsilon^n.$$

ε = true strain; n = constant.

Chapter 14: Metal processing

Forming pressure
  No friction

$$p_f = \sigma_y.$$

  Sticking friction

$$p_f = \sigma_y \left\{ 1 + \frac{(w/2) - x}{d} \right\}.$$

σ_y = yield strength; w = width of forging die; x = distance from centre of die face; d = distance between dies.

Chapter 17: Ceramic strengths

Sample subjected to uniform tensile stress
  Tensile strength

$$\sigma_{TS} = \frac{K_c}{\sqrt{\pi a_m}}.$$

K_c = fracture toughness; a_m = size of widest microcrack (crack width for surface crack; crack half-width for buried crack).
  Modulus of rupture

$$\sigma_r = \frac{6M_r}{bd^2}.$$

M_r = bending moment to cause rupture; b = width of beam; d = depth of beam.
  Compressive strength

$$\sigma_c \approx 15\sigma_{TS},$$

$$\sigma_c = \frac{CK_c}{\sqrt{\pi \bar{a}}}.$$

C = constant (≈ 15); $\bar{a}$ = average crack size.

Thermal shock resistance

$$\Delta T = \sigma_{TS}/E\alpha.$$

E = Young's modulus; α = linear coefficient of thermal expansion.

$$\dot{\varepsilon}_{ss} = A\sigma^n \exp(-Q/RT).$$

$\dot{\varepsilon}_{ss}$ = steady-state tensile strain rate; A, n = constants; σ = tensile stress; Q = activation energy for creep; R = universal gas constant; T = absolute temperature.

Chapter 18: Statistics of fracture

Weibull distribution

$$P_s(V) = \exp\left\{ -\frac{V}{V_0} \left(\frac{\sigma}{\sigma_0} \right)^m \right\}$$

or

$$\ln\left\{ \ln\left(\frac{1}{P_s} \right) \right\} = \ln\frac{V}{V_0} + m \ln\left(\frac{\sigma}{\sigma_0} \right).$$

P_s = survival probability of component; V = volume of component; σ = tensile stress on component; V_0 = volume of test sample; σ_0 = stress that, when applied to test sample, gives $P_s = 1/e(= 0.37)$; m = Weibull modulus.

Failure probability

$$P_f = 1 - P_s.$$

Slow crack-growth

$$\left(\frac{\sigma}{\sigma_{TS}} \right)^n = \frac{t(\text{test})}{t}.$$

σ = strength of component after time t; σ_{TS} = strength of component measured over time $t(\text{test})$; n = slow crack-growth exponent.

Chapter 19: Ceramics processing

Sintering

$$\frac{\mathrm{d}\rho}{\mathrm{d}t} = \frac{C}{a^n} \exp(-Q/RT).$$

ρ = density; t = time; C, n = constants; a = particle size; Q = activation energy for sintering; R = universal gas constant; T = absolute temperature.
Glass forming

$$\eta \propto \exp(Q/RT).$$

η = viscosity; Q = activation energy for viscous flow.

Chapter 20: Cements and concretes

$$\text{Hardening rate} \propto \exp(-Q/RT).$$

Q = activation energy for hardening reaction; R = universal gas constant; T = absolute temperature.

Chapter 23: Mechanical behaviour of polymers

Modulus: WLF shift factor

$$\log(a_T) = \frac{C_1(T_1 - T_0)}{C_2 + T_1 - T_0}.$$

C_1, C_2 = constants; T_1, T_0 = absolute temperatures.
Polymer viscosity

$$\eta_1 = \eta_0 \exp\left\{ \frac{-C_1(T_1 - T_0)}{C_2 + T_1 - T_0} \right\}.$$

Chapter 25: Composites

Unidirectional fibre composites

$$E_{c\parallel} = V_f E_f + (1 - V_f)E_m,$$

$$E_{c\perp} = \left\{ \frac{V_f}{E_F} + \frac{1 - V_f}{E_m} \right\}^{-1}.$$

$E_{c\parallel}$ = composite modulus parallel to fibres; $E_{c\perp}$ = composite modulus perpendicular to fibres; V_f = volume fraction of fibres; E_f = Young's modulus of fibres; E_m = Young's modulus of matrix.

$$\sigma_{TS} = V_f \sigma_f^f + (1 - V_f)\sigma_y^m.$$

σ_{TS} = tensile strength parallel to fibres; σ_f^f = fracture strength of fibres; σ_y^m = yield strength of matrix.

Optimum toughness

$$G_c = V_f \frac{d}{8} \frac{(\sigma_f^f)^2}{\sigma_s^m}.$$

d = fibre diameter; σ_s^m = shear strength of matrix.

Magnitudes of properties The listed properties lie, for most structural materials, in the ranges shown

Property		Metals	Ceramics	Polymers (unfoamed)	Composites (polymer matrix)
Density	$(Mg\ m^{-3})$	2 to 10	1 to 5	1 to 2	1.5 to 2.0
Young's modulus	(GPa)	50 to 200	10 to 1000	0.01 to 10	10 to 200
Yield strength	(MPa)	25 to 1500	3000 to 50,000	–	–
Tensile strength	(MPa)	50 to 2000	1 to 800	5 to 100	100 to 1000
Fracture toughness	$(MPa\ m^{1/2})$	5 to 200	0.1 to 10	0.5 to 5	20 to 50
Creep temperature	(°C)	50 to 1000	−20 to 2000	0 to 200	0 to 200

References

Ashby, M. F. *Materials selection in mechanical design*, 3rd edition. Elsevier, 2005.

Ashby, M. F. and Cebon, D. *Case studies in materials selection*. Granta Design, 1996.

Ashby, M. F. and Johnson, K. *Materials and design – the art and science of material selection in product design*. Elsevier, 2002.

Ashby, M. F. and Jones, D. R. H. *Engineering materials 1 – an introduction to properties, applications and design*, 3rd edition. Elsevier, 2005.

ASM. *Metals handbook*, 2nd desktop edition. ASM, 1999.

British Standards Institution. BS 3F 70: Specification for heavy duty braided rubber cord, 1991.

British Standards Institution. BS 7608: Code of practice for fatigue design and assessment of steel structures, 1993.

Brydson, J. A. *Plastics materials*, 7th edition. Butterworth-Heinemann, 1999.

Calladine, C. R. *Plasticity for engineers*. Ellis Horwood, 1985.

Campbell, J. *Castings*, 2nd edition. Butterworth-Heinemann, 2003.

Charles, J. A., Crane, F. A. A. and Furness, J. A. G. *Selection and use of engineering materials*, 3rd edition. Butterworth-Heinemann, 1997.

Cottrell, A. H. *An introduction to metallurgy*, 2nd edition. Arnold, 1975.

Crawford, R. J. *Plastics engineering*, 3rd edition. Butterworth-Heinemann, 1998.

Desch, H. E. *Timber, its structure, properties and utilization*, 6th edition. Macmillan, 1985.

Easterling, K. E. *Introduction to the physical metallurgy of welding*. Butterworths, 1992.

Esslinger, V., Kieselbach, R., Koller, R. and Weisse, B. The railway accident of Eschede – technical background. *Engineering Failure Analysis* 11(2004), 515–535.

Gale, W. and Totemeier, T. *Smithells reference book*, 8th edition. Elsevier, 2003.

Gibson, L. J. and Ashby, M. F. *Cellular solids*, 2nd edition. Butterworth-Heinemann, 1997.

Gordon, J. E. *Structures – or why things don't fall down*. Da Capo Press, 2003.

Gordon, J. E. *The new science of strong materials, or why you don't fall through the floor*, 2nd edition. Princeton University Press, 1988.

Hertzberg, R. W. *Deformation and fracture of engineering materials*, 4th edition. Wiley, 1996.

Honeycombe, R. W. K. and Bhadeshia, H. K. D. H. *Steels: microstructure and properties*, 2nd edition. Arnold, 1995.

Hull, D. and Clyne, T. W. *An introduction to composite materials*, 2nd edition. Cambridge University Press, 1996.

Jones, D. R. H. Analysis of a fatal bungee-jumping accident. *Engineering Failure Analysis* 11(2004), 857–872.

Jones, D. R. H. *Engineering materials 3 – materials failure analysis.* Pergamon, 1993.

Kalpakjian, S. and Schmid, S. R. *Manufacturing processes for engineering materials*, 4th edition. Addison-Wesley, 2002.

Kingery, W. D., Bowen, H. F. and Uhlmann, D. R. *Introduction to ceramics*, 2nd edition. Wiley, 1976.

Lawn, B. R. *Fracture of brittle solids*, 2nd edition. Cambridge University Press, 1993.

Lee, S-B. Fatigue failure of welded vertical member of a steel truss bridge. *Engineering Failure Analysis* 3(1996), 103–108.

Lewis, P. R. *Beautiful railway bridge of the silvery Tay.* Tempus, 2004.

Lewis, P. R. and Jones, D. R. H. *T839 forensic engineering, block 3, catastrophic failures.* The Open University, 2001.

Lewis, P. R., Reynolds, K. and Gagg, C. *Forensic materials engineering – case studies.* CRC Press, 2004.

Llewellyn, D. T. and Hudd, R. C. *Steels – metallurgy and applications*, 3rd edition. Butterworth-Heinemann, 1998.

McEvily, A. J. *Metal failures.* Wiley, 2002.

Ministry of Transport and Civil Aviation (UK). *Civil aircraft accident: report of the Court of Enquiry into the accidents to Comet G-ALYP on 10th January 1954 and Comet G-ALYY on 8th April 1954.* HMSO (UK), 1955.

Polmear, I. J. *Light alloys*, 4th edition. Elsevier, 2005.

Porter, D. A. and Easterling, K. E. *Phase transformations in metals and alloys*, 2nd edition. Chapman and Hall, 1992.

Powell, P. C. and Ingen Housz, A. J. *Engineering with polymers*, 2nd edition. Stanley Thornes, 1998.

Reed-Hill, R. E. *Physical metallurgy principles.* Van Nostrand Reinhold, 1964.

Richard, H. A., Fulland, M., Sander, M. and Kullmer, G. Fracture in a rubber-sprung railway wheel. *Engineering Failure Analysis*, in press.

Seymour, R. B. *Polymers for engineering applications.* ASM International, 1987.

Shewmon, P. G. *Diffusion in solids*, 2nd edition. TMS Publishers, 1989.

Ward, I. M. and Sweeney, J. *An introduction to the mechanical properties of solid polymers.* Wiley, 2004.

Waterman, N. A. and Ashby, M. F. *Elsevier materials selector.* Elsevier, 1991.

Withey, P. A. Fatigue failure of the de Havilland Comet 1. *Engineering Failure Analysis* 4(1997), 147–154.

Young, W. C. and Budynas, R. G. *Roark's formulas for stress and strain*, 7th edition. McGraw-Hill, 2001.

INDEX

Page numbers for figures have suffix f, those for tables have suffix t, those for equations have suffix e.